s sponsored by the NATO Science
ion of advanced scientific and technological
links between scientific communities.

nal board of publishers in conjunction with the

m Publishing Corporation
on and New York

Reidel Publishing Company
drecht and Boston

artinus Nijhoff Publishers
ordrecht/Boston/Lancaster

Springer-Verlag
Berlin/Heidelberg/New York

ied Sciences – No. 109

Fast

NATO ASI Series

Advanced Science Institutes Series

A Series presenting the results of activities
Committee, which aims at the dissemina
knowledge, with a view to strengthening

The Series is published by an internatio
NATO Scientific Affairs Division

A	Life Sciences	Plenu
B	Physics	Lon
C	Mathematical and Physical Sciences	D. Do
D	Behavioural and Social Sciences	M
E	Applied Sciences	D
F	Computer and Systems Sciences	
G	Ecological Sciences	

Series E: App

Fast Electrical and Optical Measurements

Volume II – Optical Measurements

edited by

James E. Thompson, PhD.

Chairman, Electrical Engineering Department
University of Texas at Arlington
Arlington, Texas 76019, USA

Lawrence H. Luessen

Head, Directed Energy Branch
Naval Surface Weapons Center
Dahlgren, Virginia 22448, USA

Editorial Committee:

Anthony K. Hyder, Jr., PhD.
Associate Vice President for Research
Auburn University, Auburn, Alabama, USA

Millard F. Rose, PhD.
Director, Space Power Institute
Auburn University, Auburn, Alabama, USA

Magna Kristiansen, PhD.
P.W. Horn Professor of Electrical Engineering
Texas Tech University, Lubbock, Texas, USA

Susie M. Anderson
EG&G Washington Analytical Services
Dahlgren, Virginia, USA

1986 **Martinus Nijhoff Publishers**
Dordrecht / Boston / Lancaster
Published in cooperation with NATO Scientific Affairs Division

Proceedings of the NATO Advanced Study Institute on Fast Electrical and Optical Diagnostic Principles and Techniques, Il Ciocco, Castelvecchio Pascoli, Italy, July 10-24, 1983

Library of Congress Cataloging in Publication Data

```
NATO Advanced Study Institute on Fast Electrical
    and Optical Diagnostic Principles and Techniques
    (1983 : Castelvecchio Pascoli, Italy)
    Fast electrical and optical measurements.

    (NATO ASI series. Series E, Applied sciences ;
108-109)
    "Proceedings of the NATO Advanced Study Institute
on Fast Electrical and Optical Diagnostic Principles
and Techniques, Il Ciocco, Castelvecchio Pascoli,
Italy, July 10-24, 1983"--T.p. verso.
    "Published in cooperation with NATO Scientific
Affairs Division."
    Includes index.
    1. Electric measurements--Congresses.  2. Optical
measurements--Congresses.  I. Thompson, James E.
II. Luessen, Lawrence H.  III. North Atlantic Treaty
Organization. Scientific Affairs Division.  IV. Title.
V. Series.
TK277.N37 1983        621.37          86-700
ISBN 90-247-3296-4 (set)
ISBN 90-247-3294-8 (v. 1)
ISBN 90-247-3295-6 (v. 2)
```

ISBN 90-247-2689-1 (series)
ISBN 90-247-3296-4 (set)

Distributors for the United States and Canada: Kluwer Academic Publishers, 190 Old Derby Street, Hingham, MA 02043, USA

Distributors for the UK and Ireland: Kluwer Academic Publishers, MTP Press Ltd, Falcon House, Queen Square, Lancaster LA1 1RN, UK

Distributors for all other countries: Kluwer Academic Publishers Group, Distribution Center, P.O. Box 322, 3300 AH Dordrecht, The Netherlands

PREFACE

An Advanced Study Institute on Fast Electrical and Optical Diagnostic Principles and Techniques was held at Il Ciocco, Castelvecchio Pascoli, Italy, 10-24 July 1983. This publication is the Proceedings from that Institute.

The Institute was attended by ninety-seven participants representing the United States, West Germany, the United Kingdom, Switzerland, Norway, the Netherlands, Italy, and France.

The objective of the Institute was to provide a broad but comprehensive presentation of the various measurement and analysis techniques that can be employed to investigate fast physical events, nominally in the sub-microsecond regime. This requires both an understanding of the basic principles underlying the diagnostic employed and its limitations, and a knowledge of the practical techniques available to obtain reliable and repeatable data. This Institute was thus structured to begin tutorially, followed by more practical techniques, demonstrations, and discussions.

The Institute was divided into the following major sections: (1) Overview of Applications and Needs; (2) Voltage and Current Measurements; (3) Data Acquisition; (4) Grounding and Shielding; (5) Fast Photography; (6) Refractive Index Measurements; (7) X-ray Diagnostics; (8) Spectroscopy; and (9) Active Optical Techniques. This Proceeding has been divided into two separate volumes. Volume 1, **Current and Voltage Measurements**, includes Sections (1) through (4) above; Volume 2, **Optical Measurements**, includes Sections (5) through (9).

In addition, three sessions were made up of presentations contributed by participants which summarized various research efforts employing high-speed diagnostics; these presentations have not been included in this Proceedings. Several companies also displayed and demonstrated high-speed diagnostic equipment.

We are grateful to a number of organizations for providing the financial assistance that made this Institute possible. Foremost is the NATO Scientific Affairs Division, which provided the single most major contribution for the Institute. In addition, the following US sources made contributions: the Naval

Surface Weapons Center, Air Force Office of Scientific Research, Office of Naval Research, Air Force Wright Aeronautical Laboratories, Army Electronics Research and Development Command, and Naval Research Laboratory.

We would also like to thank the staff of Il Ciocco, particularly Gian Piero Giannotti, its manager, and Bruno Giannasi, the Hotel's Conference Coordinator and the main reason for the on-site success of the Institute. The phrase "Ask Bruno" is now firmly imbedded in the memory of all in attendance for those two wonderful weeks in Tuscany. Our thanks to Dr. Mario di Lullo and Dr. Craig Sinclair, the present Directors
of the NATO ASI Program, and Dr. Tilo Kester and his wife Barbara, of the Publications Coordination Office, for their suggestions, patience, and overall help.

Particular thanks to our Associate Editors who willingly volunteered to assist us in reviewing and editing the manuscripts: Anthony Hyder of Auburn University, Kris Kristiansen of Texas Tech University, and Frank Rose, formerly of the Naval Surface Weapons Center, and now keeping Dr. Hyder company at Auburn University. To our Organizing Committee, lecturers, and participants - we couldn't have done it without you. And a special thanks to our wives Lynn L. and Elizabeth T., for their assistance during the Institute; and to Linda Maynard of the Universtiy of South Carolina's Office of Continuing Engineering Education, for her efforts before, during, and following the Institute.

And finally, our appreciation to those most responsible for the actual production of this Proceedings. First, to the EG&G Washington Analytical Services Center Office at Dahlgren, Virginia, which had the task of centrally retyping every lecturer's manuscript and producing a camera-ready document for delivery to the publisher. To Susie M. Anderson, the Supervisor for Word Processing, and Christie K. Wood - thank you for a job well done. Second, to Martinus Nijhoff Publishers, especially Henny Hoogervorst, who had to contend with numerous delays in the delivery of the final manuscript, but never lost her sense of purpose or humor during the past two years.

Lawrence H. Luessen
Naval Surface Weapons Center
Dahlgren, Virginia

James E. Thompson
University of Texas
Arlington, Texas

September 1985

VOLUME 1: CURRENT AND VOLTAGE MEASUREMENTS

OVERVIEW OF APPLICATIONS AND NEEDS

Overview of Applications and Needs   1
  M. F. Rose, A. K. Hyder, M. Kristiansen

VOLTAGE AND CURRENT MEASUREMENTS

Electro-Optical Measurement Techniques   5
  R. Hebner

Electro-Optical and Magneto-Optical Studies of
    Cold Cathode Electron Beam Gun Discharges   27
  M. Hugenschmidt

Fiber Optic Magnetic Field and Current Sensors   41
  G. Chandler

An Electro-Optical Technique for Measuring High
    Frequency Free-Space Electric Fields   57
  J. Chang and C. N. Vittitoe

Electromagnetic Sensors and Measurement Techniques   73
  C. E. Baum

Ultrafast Electral Voltage and Current Monitors   145
  W. Pfeiffer

High Speed Electric Field and Voltage Measurements   175
  M. S. DiCapua

High Speed Magnetic Field and Current Measurements   223
  M. S. DiCapua

Nuclear Reaction Diagnostics for Intense Particle
     Beam Measurements   263
  R. J. Leeper

Particle Analyzer Diagnostics for Intense Particle
     Beam Measurements   317
  R. J. Leeper, J. R. Lee, L. Wissell, D. J. Johnson,
  and W. A. Stygar

DATA ACQUISITION

Software Correction of Measured Pulse Data   351
  N. J. Nahman

Fiber Optic Links for Data Acquisition, Communication,
     and Control   419
  G. Chandler

A High Speed Multi-Channel Data Recorder   437
  J. Chang, J. Foesch, and C. Martinez

An Iterative Deconvolution Algorithm for the
     Reconstruction of High-Voltage Impulses
     Distorted by the Measuring System   445
  K. Schon

Test Methods for the Dynamic Performance of Fast
     Digital Recorders   453
  K. Schon

GROUNDING AND SHIELDING

Electromagnetic Topology for the Analysis and Design
     of Complex Electromagnetic Systems   467
  C. E. Baum

Basic Principles of Grounding and Shielding with
     Respect to Equivalent Circuits   549
  J. Wiesinger

System Design: Practical Shielding and Gounding
     Techniques Based on Electromagnetic Topology   567
  W. Graf

A Systematic, Practical Approach to the Design of
     Shielded Enclosures for Data Acquisition and
     Control   585
  G. Chandler

CONTENTS

VOLUME 2: OPTICAL MEASUREMENTS

FAST PHOTOGRAPHY

Advances in High Speed Photography: 1972 - 1982   595
    J. S. Courtney-Pratt

Recent Techniques for High-Speed Photography and
    Low-Level Image Recording   609
    W. Pfeiffer

Laser Photographic and Cinematographic Applications
    to the Investigation of Transient Phenomena . . . .   643 <
    M. Hugenschmidt

REFRACTIVE INDEX MEASUREMENTS

Concepts and Illustrations of Optical Probing
    Diagnostics for Laser-Produced Plasmas   691
    J. A. Stamper, E. A. McLean, S. P. Okenschain,
    and B. H. Ripin

Sub-Nanosecond, Four-Frame, Holographic Interferometry
    Diagnostics . . . .   729 <
    G. Allen, H. P. Davis, L. P. Mix, and J. Chang

Laser Interferometry: Streaked Shadowgraphy and
    Schlieren Imaging . . . .   743 <
    C. Popovics and R. Benattar

Moire-Schlieren, Time-Differential Interferometry,
    and Enhanced Sensitivity of Faraday Rotation
    Measurements   771
    P. R. Forman

Twenty-Picosecond Pulsed UV Holographic Interferometry
    of Laser-Induced Plasmas   789
    G. E. Busch

X-RAY DIAGNOSTICS

Nondispersive X-Ray Diagnostics of Short-Lived
    Plasmas   795
    R. H. Day

High Energy X-Ray Diagnostics of Short-Lived Plasmas . . .   827
    R. H. Day

Fiber Optics in X-Ray Diagnostics Applications   835
    R. H. Day

Flash Radiography   845
    F. Jamet

3-ns Flash X-Radiography   863
    J. Chang

SPECTROSCOPY

Principles of Plasma Spectroscopy   885
    H. R. Griem

Spectroscopy of Laser-Produced Plasmas   911
    G. Tondello

ACTIVE OPTICAL TECHNIQUES

Laser Induced Fluorescence Techniques   951
    S. J. Davis

Multiphoton Techniques for the Detection of Atoms   971
    W. K. Bischel

Coherent Anti-Stokes Raman Scattering   1001
    J. J. Valentini

Thomson Scattering Diagnostic for Intense Relativistic
      Electron Beam Experiment   1023
    G. R. Allen, H. P. Davis, and J. Chang

Appendix A: Ranking of Diagnostic Techniques   1037
    A. K. Hyder, M. F. Rose, M. Kristiansen

Participants   1053

Index   I

FAST PHOTOGRAPHY

ADVANCES IN HIGH-SPEED PHOTOGRAPHY: 1972-1982

J. S. Courtney-Pratt

American Bell Inc.
Holmdel, NJ
USA

1. INTRODUCTION

The variety, range and precision of methods available for photographic recording of fast phenomena have been increasing steadily. The capabilities of the newer techniques are considered, classifying the methods by the kind of record obtained. Descriptions of experimental techniques and apparatus, and illustrations, are given in an earlier article entitled, "A Review of the Methods of High-Speed Photography," published in Reports on Progress in Physics in 1957; [1*] and in "Advances in High Speed Photography 1957-1972" published in the Proceedings of the Tenth International Congress on High Speed Photography (HSP10)[116*] and also in JSMPTE 82 167-175 (1973) [117*].

This present paper is in the nature of a survey of the limits to which the various techniques have been pressed as compared to the limits attained, or reported in the open literature, at the date of the reviews 10 and 25 years ago. There are a number of recent books and articles which also provide excellent surveys and impressive bibliographies [129-138].

---

* Reference numbers below 115 are the same as those given in the earlier review 116, 117.

2.   STREAK PHOTOGRAPHY LIMITS

Streak records with drum cameras can give a time resolution of 5×10^{-9} s [2,3]. Rotating mirror streak cameras with a single reflection [15] at present approach 10^{-9} s and may, with multiple reflections, achieve 10^{-10} s. The Schardin limit [4] for presently available rotor materials is 0.25×10^{-9} s, but this is predicted assuming a single reflection of the light beam from the rotor and can be surpassed if the camera is designed to take advantage of multiple reflections. Deflecting image converters go much further: 5×10^{-13} s [5-12,115,125,139-143] (see Table 1).

3.   SINGLE FRAME PHOTOGRAPHY

There has been a notable increase in the last ten years in the use of image converter tube cameras of many kinds, particularly for the study of laser pulse structure and for investigations of electrical breakdown.

Deflecting image converter tube cameras are now also being used for studies at UV and shorter wavelengths - though the time resolution is somewhat longer than for visible light studies [123]. Twenty-five percent of all papers presented at recent HSP Congresses involve ICTs.

Single flashes of light, bright enough for silhouette recording, can be as short as 3×10^{-14} s [118,137,138] and similarly short for reflected-light recording of small near objects, about 10^{-12} s for a field of view a meter square or more. The very short flashes just mentioned are laser flashes. The availability in many labs of picosecond laser pulses is one of the significant advances of the decade. The power in the laser flash can be very high, but it should be noted that the integrated energy in the flash is always less than the energy in the flashlamps or primary lasers that pump the laser. There has been a steady advance in the design of open sparks and of gas discharge tubes and associated equipment, though this light output is rarely briefer than $N/10^{7}$ s where N is the stored energy in joules [16-26,129-131].

Electrically driven Kerr cells can operate with an exposure time of 5×10^{-10} s. Optically driven Kerr cells can give exposures of a few picoseconds [32]. Simple image converter tubes

Table 1.   Characteristics of streak cameras.

Streak Cameras	Time resolution, s		
	<1957	<1972	<1982
Streak record, drum camera	10^{-7}	5×10^{-9}	5×10^{-9}
Streak record, rotating mirror, single rotor	10^{-8}	2×10^{-9}	10^{-9}
Streak record, rotating mirror, multiple reflection	0.25×10^{-9}	10^{-10}	10^{-10}
"Schardin Limit"	0.25×10^{-9}	0.25×10^{-9}	0.25×10^{-9}
Deflecting image converter tubes	$10^{10}-10^{-11}$	$(2 \text{ or } 3) \times 10^{-12}$	5×10^{-13}

can work at 10^{-10} s and with greater light transmission than Kerr cells [7,30,31].

Image tubes with collimating microchannel plates can allow recording at 10^{-12} s. This seems to be one of the most significant new developments in electronic cameras [121].

Table 2 summarizes the methods for taking single shots, and the exposure times possible.

4.   MULTIPLE EXPOSURE PHOTOGRAPHY

Several pictures can easily be taken at short intervals superimposed on the one plate [33-35]. With conventional flash or spark equipment the maximum rate and length of sequence depend on power dissipated and the deionization time of the spark gap or series quenching gap. Minimum time between flashes is rarely shorter than $N/10^4$ s where N is the stored energy in joules/ flash. Image separation is achieved by movement of the object itself. Strobe equipment can operate up to 300,000 p.p.s.

Table 2. Methods and times of taking single shots

Single exposures	Exposure time, s		
	<1957	<1972	<1982
Single flashes of light, silhouette recording	10^{-8}	10^{-11}	3×10^{-14}
Single flashes of light, reflected light, near objects	10^{-7}	2×10^{-10}	3×10^{-14}
Single flashes of light, reflected light, 1 m^2	10^{-6}	$1\text{-}10 \times 10^{-9}$	10^{-12}
Kerr cells, electrically driven	10^{-8}	5×10^{-10}	5×10^{-10}
Kerr cells, optically driven	-	5×10^{-12}	5×10^{-12}
Simple image converter tubes	10^{-9}	$10^{-10} - 10^{-9}$	10^{-10}
ICTs with collimating microchannel plates	-	-	10^{-12}

(pictures per second). Repetitive laser flashes can be much briefer - down to fractional picoseconds, and repetition rates may exceed 10^8 p.p.s.

Frames may be separated by intermittent film movement, at rates less than 600 p.p.s.; or, for speeds up to 10^5 p.p.s., by using continuously moving film and short exposures [36,37]. Short sequences of separate pictures have been taken at rates above 10^8 p.p.s. using a repetitively pulsed laser to give exposure times of a few picoseconds and separation of the frames by using an image converter tube in the streak mode [7,13]. Where exposures are a significant proportion of the interframe interval, one must use some form of image movement compensation. Rotating prism cameras continue to be used by more people than any other class of high-speed camera. The classic work was done

by John Waddell and W. Herriott. There are now many makes on the
market. Maximum rates for full frame records have not increased
much, but spatial resolution, in line pairs per frame height, has
doubled [39,135]. With smaller pictures of lower resolution,
higher speeds are possible. Polaroid has brought out a framing
camera for rates up to 300 p.p.s. and with 'instant' processing.

For higher speeds still, one uses effectively separate
cameras exposed in succession by mechanical means (such as a
rotating slotted disc) up to 3×10^5 p.p.s. [40,41] or by optical means (such as the use of a rotating mirror) up to 5×10^7
p.p.s. [42-54,136]. There are designs of such rotating mirror
framing cameras from several countries: England, France, Japan,
USA and USSR. Electronic means (such as phased shutters or
phased spark sources) allow rates up to 10^8 p.p.s. [55-60], as
for example in the famous Cranz-Schardin method. Multiple exposures can also be recorded with an image converter tube at rates
up to 6×10^8 p.p.s., as for example with the IMACON camera [61,
64].

Table 3 summarizes methods of multiple exposure recording
with repetition rates and exposure times per frame.

5. MISCELLANEOUS TECHNIQUES

Comparably high (and even higher) speeds, with less elabo-
rate equipment, are made possible by image dissection [65,71].
See Table 4. Simple dissection plates, with clear lines or holes
in an opaque ground plate, allow recording rates of 10^5 to 10^6
p.p.s., but with low throughput of light [66]. Using a simple
slit dissection plate and a rotating mirror camera to traverse
the image elements, rates of 10^8 p.p.s. have been achieved for
brilliant objects [67-69]. Dissection by means of lenticular
plates allows in some cases, a considerably greater throughput of
light [70,71]. Aperture scanning cameras have been built that
can take 3000 pictures at 10^6 p.p.s. [72]. Lenticular plate
image dissection with mechanical plate traverse, has been applied
to cinemicrography so that a series of 300 pictures can easily be
taken at 10^5 p.p.s., at magnifications up to 2000 x [73-73]. Al-
ternatively, the use of fiber light guides for dissection allows
long series of pictures, at low resolution, at 10^5 p.p.s. [76] or
shorter series at 10^6 p.p.s. [77]. Using tapered SELFOC fibers

Table 3. Methods of multiple exposure recording.

Methods	<1957 Exposure time,s (t)	<1957 Repetition rate,p.p.s. (n)	<1972 Exposure time,s (t)	<1972 Repetition rate,p.p.s. (n)	<1982 Exposure time,s (t)	<1982 Repetition rate,p.p.s. (n)
Repetitive flashes, electrical discharges, superposed pictures	10^{-7}	2.5×10^4	5×10^{-9}	3×10^5	5×10^{-9}	3×10^5
Repetitive flashes, laser light	-	-	$(2 \text{ or } 3) \times 10^{-12}$	10^8	3×10^{-14}	10^8
Frame separation by intermittent film movement	$1/2n - 1/100n$	300	$1/2n - 1/100n$	600	$1/2n \text{ to } 1/100n$	600
Frame separation with image motion compensation by rotating prism	$1/n - 1/100n$	10^4	$1/n - 1/100n$ (but resolution has doubled)	10^4	$1/n \text{ to } 1/100n$	10^4
Frame separation by continuous film movement with short exposures	$1/100n > t > 10^{-7}$	10^4	$1/100n > t > 10^{-11}$	10^5	$1/100n > t > 3.10^{-14}$	10^6
Frame separation by pulsed laser illumination and image converter tube continuous sweep	-	-	$(2 \text{ or } 3) \times 10^{-12}$	10^8	3×10^{-14}	10^{10}
'Separate cameras,' or separate lenses, and rotating slotted disc	$10/n > t > 10^{-6}$	10^5	$10/n > t > 10^{-6}$	3×10^5	$10/n > t > 10^{-6}$	3×10^5
'Separate cameras' and rotating mirrors	$1/n - 1/10n$	10^7	$1/n - 1/10n$	5×10^7	$1/n \text{ to } 1/10n$	5×10^7
'Separate cameras' and phased shutters	$1/n > t > 10^{-8}$	10^7	$1/n > t > 10^{-9}$	10^8	$1/n > t > 10^{-9}$	10^8
'Separate cameras' and phased spark sources	$1/n > t > 10^{-7}$	10^7	$1/n > t > 5 \times 10^{-9}$	10^8	$1/n > t > 5 \times 10^{-9}$	10^8
Repetitively pulsed and swept image converter tube cameras	$1/3n > t > 5 \times 10^{-7}$	2.5×10^5	$1/3n > t > 5 \times 10^{-10}$	6×10^8	$1/3n > t > 5 \times 10^{-10}$	6×10^8

Table 4.   Image dissection cameras.

Type	Repetition rate, p.p.s.		
	<1957	<1972	<1982
Clear lines or holes in opaque plate	$10^5 - 10^6$	$10^5 - 10^6$	10^6
Slit plate and rotating mirror	10^8	10^8	10^8
Lenticular plate, aperture scanning	2.5×10^5	10^6	10^6
Lenticular plate with mechanical traverse (including cinemicrography up to 2000 X)	10^5	10^5	10^5
Fiberoptics dissection, long series, low resolution	10^5	10^5	10^5
Fiberoptics, short series, resolution of -200 samples per frame width	-	10^6	10^{10}
Lenticular plate, rotating mirror,	-10^8	10^9	10^9
Lenticular plate or slit plate and image converter tube	$>10^9$	$>10^9$	$10^{10}-10^{12}$

to give small spots, rates $\sim 10^{10}$ p.p.s. are claimed [122].   In the U.S.S.R., image dissection cameras have been extensively developed and models are available to take pictures at rates exceeding 10^9 p.p.s. [78-84].   The combination of image dissection and deflecting image converter provides means for taking a series of 50 or more pictures at rates in excess of 10^{10} p.p.s. and with a direct possibility of an increase to 10^{12} p.p.s. [84,85,144].

Schlieren studies have continued steadily. There is a good bibliography in Reference [126]. The introduction of lasers has allowed one to take interferograms either directly or using holographic techniques with short exposures and at short intervals [86-95]. Again, because of their monochromatic nature, one can study detail of strongly self-luminous events [96]. X-rays can be used at nanosecond exposures and submicrosecond intervals for penetration of several inches of steel or or to take records through smoke or flash [97-106,146]. Picosecond flashes are now possible [145]. This reference provides a good survey of the characteristics of flash x-ray sources [145]. X-ray streak cameras can achieve subpicosecond resolution [123]. X-ray microscopes have a resolution of a few microns [147, 148]. There has been a marked increase in the number of papers dealing with x-ray studies. For example, ~20 papers will be given at HSP15. Multiple exposures with x-rays are also possible in combination with ICT [127]. β-rays provide equally short exposures for enhancement of contrast in specialized cases [107]. "Memory" tubes and intensifier tubes have widened the range of phenomena that can be studied, eased synchronization problems, and lowered the requirements for provision of adequate illumination of brief events [108-112]. The use of intensifier tubes (single stage, multistage and microchannel plate), has become much more widespread in recent years [119,141,142].

Some years ago, an experimental arrangement using the time of flight of a light pulse, was proposed to provide interframe intervals in the picosecond range. Similar methods are proliferating [116,124]. Particularly noteworthy is the holographic time of flight recorded by Nils Abramson at 30×10^9 p.p.s. [128]. With the shorter pulses described by C. V. Shank [118] one could increase even this to $\sim 10^{13}$ p.p.s.

A few of these more uncommon methods are listed in Table 5.

One of the most notable changes in the field of high speed recording in the last decade has been the introduction and rapid spread of video techniques, using both video tapes and video discs. There is a good discussion of the relative merits of video and film in Reference [157]. Resolutions that are customary are ~128, ~256, ~512, and up to 1000 scan lines/frame. Frame rates vary from 30 p.p.s., at up to 1000 line resolution, to about 2000 p.p.s., with a resolution of ~200 lines (or several times faster with fractional frame size). Exposure times with coupled flash can be a microsecond [120].

Table 5. Miscellaneous methods.

	Exposure time, s		
	<1957	<1972	<1982
X-rays for penetration of several inches of steel	10^{-6}	10^{-8}	10^{-8}
X-rays for penetration of flash and smoke	10^{-7}	10^{-9}	10^{-11}
B-rays; or superradiance in the visible	-	10^{-9}	10^{-9}
Intensifiers, memory tubes, holographic techniques, etc.: Rates and times as described but allowing greater ease in recording at low light levels, simplification of synchronization, and the recording of "before and after" interferograms.			

Using one or another of the photographic or video techniques, photographs may be taken of most macroscopic phenomena that are not too remote, though synchronization problems are sometimes of overriding significance. The difficulties are greater when the interest is in fine spatial detail. These problems are accentuated in cinemicrography where the image moves many times faster than the event [113,114,133,134]. In such cases, one must take advantage of the most advanced techniques, and there is urgent need for the continued development of better methods.

6.   RECENT RESULTS

Attention is drawn to the recently published Proceedings of the 14th International Congress on High Speed Photography and Photonics which was held from October 19-24, 1980, in Moscow. Apart from the contributed papers, there are 15 invited papers which provide extensive surveys of the different branches of high speed photography. The six papers by the Russian authors should be emphasized.

The opening paper by B. M. Stepanov [149] has a particularly interesting description of a variety of infrared recording techniques. He also draws attention to the extensive use of image converter tubes and points out that individual photoelectrons are recordable and that fast ICTs are, for all practical purposes, noiseless (in contrast to photomultipliers), as the statistical likelihood of a dark current electron occuring from any given image element, within the time of the recording, is negligible. He says that the time resolution of an image converter tube, with a circular time base, can be as small as 3×10^{-14} s. This claim is repeated by S. D. Fanchenko [15]. Indeed, Fanchenko states his present estimate of the limit of resolution of a swept image converter tube will be 10^{-14} s, as he predicted in 1956.

The Proceedings of the 15th International Congress on High Speed Photography and Photonics have recently been published by the SPIE under the able editorship of Lincoln L. Endelman. There are about a dozen invited review papers in addition to a large number of contributed papers. There is a good survey by H. F. Swift [151] with an extensive discussion of the many advances in the whole range of high-speed cameras.

In the paper by Schelev [152] there is again a brief description of the Russian interest in infrared recording.

There is an excellent review by M. Hugenschmidt [153] of high speed photography in the Federal Republic of Germany.

A noticeable advance in videography is the availability from NAC of color videography at 200 frames/second [154].

Dr. H. E. Edgerton has presented a brief description of a wide variety of flash sources [155].

In closing, I might mention a new use of high speed photography, not used to determine the change of a scene as time passes, but to record the depth structure of an image, by recording the difference in the time of return of light reflected from that structure [156].

I am sure there are many other contributions of outstanding merit that by inadvertence I have missed. I would be grateful to hear of them from those working in the field.

REFERENCES

HSPN is used throughout the references as an abbreviation for The Proceedings of the nth Congress on High Speed Photography. Reference numbers below 115 are the same as those given in an earlier review [116,117].

116. Courtney-Pratt, J. S., Advances in High Speed Photography 1957-1972 HSP10 pp. 59-63 (1972).

117. Courtney-Pratt, J. S., Advances in High Speed Photography 1957-1972, JSMPTE 82, pp. 167-175 (1973).

118. Shank, C. V., Private Communication. See also Proceedings of SPIE Conference on Picosecond Lasers and Applications; January 26-27, 1982.

119. Hadland, J., Haynes, K. A. F., Helbrough, K. and Huston, A. E., "Two Special Purpose Cameras Using Microchannel-plate Intensifiers, HSP 13, pp. 465-467.

120. Hyzer, W. G., "V200 Color High-speed Video System, and The Spin Physics SP-2000 Motion Analysis Systems", Photomethods pp. 40, 41 and 47, 48 (May 1981).

121. Lieber, A. J. and Sutphin, H. D., "Picosecond Framing Camera Using a Passive Microchannel Plate", Applied Optics 18, pp. 745, 746, (1979).

122. Kung-Tsu-Tung, "Applications of GRIN Fiber (SELFOC fiber) to Speed Raster Photography", HSP 13, pp. 812-817.

123. Lieber, A. J., Sutphin, H. D., Webb, C. B. and Williams, A. H., "Subpicosecond X-ray Streak Camera Development for Laser Fusion Diagnsotics", HSP 11, pp. 194-199.

124. Courtney-Pratt, J. S. and Rentzepis, P. M., "Picosecond Photography and Time Resolved Spectrography", JSMPTE 84, pp. 478-480, 1975.

125. Schiller, N. H., et al., An Ultrafast Streak Camera System, Optical Spectra, pp. 55-63 (June 1980).

126. Davies, T. P., "Schlieren Photography Short Bibliography and Review", Optics and Laser Technology, pp. 37-42 (February 1981).

127. Bracher, R. J. and Huston, A. E., "High Speed Cineradiography of Projectiles", HSP12, pp. 532, 533.

128. Abramson, N., "Light in Flight Recording by Holography," SPIE 1980 and SHPP Newsletter 1 [1], 11 (1981).

129. Früngel, F., High Speed Pulse Techniques Volumes I to IV, Academic Press (1965-1980).

130. Edgerton, H. E. and Killian, J. R., Moments of Vision, M.I.T. Press (1979).

131. Edgerton, H. E., Electronic Flash, Strobe, 2nd Edition, M.I.T. Press (1979).

132. Dubovik, A., The Photographic Recording of High Speed Processes, John Wiley & Sons (1981).

133. Schall, R., "High Speed Photomicrography", HSPP Newsletter 1 [1], 10 (1981).

134. Chaudhri, M. M., "High Speed Photomicrography and Electron Microscopy", Photomethods 24, [11], 45 (1981).

135. Hyzer, W. G., "Techniques and Applications of Intermittent and Rotating Prism Cameras," HSP13, pp. 49-56.

136. Nebeker, S. J., "Techniques and Applications of Rotating Mirror Cameras," HSP13, pp. 57-67.

137. Edgerton, H. E., "Techniques and Applications of Xenon Flash", HSP13, pp. 27-36.

138. Früngel, F., "Repetitive Submicrosecond Light and X-ray Flash Techniques"..., HSP13, pp. 37-48.

139. Yamanaka, C. and Yamanaka, T., "Development of Picosecond Photonics for Laser Fusion Research," HSP13, pp. 123-129.

140. Bradley, D. J., "Recent Developments in Picosecond Photochronoscopy," HSP13, pp. 130-141.

141. Schelev, M. Y., "Image Converter Diagnostics of Laser and Laser Plasma in Pico-femto Second Region," HSP13, pp. 142-150.

142. Hadland, R., Techniques and Applications of Image Converter Tubes," HSP13, pp. 151-161.

143. Niu, H., Sibbett, W. and Baggs, M. R., "Theoretical Evaluation of the Temporal and Spatial Resolutions of Photochron Streak Image Tubes," Rev. Sci. Instrum. 53, pp. 563-569 (1982).

144. Niu, H., Chao, T. and Sibbett, W., "Picosecond Framing Technique Using a Conventional Streak Camera," Rev. Sci. Instrum. 52, pp. 1190, 1191 (1981).

145. Nagel, D. J. and Dozier, C. M., "Characteristics of Flash X-ray Sources", HSP12, pp. 132-139.

146. "Livermore Flash-Radiography Facility," Physics Today 35, 21 (June 1982).

147. Fleurot, N., Gex, J. P., Lamy, M., Quinnessiere and Sauneuf, R., HSP12, pp. 200-206.

148. Attwood, D. T., "Time Resolved X-ray Pinhole Photography of Compressed Laser Fusion Targets," HSP12, pp. 325-332.

149. Stepanov, B. M., "Physics and Techniques of Transient Event Measurement and Recording," HSP14, pp. 17-24.

150. Fanchenko, S. D., "Electron-Optical Photography Trends," HSP14, pp. 24-37.

151. Swift, H. F., "Current and Future Activities in High Speed Photography and Photonics," HSP15, pp. 8-14.

152. Schelev, M. Y., "New Trends in Picosecond Photonics," HSP15, pp. 75-82.

153. Hugenschmidt, M., "High Speed Photography in the Federal Republic of Germany," HSP15, pp. 59-66.

154. Yamamoto, T., "Advanced Techniques in Applications of High Speed Videography," HSP15, pp. 353-358.

155. Edgerton, H. E., "Exposure Time: It Can Be Important," HSP15, pp. 67-76.

156. Courtney-Pratt, J. S., "High Speed Photography to Provide a Three-Dimensional View," HSP15, pp. 254-259.

157. Photomethods, 25, 19-30 (1982).

RECENT TECHNIQUES FOR HIGH-SPEED PHOTOGRAPHY AND LOW-LEVEL IMAGE RECORDING

W. Pfeiffer

Technische Hochschule
6100 Darmstadt, Schloßgraben 1,
FRG

1. INTRODUCTION

As transient physical phenomena generally are connected with radiation, they may be analyzed by means of high-speed-photography in the nanosecond or, with some restrictions, even in the pico-second range. The most important requirements which have to be met by high-speed cameras are short exposure time, high spatial resolution, low geometrical distortion, wide spectral input sensitivity and precise triggering with short delay. The shutter ratio becomes especially important for recording transients with extremely high dynamics. Generally, short exposure times and/or weak input luminosity will require additional amplification of the recorded images.

Two important innovations have greatly influenced the field of high-speed-photography. One of these is the development of gatable image intensifiers with proximity focus. Compared with former tubes, used for high-speed streak or framing photography, these have zero distortion, high resolution and high gain. Also, an extraordinary high shutter ratio may be achieved. The fact that only a single frame is possible with such tubes seems to be only a minor problem.

The second improvement has become possible through the use of low-light level (LLL) video techniques. Using these tech- niques the output luminosity of the gated image intensifiers can be recorded with high sensitivity and evaluated with high pre- cision. However, for use in scientific fields, the standard LLL- cameras should be modified. Low temperature operation and slow- scan read out will allow many improvements. Additionally, the

silicon-intensified-target (SIT) vidicons should be replaced by
proximity-focus image intensifiers and solid state image devices.

2.   EXPERIMENTAL METHODS

2.1   Requirements for High-Speed Photography in Insulation
      Research.

      Pressurized insulating gases with high specific electrical
strength are increasingly used [1].   Under high stress, predis-
charges may develop, within small gas volumes, with extraordinary
high speed (Fig. 1).   Typical for these processes, especially in
the streamer phase, are luminosity variations of many orders of
magnitude within some nanoseconds (Fig. 2) [2].

      The very first predischarge phenomena are of special inter-
est but are somewhat obscured since breakdown occurs some 10 ns
later and usually cannot be prevented.   Consequently, the shutter
ratio is of extreme importance.   Additionally, high time resolution is necessary since the luminous phenomena have to be correlated precisely with the electrical phenomena.

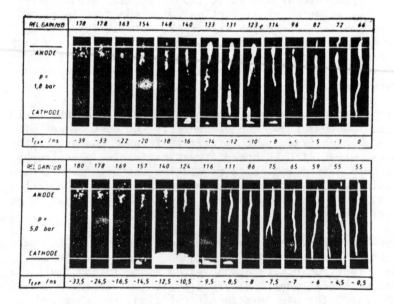

Fig. 1.   Spatial and temporal resolved predischarge phenomena in
          compressed nitrogen.

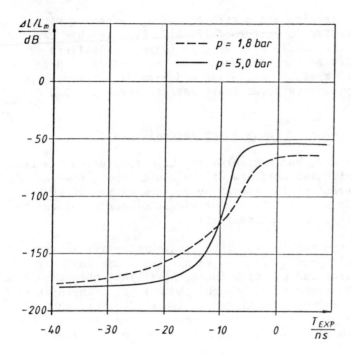

Fig. 2.   Temporal resolved luminosity during the predischarge
           development.

2.2 Selection of Suitable High-Speed Photographic Methods.

The highest time resolution, in the ps-range, is possible
by means of laser techniques (striation or interference) [3].
These methods, however, are not suitable for processes with high
dynamics. The techniques employed must have a very high shutter
ratio and high luminous gain [4].

With streak techniques, a very high time resolution can be
achieved. However, in our case, the dynamic range of a streak
camera will be exceeded for streak lengths of a few ns. This
would also be true for framing cameras, which usually have some
ns framing capability [6]. Thus, in each case, the complete
phenomenon has to be reconstructed from several separate measurements. Additionally, for spatially scattering processes, it is
rather troublesome to focus the phenomena on the input slit of a
streak-camera. A further disadvantage of streak cameras is the
fact that they do not allow spatial and spectral resolution.
Moreover, in the ns-range and below, both conventional high-speed
streak and framing cameras will have rather poor image reproduc-
tion.

Much better results are possible if single-framing with
minimal exposure time is performed [7, 8]. Even in this case
multiple frames are possible if several image intensifiers or
several projection systems are used. In the ns-range single-
framing with high spatial resolution and zero distortion is possible if gated proximity-focus image intensifiers are used [9].

2.3 Gating of Proximity Focus Image Intensifiers.

Low gain ($\sim$ 20), but extraordinary high resolution, is characteristic of image intensifiers of the diode type. However, as
in the gated operation, the operating voltage of 5 - 10 kV has
to be pulsed. These tubes have most recently been displaced by
microchannel plate image intensifiers.

Meanwhile, the technology of the image intensifier diodes
has been greatly improved (Fig. 3, Proxifier). Due to their
simple construction and the high focussing fields, they have some
advantages in the field of high-speed photography. The highest
resolution ($\sim$ 60 lp/mm) is achieved if tubes with very short
distance between photocathode and luminous screen are available
(1,5 mm). For these tubes, a time resolution in the nanosecond

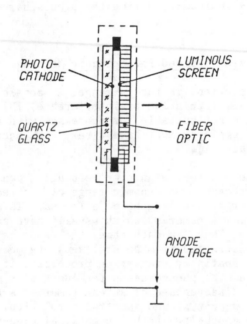

Fig. 3.   Image intensifier diode (PROXIFIER).

range is possible since their operating voltage, which has to
be pulsed, is below 10 kV. Of course, the generation of steep
high voltage pulses across a 30 - 40 pF load is a severe problem
(Fig. 4), especially if low trigger delay and negligible jitter
are required. However, with transistorized Marx banks, these
conditions can be met for gating times on the order of 5 ns
(Fig. 5) [10]. The trigger delay is below 20 ns with negligible
jitter.

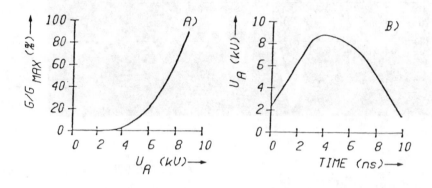

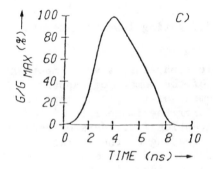

Fig. 4.   Gating of image intensifier diodes
          a) Transfer characteristic, b) Gating impulse,
          c) Opening characteristic.

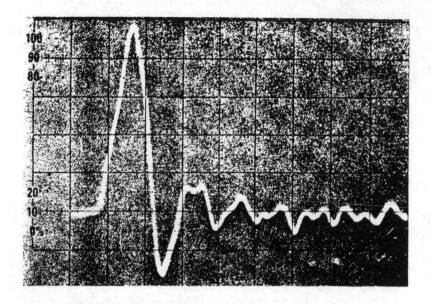

Y: 1.4 kU/DIU
T: 10 ns/DIU

Fig. 5.   Output voltage of the loaded gating impulse generator.

For all high-speed photographic apparatus, a very important requirement is the exact knowledge of the moment of exposure. This demands a very precise measurement of the gating impulse directly at the gating electrodes. For the image intensifier diode this may be a problem, as conventional dividers of very short rise time may cause excessive loading of the gating impulse generator. Additionally, for this and some other reasons, it would be very desirable to have a continuously coaxial arrangement (Fig. 6). In this case, an integrated, coaxial capacitive divider can be used [11].

Without additional amplification such a high-speed camera will allow measurements in the early spark phase. For 5 ns gating a dynamic resolution of more than 28 1p/mm can be achieved (Fig. 7), which seems to be limited by the measuring arrangement. This is far better than the performance of any other high-speed camera [12]. Of course, the effective amplification can be increased, if two further, non gated, image intensifier diodes are

Fig. 6a. Nanosecond gated image intensifier diode (NANOGATE)
          gatable image intensifier diode with coaxial elec-
          trodes.

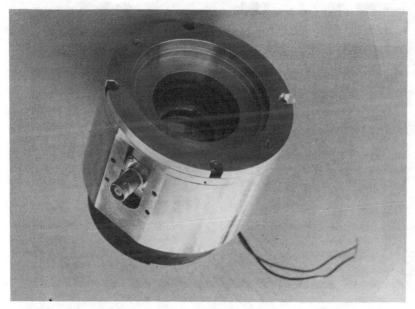

Fig. 6b. Nanosecond gated image intensifier diode (NANOGATE),
          image intensifier housing.

Fig. 6c.   Nanosecond gated image intensifier diode (NANOGATE).

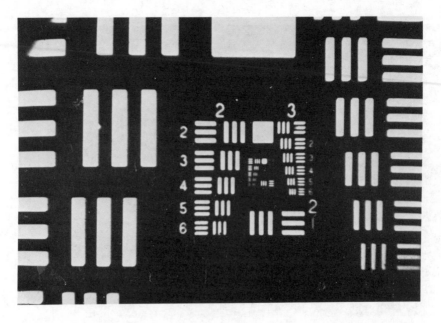

Fig. 7.   Dynamic resolution of a 5 ns gated image intensifier
          diode (limiting resolution 4/6 = 28 linepairs per
          millimeter).

cascaded (Fig. 8). The total resolution is slightly degraded by
this but, with more than 20 lp/mm, is still superior to any other
system.

As already mentioned, the shutter ratio of gated image intensifiers is of great importance as it will limit the dynamics
which can be allowed for the phenomena under investigation.
Presently this is better than 10^6 (Fig. 9) and further improvements are going on. An additional benefit of image intensifier
diodes with high operating fields is the field enhanced sensitivity in the long wavelength region. As a result, a very homogeneous input characteristic between 200 ... 800 nm is achieved
(Fig. 10).

High gain and multiple gating capabilities are the reasons
that image intensifiers with microchannel plates are mainly used
in the field of high-speed single-framing photography (Fig. 11)
[13]. Due to double proximity focussing and the limited resolution of the microchannel-plate itself, the maximal resolution is

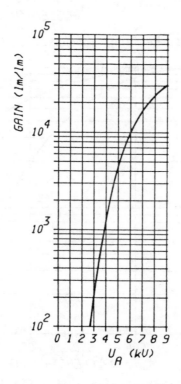

Fig. 8. Luminous gain of a three-stage image intensifier cascade.

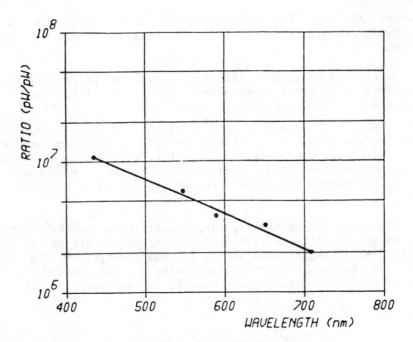

Fig. 9.   Shutter ratio of the image intensifier diode.

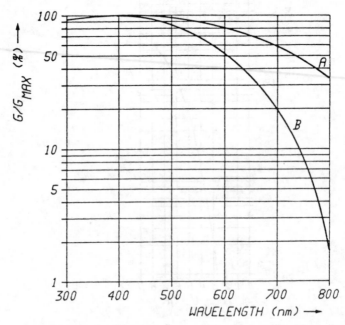

Fig. 10.   Spectral input characteristic of a field intensified
           photo-cathode (A) compared with a standard S20 charac-
           teristic (B).

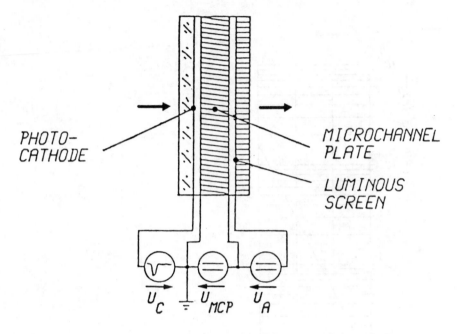

Fig. 11. Image intensifier with microchannel plate (MCP).

in the order of 25 lp/mm. The luminous gain of such tubes is better than 10^4 (Fig. 12). It is also possible to use cascaded microchannelplates (Fig. 13). In this case the luminous gain will be on the order of 10^6 and single photon sensitivity may be achieved. The resolution, however, will only be 15 lp/mm.

Generally MCP-tubes can be gated in three different ways [14]. However, it is not useful to pulse the high voltage (5 kV) between MCP-output and luminous screen. Neglecting the problems of generating such pulses, the main disadvantage is possible saturation of the non gated MCP. This will not occur if the MCP is gated. However, it may not be a problem if, for gating in the nanosecond range, the correct voltage distribution in the channel itself can be reached. Additionally, the high input capacitance of the MCP will cause severe loading of the pulser. Across the MCP the precise voltage distribution is a very severe problem since the gain is proportional to the 10th power of the voltage. Therefore, for fast gating, hollow effects are likely to occur.

The most important benefit of the MCP-tubes becomes effective if they are gated with a rather low voltage between photocathode and MCP-input (Fig. 14). About 50 V are sufficient for 50% gating. The generation of such low amplitude pulses across

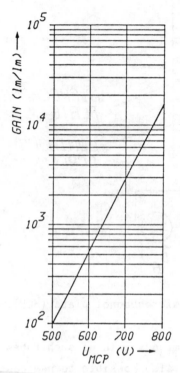

Fig. 12.   Luminous gain of a MCP-tube.

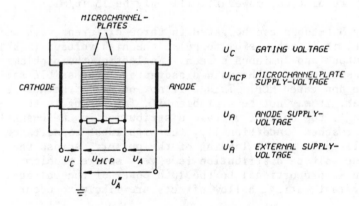

Fig. 13.   Dual microchannel plate image intensifier tube.

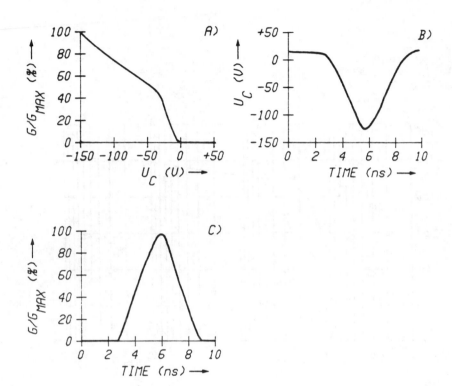

Fig. 14.   Gating of image intensifiers with MCP,
           a) transfer characteristic, b) gating impulse,
           c) opening characteristic.

a 30 ... 50 pF load is rather simple. However, due to the typical gating characteristic of such tubes, one has to be very careful to avoid multiple exposure. As it is nearly impossible to
have a truly aperiodic pulse form, a dc-blocking of the tube with
20 ... 50 V bias is inevitable. In addition, it has to be kept
in mind that gating-on the tube, which will only require some few
volts (Fig. 15), will not provide good resolution. Good resolution will require at least 30 V. Therefore, for short exposure
times, good resolution will only be obtained if extraordinary
fast rising pulses are used. Otherwise, especially in case of
rather low amplitude gating pulses, the dynamic resolution may be
reduced drastically (< 4 lp/mm at 4 ns gate) [15]. In this case,
the spatial inhomogeneity of the exposure will also become ap-
parant.

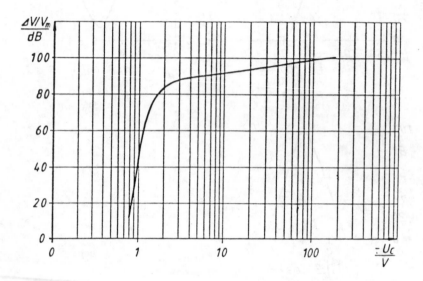

Fig. 15. Gating characteristic of the dual MCP-tube.

Two effects will mainly limit the shortest exposure times.
Within the tube, the propagation of the gating pulse is limited
by the distributed RC behavior in the active region. With approximately 10 pF and surface resistances below 50 $\Omega/\square$, exposure
times of 1 ns should be possible. Additionally, the voltage at
the tube is influenced by a total capacitance on the order of
30 pF and the output impedance of the pulser. The pulser impedance should be very low, less than 10 Ω.

For ultrafast-gating, the measurement of the gating pulse
directly across the tube is of great importance. However, for
low amplitudes, this is not difficult, even for 1 ns gating (Fig.
16). The pulsers used here have rather short trigger delays.
They are generally less than 5 ns, where the jitter can be neglected (< 50 ps).

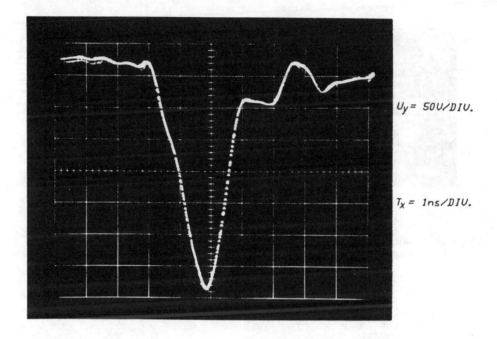

$U_y = 50 U/DIV.$

$T_x = 1 ns/DIV.$

Fig. 16.   Gating impulse at the image intensifier (dual MCP).

With gated single, MCP-tubes early predischarge phenomena
can be investigated. Gated, dual MCP-tubes will provide single
photon sensitivity (Fig. 17). By low temperature operation of
such image recording systems (Fig. 18), the transfer characteristics can be further improved. However, we should realize that,
for such high gains and the very short exposure time, the number
of recorded photons per mm^2 may be very small. Of course this
will severely degrade the effective resolution (Fig. 19). In
case of sufficient input photons, a single MCP-tube with 5 ns
gate will provide a resolution of 20 lp/mm (Fig. 20) and a dual
MCP-tube with 1 ns gate will have a resolution of about 10 lp/mm
(Fig. 21). A very important feature of gated MCP-tubes is the
extraordinarily high shutter ratio which may be up to 10^8 (Fig.
22).

In summary, we see that gated MCP-tubes are presently a very
effective tool in the field of high speed framing photography
which can be simply applied. However, in some cases, image intensifier diodes may also be very advantageous.

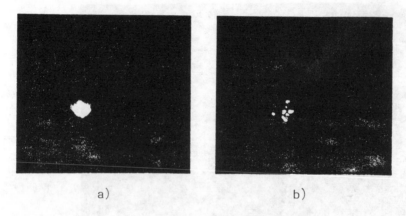

a)   b)

Fig. 17.   Demonstration of single-photon-sensitivity of the high
            speed image recording system.

            a) response for 10^8 photons/frame at the input,
            b) response for 40 photons/frame at the input; accord-
               ing to the quantum efficiency ($\sim$ 20 %) 8 luminous
               spots are recorded.

Fig. 18.   Low temperature operated image recording system.

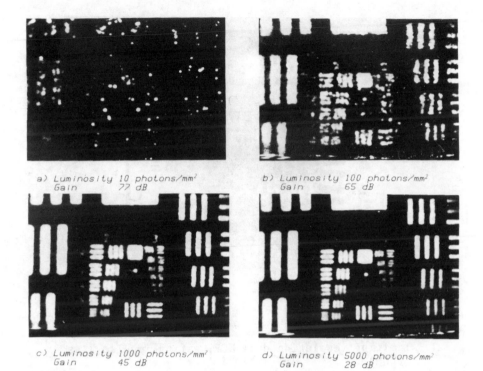

a) Luminosity 10 photons/mm²
   Gain   77 dB

b) Luminosity 100 photons/mm²
   Gain   65 dB

c) Luminosity 1000 photons/mm²
   Gain   45 dB

d) Luminosity 5000 photons/mm²
   Gain   28 dB

Fig. 19.   Dynamic resolution of a 1 ns gated dual MCP intensifier
           for different luminosity and gain.

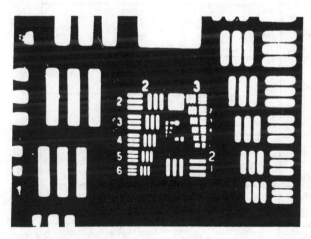

Fig. 20.   Dynamic Resolution of a 5 ns gated MCP intensifier
           (limiting resolution 4/1 = 16 linepairs per milli-
           meter).

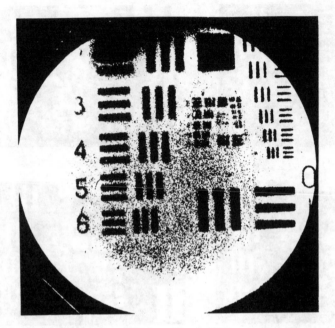

Fig. 21.   Limiting resolution dual MCP, gating impulse 1 ns
(FWHM).

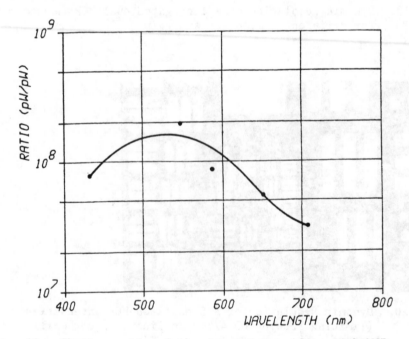

Fig. 22.   Shutter ratio of the image intensifier with MCP.

2.4 Image Recording and Evaluation.

The weak luminous phenomenon at the output of the nanosecond gated image intensifiers usually has to be amplified before it can be recorded easily. Both requirements can be met with video techniques. High sensivity is attained if low-light-level video cameras are used. Since the vidicons can act as an intermediate store for the very short afterglow of the luminous screen of the image intensifier, the recording can be done using video techniques. Either video tapes or video frame stores can now act as a permanent storage media (Fig. 23). Additionally, the video-recording allows rather precise measurement techniques and, with some preconditions, the amplitude of the video signal can be

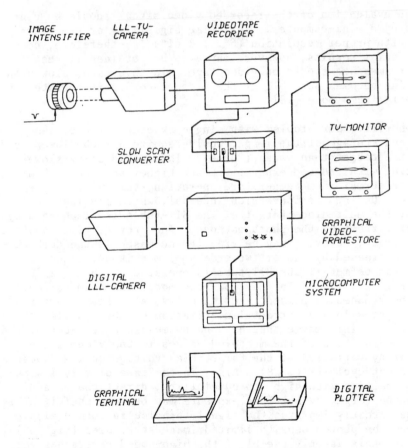

Fig. 23. Image processing and evaluation.

calibrated with respect to the input luminosity [16]. The transfer characteristics of such a system can be corrected by a mini-
computer which allows further image processing (Fig. 24) and
multiple displays (Fig. 25).

Presently, most of the low-light-level cameras are operated
with silicon-intensified-target (SIT)-vidicons (Fig. 26) [17].
Even ISIT-cameras have been proposed, where an additional image
intensifier of the 1st generation is inserted. The benefit of
the 10 times higher sensitivity, however, is offset by lower resolution and severly increased geometrical distortion. Therefore,
such ISIT-cameras are not regarded as useful. In our field of
application of a single, rather short exposure, it must be considered that the transfer characteristics of a SIT-camera may
greatly differ from that for continuous input illumination.

The evaluation of the recorded video signal should be controlled by a mini-computer. Real time digitizing of the video
signal with proper resolution (6 ... 8 bit) is, therefore, necessary. Useful results, however, may also be obtained if the
single frames are stored on video tape. In this case, slow-scan
digitizing becomes possible if the video recorder is operated in
the single-frame mode [18].

Generally, our problem with single exposure of the video
camera and storage of the image, would require a much slower read
out of the target than usual (20 ms). This would allow slower
digitizing with reduced expenditure and higher accuracy. However, for normal room temperature operation, this will not be
possible. During the conversion interval, which may last some
seconds, the multidiode target of the SIT-tubes is discharged by
the dark-current of the photo cathode. The storage time constant
of the target, however, may be greatly increased if the dark current is decreased by one or two orders of magnitude. These conditions can be met if the SIT-tube is operated at about -20°C to
-30°C (Fig. 27) [19]. In the slow-scan mode, the beam current
has to be reduced or pulsed. Additionally, slow-scan operation
of SIT-tubes allows a perfect linearization of the transfer characteristics. This may either be done by multiple read-out of the
target and addition of the measured values in the video frame
store, or by variation of the target bias voltage during exposure
and read out cycle (Fig. 28). Also, in the case of very low input luminosities with high reproducibility, direct charge addition on the target by multiple exposure may be very useful. This
technique greatly improves the signal-to-noise ratio and also
overcomes the photon density limited image structures (Fig. 29).
Of course, this is only useful if the high-speed camera has very
low noise, which may require low temperature operation of the
image intensifier.

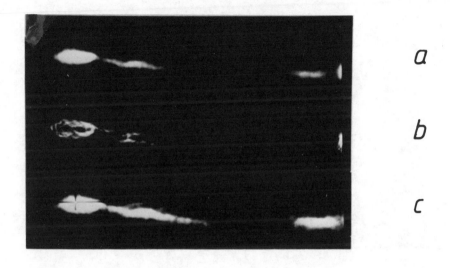

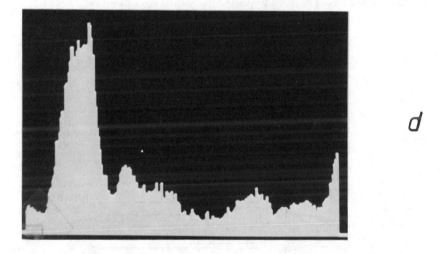

Fig. 24.   Applications of digital image processing
a) original record,
b) gradient display,
c) contrast enhancement
d) display of axial luminosity distribution
(marked line)

Fig. 25.   Spatial resolved luminosity during predischarge
           development.

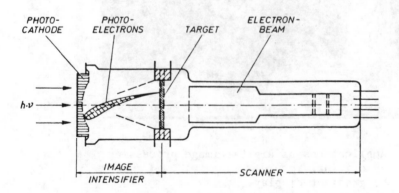

Fig. 26.   SIT - vidicon.

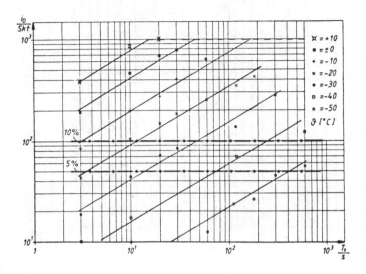

Fig. 27.   Storage time of the silicon target vidicon versus
           temperature.

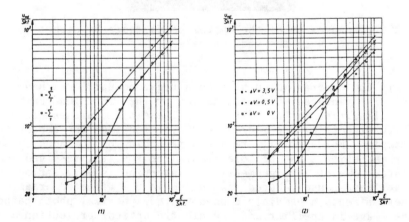

Fig. 28.   Linearization of the transfer characteristic of the
           SIT-vidicon by multiple read-out (1) or variation of
           the target voltage (2).

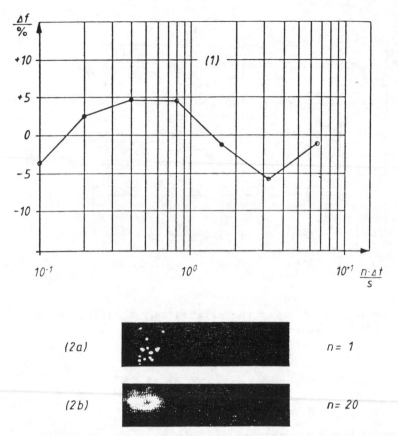

Fig. 29.   Improvement of the system performance by direct integration on the vidicon target.

     All LLL-cameras which have been regarded until now have some
common disadvantages which are caused by the first generation
image intensifier in the SIT-tubes.   These difficulties are especially vignetting and increased distortion from the center to
the margin.   For spatially resolved recording of luminous phenomena, these effects may be tolerated.   For spectral resolved measurements, however, the geometrical distortion is very serious and
will at least cause reduced resolution in the affected areas.
As these effects are mainly caused by the spherical photo cathode
of the image intensifier and the electro-optical projection
system, significant improvements are possible if a second generation image intensifier is coupled with a regular vidicon or another camera tube (newicon) (Fig. 30).   Thus, a low distortion
LLL-camera can be realized which has UV-sensivity and good resolution.

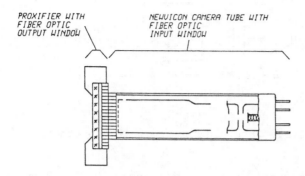

Fig. 30. Low light level camera tube with proximity focus image
          intensifier.

     Zero distortion is obtained if the vidicon is also replaced
by a solid state image device (Fig. 31). Presently the most
usual matrix elements are of the CCD (charge coupled device)
type [20]. Compared with conventional camera tubes, they have
low noise, which may be further improved by low temperature
operation. A dynamic range of better than 10^3 seems to be possible. Both control electronics and equipment for low temperature
operation are much simpler than in conventional cameras. A problem may arise from the fiber-optic coupling of image intensifier
and CCD-device as the latter generally do not provide fiber-optic
input. In each case there will be a loss of resolution as the
diameter of image intensifiers (> 18 mm) usually will greatly exceed the active area of image devices (6,6 x 8,8 mm). This could
only be improved if a tapering fiber-optic could be used. However, it seems to be more favorable to use high-aperture, tandem
optics which will provide a light transmission of about 20 %.

2.5 Experimental Set-up for the Analysis of Spontaneous Break-
     down Phenomena

     Electrical breakdown for DC or AC voltage stress is a random
process which cannot be predicted temporally. Typically, the
radiation phenomena has an extraordinarily high dynamic range.
High-speed cameras have to be triggered by the breakdown process
itself. The phenomena which are of interest from, the point of

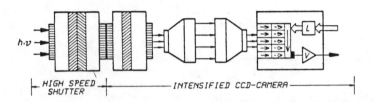

Fig. 31.   High-speed image recording system with CCD-camera.

view of insulating behavior, however, will occur 20 ... 50 ns before this event.   Therefore, the short trigger delay of the high-
speed camera alone will not be sufficient.   An optical delay path
with sufficient transit time, wide spectral transfer characteristic, and low attenuation will be necessary [21].   Such an experimental set-up will also allow multiple-frame operation if several
delay paths, with different transit times are used (Fig. 32).
With a similar arrangement, even three dimensional records, at
one moment of exposure, are possible [5].

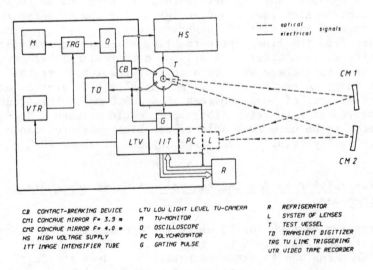

CB CONTACT-BREAKING DEVICE	LTV LOW LIGHT LEVEL TV-CAMERA	R REFRIGERATOR
CM1 CONCAVE MIRROR F= 3.9 m	M TV-MONITOR	L SYSTEM OF LENSES
CM2 CONCAVE MIRROR F= 4.0 m	O OSCILLOSCOPE	T TEST VESSEL
HS HIGH VOLTAGE SUPPLY	PC POLYCHROMATOR	TD TRANSIENT DIGITIZER
ITT IMAGE INTENSIFIER TUBE	G GATING PULSE	TRG TV LINE TRIGGERING
		VTR VIDEO TAPE RECORDER

Fig. 32.   Test arrangement for spatial or spectral resolved recording of predischarge phenomena.

An essential condition for the evaluation of the recorded optical phenomena is the exact correlation with the electrical phenomena of the experiment. The time uncertainty must be less than the exposure time and the framing interval. This will require precise voltage or current measurements [12] and a homogenous coaxial test vessel (Fig. 33). In this way double-framing records, with approximately 1 ns exposure time and 1 ns framing interval, were performed which could be correlated with the electrical phenomena with an error well below 0.5 ns.

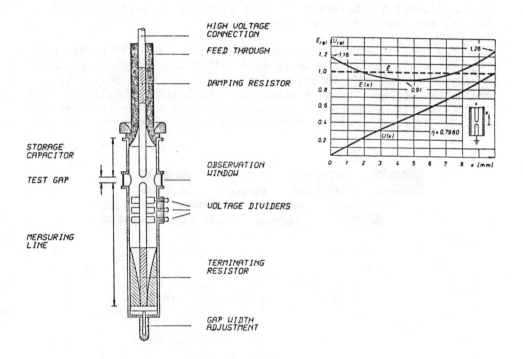

Fig. 33.   Test vessel and the axial field distribution in the test gap.

3.   APPLICATIONS

3.1 Application of the High-speed Image Recording System for the Spatial, Temporal and Spectral Resolved Analysis of Pre-breakdown Phenomena in Compressed Gases

By inserting a grating spectrograph in front of the gated image intensifier, spectral resolution can be obtained. A great advantage, when compared with optical multi-channel analysers is the fact that such a system will still provide spatial resolution in one dimension. Moreover, it is superior as far as time resolution and sensitivity are concerned.

The broad spectral input characteristic, from 270 to 680 nm, requires eight separate measurements with a bandwidth of approximately 50 nm in order to get a high resolution (Fig. 34). The evaluation of the intensity measurements, including amplitude correction according to the wavelength dependent transfer characteristic (dotted line), is performed by a mini-computer.

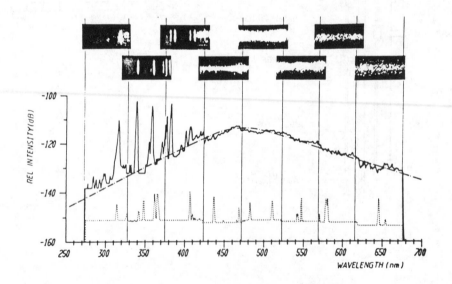

Fig. 34.   High speed spectrogram, video records from the different grating positions (straight line) and the reference spectrogram of a mercury-cadmium lamp (dotted line).

These experimental techniques require many separate measurements. In case of spatial resolved luminous phenomena, the predischarge development has to be reconstructed from different
records at different exposure times. Short time spectrograms,
even for one moment of exposure, require eight separate measurements. This means that these techniques will require some
reproducibility of the phenomena under investigation. An additional problem is the extraordinary high dynamics of the luminous
phenomena. For instance, in compressed SF_6, the dynamics does

not allow conventional high-speed techniques to be used (Fig.
35). Even for streak lengths of about 1 ns, or similar framing
intervals, the dynamic range may be exceeded during measurements
in the streamer phase. Also, during measurements of first predischarge phenomena, the shutter ratio will not be sufficient.

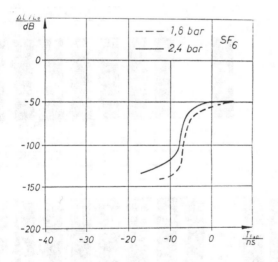

Fig. 35.   Temporal resolved luminosity during the predischarge
           development in SF_6.

The spatial and temporal resolved records of the predischarge development in SF_6 (Fig. 36) give some idea of these extreme difficulties. Due to the rather high spatial scatter of the predischarge phenomena in SF_6, no reliable high-speed spectrograms were previously available. In N_2, however, very detailed spectral resolved measurements have been possible (Fig. 37 - 38). The moments of exposure do exactly coincide with those of the luminous phenomena shown in Fig. 1. However, due to the limited aperture of the spectrograph (f = 5) and its grating efficiency ($\eta \sim 60$ %), the very first luminous phenomena could not be resolved spectrally.

4.   SUMMARY

Proximity focus image intensifiers and new video recording techniques allow two-dimensional image recording with 1 ns exposure time and single photon sensitivity. The performance can be further improved if a modified video tehcnique, combined with low temperature operation of the electro-optical components, is applied.

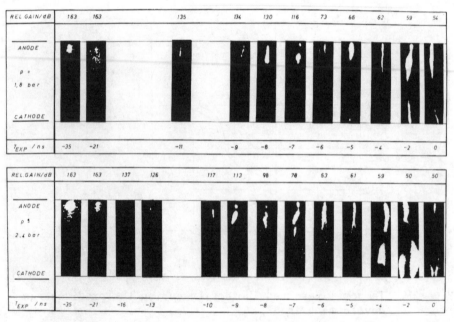

Reconstruction of predischarge development in SF_6

Fig. 36.   Reconstruction of predischarge development in SF_6.

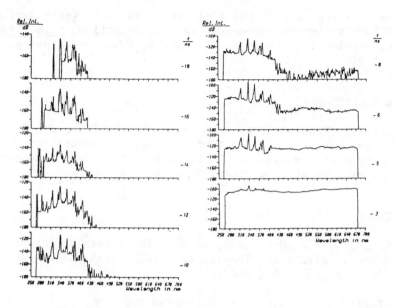

Fig. 37. Corrected N_2 spectrograms at selected exposure times, pressure p = 1,8 bar.

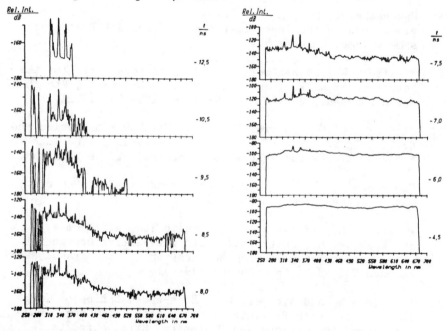

Fig. 38. Corrected N_2 spectrograms at selected exposure times, pressure p = 5 bar.

New developments in the field of solid state image devices will allow further improvements. Also a reduction of the exposure time below 1 ns seems to be possible from the technical point of view.

5.   ACKNOWLEDGEMENT

Contributions of B. Aulbach, B. Lieberoth-Leden and D. Wittmer are appreciated. Part of the work was supported by the German Research Association. Gatable Proxifier components were provided by a grant from Proxitronic Inc.

REFERENCES

1. Cooke, C. M. and Cookson, A. H., "The Nature and Practice of Gases as Electrical Insulators", IEEE Trans. Elec. Ins. Vol. EI-13, p. 239-248 (1978).

2. Pfeiffer, W., "Breakdown Mechanism and Time Development of Discharges in Compressed Insulating Gases", Nuclear Instruments a. Methods Vol. 220, p. 63-72 (1984).

3. Hugenschmidt, M., "Laser Photographic and Cinematophic Applications in the Investigations of Transient Phenomena", same volume, Chapter 5

4. Pfeiffer, W. and Wittmer, D., "Image Intensifiers for the Detection of Weak, Transient Luminous Phenomena", 9th IMEKO World Congress, Berlin, Vol V/II, p. 286-295 (1982).

5. Giesselmann, M. and Pfeiffer, W., "Flashover Mechanism and Dielectric Strength of Gas/Solid Interfaces", 1984 IEEE Int. Symp. on Electrical Insulation, Montreal, to be published.

6. Courtney-Pratt, J. S., "Research Trends in Fast Photography", Same volume, Chapter 5.

7. Price, R. H., Wiedwald, J. D., Kalibjian, R., Thomas, S. W., and Cook, W. M., "Ultrafast Gated Intensifier Design for Laser Fusion X-ray Framing Applications", IEEE Trans. Nucl. Science Vol NS-31, p. 504-508 (1984).

8. Pfeiffer, W. and Wittmer, D., "High-speed Gating of Image Intensifiers with Proximity Focus", 16th Int. Congr. on High Speed Photography and Photonics, Strasbourg, to be published.

9. Pfeiffer, W., Aulback, B., Lieberoth-Leden, B., and Wittmer, D., "Fast Photographic Diagnostics Applied to Breakdown Phenomena", 4th IEEE Pulsed Power Conference, Albuquerque, p. 282-288 (1983).

10. Pfeiffer, W. and Wittmer, D., "High Resolution Nanosecond Gated Image Intensifier Diode", 15th Int. Congr. on High Speed Photography and Photonics, San Diego, p. 260-266 (1982).

11. Pfeiffer, W., "Ultra-High-Speed Methods of Measurement for the Investigation of Breakdown Development in Gases", IEEE Trans. Instrum. a. Meas. Vol. IM-26, p. 367-372 (1977).

12. Yates, G. J., King, N. S. P., Jaramillo, S. A., Ogle, J. W., Noel, B. W. and Thayer, N. N., "Image Shutters: Gated Proximity-Focused Microchannel-Plate (MCP) Wafer Tubes Versus Gated iSlicon Intensified Target (SIT) Vidicons", 15th Int. Congr. on High Speed Photography and Photonics, San Diego, p. 422-433 (1982).

13. Pfeiffer, W., "Ultra-Kurzzeitkamera mit hoher Lichtverstärkung", Fernseh- und Kino-Technik Vol. 32, p. 467-470 (1978).

14. Lundy, A. S. and Iverson, A. E., "Ultrafast Gating of Proximity Focused Microchannel-Plate Intensifiers", 15th Int. Congr. on High Speed Photography and Photonics, San Diego, p. 178-189 (1982).

15. Yates, G. J., King, N. S. P., Jaramillo, S. A., Noel, B. W., Gobby, P. L., Aeby, I., and Detch, J. L., "Nanosecond Image Shuttering Studies at Los Alamos National Laboratory", IEEE Trans. Nucl. Science Vol. NS-31, p. 484-489 (1984).

16. Liesegang, G., and Smith P., "Vidicon Characteristics Under Continuous and Pulsed Illumination", Applied Optics Vol. 21, p. 1437-1444 (1982).

17. Crowell, M. H. and Labunda, E. F., "The Silicon Array Camera Tube", BSTJ Vol. 48, p. 1481-1528 (1969).

18. Aulbach, B., "Verfahren zur A/D-Umsetzung von Fernsehbildern", Elektronik H. 24, p. 63-68 (1982).

19. Honeycutt, R. K. and Burkhead, M. S., "Adaption of a Commercial Silicon Vidicon Detector System for Astronomical Spectroscopy", Astron. Obs. with Television-Type Sensors, Univ. British Columbia, p. 229-235 (1973).

20. Torr, M. R. and Devlin, J., "Intensified Charge Coupled
     Devices for Use as a Spaceborn Spectrographic Image-Plane
     Detector System", Applied Optics Vol. 21, p. 3091-3108
     (1982).

21. Pfeiffer, W. and Schmitz, W., "High Speed Electrical and
     Optical Investigation of Discharge Development in SF_6",
     Fifth Int. Conf. on Gas Discharges, Liverpool, IEE Publ.
     165, p. 328-331 (1978).

LASER PHOTOGRAPHIC AND CINEMATOGRAPHIC APPLICATIONS TO THE INVESTIGATION OF TRANSIENT PHENOMENA

M. Hugenschmidt

Deutsch-Französisches Forschungsinstitut
Saint-Louis,
France

1. INTRODUCTION

Due to the high peak powers and short duration pulses the range and precision in high speed photography is considerably increased by the use of lasers [1]. Besides conventional laser-photographic techniques, including shadowgraphy, schlieren-techniques or interferometry, specially developed coherent optical techniques such as holography or speckle photography are in use. Single flash exposure, as well as repetitive exposure by trains of pulses in cinematographic applications, may be adapted to any optical problem, both for silhouette recording and for reflected light recording.

In the past, solid-state lasers in various modes of operation have mainly been applied to photographic diagnostics. These lasers are still in use today, especially for holographic or picosecond photographic recording. Dye lasers, however, are frequently used as well. They allow the frequency to be precisely tuned. The various dyes available cover the whole visible spectral range from the near UV to the near IR. Furthermore, gas lasers became increasingly important. Most of these lasers are transversely excited by electrical discharges of the blumlein type. These lasers are rugged and less expensive. Moreover, they provide flexible pumping sources for dye lasers. Another relatively new, promising type of laser is the copper vapour laser which was developed for isotope separation applications. Among the numerous other diagnostic applications, these lasers can also be useful, for example, for real time holographic measurements, as they provide trains of pulses, with high repetition rates, up to several kHz and pulse halfwidths of several tens of

ns, even at relatively high mean powers of 10 to 100 W. For the
investigation of relatively small objects, with a small field of
view, semiconductor laser diodes, mostly emitting in the near IR
around 900 nm, may provide a proper light source for fast photographic equipment.

The present paper gives a review on some of the most important properties of laser light sources, with respect to their
special application in photography. Moreover, examples are given
that demonstrate their applicability and usefulness in investigating rapid transient processes. With a few exceptions, most of
the examples given are concerned with studies of the interaction
of powerful laser pulses with matter.

2. THEORETICAL CONSIDERATIONS

In general, the spatial resolution of imaging system is described by the Raleigh [2] criterion. According to diffraction
theory, the minimum detectable angular separation, given for an
incident beam of diameter D, can be calculated by $\theta = 1.22\ \lambda/D$,
where λ is the wavelength of the source under consideration. The
maximum information storage capacity is obtained if the full
range of spatial frequencies lies within the spatial resolution
power of the film. Commerically available photographic materials, with various spatial resolutions, cover a large range of
spectral sensitivities from the UV to the near IR. Special materials have been developed both for conventional laser photography
and for holography. Their spatial resolution can best be described by the modulation transfer which can also be applied to
any optical element or system of elements. The modulation transfer function gives the fringe contrast obtained, for example, in
a two beam interferometric experiment. If maximum and minimum
intensities thus obtained are termed I_{max} and I_{min} the contrast,
or visibility, is determined by the following equation,

$$M = (I_{max} - I_{min}) / (I_{max} + I_{min}) .$$

It has to be taken into account that M refers to the contrast
ratio, after processing of the films, as the optical density of
the materials is subject to changes due to the photo-chemical
processes involved. M is thereby strongly dependent upon the
spatial frequencies, e.g., the number of fringes per mm, f_x, and
f_y. It should be pointed out that the high resolutions achieved
with photographic materials, which are of the order of several
hundreds to thousands of lines per mm, are largely different from
those obtained with photon-optical instrumentation such as electro-optical shutters or image converters. Typical values of

these devices are only several tens of lines per mm. Photographic materials thus allow extremely high information densities to
be stored [3].

    As an example, let us consider a commercial plate (Agfa/10
E 75) which has been developed for ruby laser applications in
holography. Its resolution is 2,800 lines per mm, corresponding
to $7.84 \cdot 10^6$ resolvable image points per mm^2. The energy density needed for exposure of these plates is 50 erg/cm^2, resulting
in an overall energy of $0.64 \cdot 10^{-4}$ per Bit. If ruby lasers are
supplied ($\lambda = 0.6943$ μm, $h\nu = 2.86 \cdot 10^{19}$, $J = 1.78$ eV) $2.24 \cdot 10^4$ photons will be required per Bit. Twenty-nanosecond duration, giant
pulses, with energies of 10 mJ, which can be generated by mono-
mode ruby laser oscillators without additional amplifier stages
are capable of exposing an area up to 2,000 cm^2. Standard plates
(9 x 12 cm^2) thus only need energies of 0.54 mJ.

    Similar considerations apply to other optical storage media.
As photochemical reactions are not reversible, stored information
cannot be erased. Special techniques have been developed, however, especially in the field of holographic recording, for example, by the use of photo-thermoplastic materials. Thereby
resolutions of several hundreds of linepairs per mm^2 and sensitivities of the order of 100 erg/cm^2 were attained. These materials can be handled easily, as, for example, the recorded
holograms are developed without removing the plate within a few
tens of seconds. This technique allows one to eliminate rather
complicated realignment of holograms in real time holographic
experiments.

    The question arises, however, as to whether the whole spec-
trum of spatial frequencies can effectively be used because some
real limitations, in laser photography, will be imposed by the
speckle. These speckles are formed by spatial interference phenomena, due to the high degree of coherence. They lead to randomly distributed intensity patterns which appear on each photograph, thus giving some type of background noise.

    In some special applications, however, the mean diameter of
the speckles is used as a spatial carrier frequency, modulating
the lower frequency spectrum of the images. Spatial frequency
filtering, corresponding to temporal demodulation processes,
then allows the original image information to be reattained.
Applications of this technique in the field of fast photography
will be given in Section 7.

3. BASIC PROPERTIES OF LASERS USED FOR FAST PHOTOGRAPHY

Lasers generate and amplify electro-magnetic waves at optical frequencies scanning a broad spectral range. This section
summarizes briefly some fundamental properties, with special emphasis to pulsed lasers that are used in high-speed photography.
Laser emission has been reported for a large number of neutral
and ionized atoms or molecules in dielectric or semiconducting
solids, in liquids and in gases [4] [5] [6]. The activation of
the laser materials is achieved by different pumping mechanisms,
depending upon the type of the laser. This includes optical
excitation by intense light sources, whereby both thermal sources
and primary laser sources are applied, as well as direct electrical excitation in semiconductors or in gas discharges. Special
techniques may thereby be used, including electron beam injection, x-ray ionization or thermal heating, followed by rapid gas-
dynamic cooling such as employed for the excitation of high power
infrared lasers.

3.1 Spectral Distribution of the Radiation

Figure 1 schematically shows the spectral distribution of
a laser output. Corresponding to the transition involved, the
the envelope of the gain curve $g(v)$, centered around y_o, is usually mathematically approximated by a Gaussian or Lorentzian
profile. This gain curve is characterized by a halfwidth δv.
Due to the stimulated emission, δv is considerably smaller than
the fluorescence linewidth of the transition. The feedback,
given by a Fabry-Perot-type or ring-type resonator provides the
actual oscillating frequency spectrum which is largely determined
by the geometrical configuration of the cavity. As the field
strengths are mainly transverse to the direction of propagation,
the modes are termed transverse electromagnetic modes TEM_{nmq},
where m and n are the number of intensity modes in the far field,
thus characterizing the transverse power distribution [7] [8].
As the cavity lengths L of all lasers largely exceed λ_{mnq} (which
means $L \gg \lambda_{mnq}$), the integer number q, describing the longitudinal mode order, is usually very large so that q is omitted. In
the spectral representation, the transverse modes are centered
symmetrically around all the longitudinal modes. For simplicity,
in Fig. 1, this is shown only for two adjacent longitudinal
modes.

In special cases the spectra include even larger numbers of
lines, corresponding to the transitions between inverted states.
This is particularly found in molecular lasers, due to the population inversion achievable in various rotational/vibrational

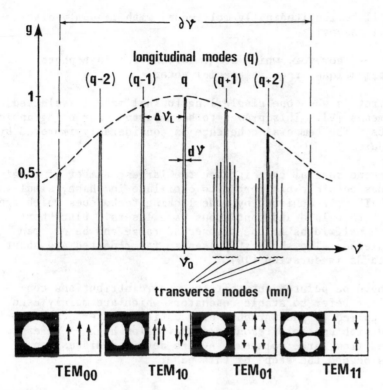

Fig. 1.   Laser spectral distribution.

levels.   Broadening mechanisms may lead to an overlap of the individual lines allowing large frequency bands to be tuned continuously.   The most important group of tunable lasers, however, are
the dye lasers.   Broad fluorescence spectra, up to several hundreds of Å, are achieved with these lasers.   They are tuned by
applying dispersive optical elements such as prisms or gratings
inside the laser cavity.   Tunable lasers are mainly applied to
spectroscopy.   In high-speed photography they may also be of
great interest, especially in interferometric experiments including classical interferometry or holographic interferometry.
By tuning the laser line to a resonance line, for example, the
fringe shifts yield information on the partial densities of the
excited molecular states involved.

In summarizing the main features of the spectral distribution of laser radiation, it can be stated that large numbers of
different longitudinal and transverse modes contribute to the
emission.   Under normal conditions, even in the case of transverse fundamental mode of operation ($m = 0$, $n = 0$) TEM_{oo}, the

emission will be longitudinally multimode, with randomly distributed initial phases.

Two cases, however, which are important for fast photographic applications, require special note.

The first is when one single longitudinal mode is selected
by proper means [9]. This proves to be important for holographic
measurements. The temporal coherence is considerably improved by
this procedure.

The second special case is when the largest number of longitudinal modes possible are forced to oscillate in phase, simultaneously. This is achieved by mode locking techniques which are
applied to both pulsed and continuous wave lasers. Ultrashort
pulses, with halfwidths in the picosecond range and below, can
then be achieved. The shortest pulses so far reported are about
30 femtoseconds in duration [10].

It should be pointed out that the mode distributions considered so far refer to stable resonators which are mainly used
in lasers applied to photographic diagnostics. Lasers can also
be operated with unstable resonators that allow high energies to
be extracted from large volumes. A more detailed discussion,
however, is beyond the scope of this paper.

3.2 Coherence Properties

One of the most striking arguments for the use of lasers in
various experiments is given by the fact that their radiation is
largely coherent. Classically coherence is described by second
or higher order correlation effects [11]. The classical theory
can be extended to quantum mechanical formalisms as first pointed
out by Glauber [12]. In the simplest form, the coherence is referred to the correlation between two electromagnetic waves which
can be separated spatially and temporally. Following the nota-
tions of Born and Wolf, their amplitudes are described by the
complex analytical signals $V_1 = V(r_1, t_1)$ and $V_2 = V(r_2, t_2)$.
The mutual coherence function is given by the cross correlation
integral $\Gamma_{12}(\tau)$, as can be seen in Fig. 2. For simplicity, the
complex integral relation is replaced by the bracket notation
$< V_1(t + \tau) \cdot V_2(t) >$. The indices 1 and 2 refer to the vectors
r_1 and r_2. Mostly, the normalized coherence function $\gamma_{12}(\tau)$ is
used in the literature. As the absolute value $\gamma_{12}(\tau)$ is related to the fringe contrast M that can be measured interferomet-
rically, the degree of coherence of laser sources can most simply

be determined experimentally. The lower part of Fig. 2 shows
that the temporal shape of a wave $V(r,t)$ at a given point is related to the spectral distribution by a Fourier transform. As
only spectral intensities can be measured, $I(v)$ is shown which
proves to be proportional to the square of the absolute value
of the Fourier transform. In the special case of the assumed

mutual coherence function

$$\Gamma_{12}(\tau) = \lim_{T \to \infty} \frac{1}{2T} \int_{t-T}^{t+T} V(\textit{r}_1, t+\tau) \, V^*(\textit{r}_2, t) \, d\textit{r}$$

$$= <V_1(t+\tau) \cdot V_2^*(t)>$$

normalized coherence function

$$\gamma_{12}(\tau) = \frac{\Gamma_{12}(\tau)}{\sqrt{\Gamma_{11}(\tau=0) \cdot \Gamma_{22}(\tau=0)}}$$

simplified example

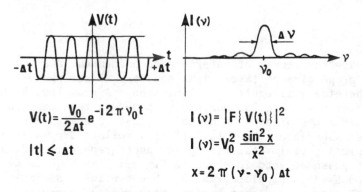

$$V(t) = \frac{V_0}{2\Delta t} e^{-i 2\pi v_0 t}$$

$$|t| \leqslant \Delta t$$

$$I(v) = |F\{V(t)\}|^2$$

$$I(v) = V_0^2 \frac{\sin^2 x}{x^2}$$

$$x = 2\pi(v - v_0)\Delta t$$

coherence time: $\Delta t = 1/(4\pi\Delta v)$

coherence length: $\Delta l = c \cdot \Delta t$

Fig. 2. Mutual coherence function and cross correlation
          function.

rectangularly shaped wave train, a $(sinx/x)^2$ distribution is obtained. In a general way it can be shown, however, that the maximum time difference Δt, the coherence time, for which well defined phase relationships are still valid is inversely proportional to the spectral width of the laser source,

$$\Delta t = 1/(4\pi\Delta v) \ .$$

Correspondingly, the coherence length is defined by $\Delta \ell = c \cdot \Delta t$, in which c is the velocity of light.

The coherence length $\Delta \ell$ will have a maximum if only a few modes or, at best, only one single longitudinal mode of the spectrum (see Fig. 1) is oscillating. The spectral width dv is then determined by the laser gain and by the reflectivities on the laser cavity.

Unfortunately, the mode spectrum, even in the pulsed mode of operation of lasers, is subject to thermally induced changes. At the same time, the mode separation varies, causing the frequencies of individual modes to be shifted. These shifts are rather small. Nevertheless a value of 3.4 MHz/ns has been measured by A.Hirth (ISL) in the case of a monomode, giant pulse, ruby laser. This frequency shift leads to an increased effective spectral width dv and, thus, to a reduction of the coherence time or coherence length, respectively. Conventional fast photography is not affected in this way but a reduced depth of the field view in holographic recording has to be taken into account.

3.3 Speckles

Speckles are the result of interference phenomena, due to the coherence properties of laser light sources. Laser photographs are, therefore, irregularly illuminated. A granular, high constrast intensity pattern is thus superimposed on the image information as some type of background noise [13]. Speckles are always obtained when laser light is diffusely reflected or transmitted. A typical experimental set-up for photographic recording of transparent phase objects is shown in Fig. 3. The objects to be investigated have to be placed in the object plane. A simple diffusely scattering transparent screen (milk glass) is mounted in the object plane. Qualitatively, it can be seen from Fig. 3 that the speckle distribution and the mean diameter of the grains mainly depend upon the diameter D_L of the aperture stop in front

of the objective lens of the imaging system. Quantitatively, these multiple interference speckles are described mathematically by a two-dimensional intensity autocorrelation function which

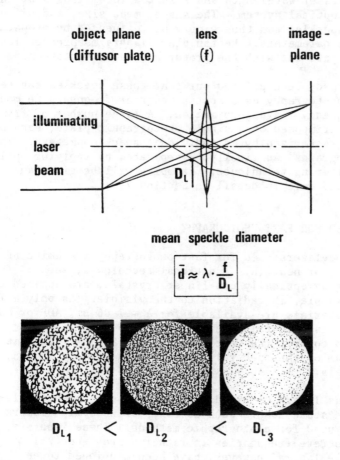

Fig. 3.   Experimental setup for photographing transparent phase
          objects.

takes into account the field contributions of all the light
waves, originating from all the scattering centers, with randomly
distributed phases. Detailed calculations show that the essential part of the autocorrelation function is proportional to the
square of the absolute value of the Fourier transform, of the intensity distribution, in the scattering plane.

For practical applications a rather simple rule can be used
for a first estimate of the mean diameter $\bar{d}$ of the speckles. In
laser photographic systems, the following equation holds

$$\bar{d} = \lambda \cdot f / D_F = \lambda \cdot F,$$

where λ is the laser wavelength and f is the focal length of the lens or of the optical system. The shape, mean size, and orientation of the speckles can thus simply be influenced by proper choice of these parameters. Rectangular diaphragms produce long shaped speckle patterns with the proper angular orientation.

It has already been pointed out that these speckles can be used in laser photography as a spatial carrier frequency to modulate lower frequency image information. A large number of frames can thus be superimposed on a single photographic plate, each one of which is characterized by an individual speckle pattern. The different photographs can finally be separated by applying optical spatial filtering techniques. Examples will be given and will be discussed more in detail in Section 7.

4. LASERS USED FOR FAST PHOTOGRAPHY

Most of the lasers used for fast photography are emitting in the UV, visible, or near IR. Additional techniques, such as frequency doubling by optically non-linear crystals, are applied to convert, for example, IR radiation to the visible. As only a few photographic materials are available for $\lambda = 1.06$ μm, ADP or KDP, or other non-linear crystals are often used in neodymium-YAG or glass lasers to convert the wavelength to $\lambda = 533$ nm. This is near the maximum of the spectral sensitivities of many photographic materials.

Neutral gas lasers, ionized gas lasers, semi-conductor lasers, and solid state lasers such as ruby- or neodymium-lasers are frequently used for photographic methods. These lasers are operated at discrete frequencies in rather narrow spectral ranges. The dye lasers, however, have been mentioned to be characterized by a larger band of fluorescence. These tunable lasers, which are also of great interest for use in fast photography, have been studied extensively during the last few years.

The main features of laser light sources are the narrow linewidths achievable down to 0.01 $\overset{\circ}{A}$ or less, the high peak powers in the pulsed mode (up to the GW-range), and the high degree of coherence. Moreover, most of the lasers can be operated continuously or in different pulsed modes, covering time scales from the μs to the sub-ps range. In the following chapters, the various modes of operation shall be discussed briefly.

4.1 Long Pulses in the μs-Range and Above

Continuous wave emission, which is most important for many scientific and technical applications of lasers in metrology or in high power applications such as in material processing, shall be excluded from the following considerations.

Pulses of several tens of μs, or even longer pulses, in the visible or near infrared, with peak powers of several tens to hundreds of kW, can be attained in some cases. Most solid state laser pulses, however, are usually statistically modulated, producing irregular spikes. Trains of periodic, regularly shaped pulses, the so-called relaxation pulses, can be produced if longitudinal, single mode operation is achieved. For photographic diagnostic purposes, both of these modes of operation are not well suited.

Long pulse operation, up to about 100 μs, has been realized with dye lasers and with neodymium-YAG lasers. The largely smooth envelope of these pulses is obtained by the randomly distributed phases of all the oscillating longitudinal modes. Among the many applications, their use as monochromatic background illuminating source in photographic high speed camera equipment, should be emphasized.

Moreover, these long pulses can easily be modulated by a special technique investigated by A. Hirth [14]. In this case, periodic trains of pulses are generated at repetition rates, typically some hundreds of kHz, and pulse durations of about 200 ns. For example, such series of pulses can be used in stroboscopic measurements.

If single pulses, with halfwidths of a few μs, are required, flashlamp pumped dye lasers provide the best solution. Again, these temporally smooth pulses with high energies and peak powers can cover a large spectral range by suitably choosing the dyes.

4.2 Giant Pulses

Short pulses, with durations of several ns, are obtainable by q-switching the laser cavity or by gain-switching the active laser medium.

The best known and most used lasers that use q-switching for fast pulses, are the ruby and the neodymium solid state lasers. By this technique some type of (mechanical, electro-optical or saturable absorber) shutter is used inside the laser cavity to suppress optical feedback, during the pumping process. Nearly all of the stored energy can then be delivered to a single

pulse by rapidly establishing a high q-value for the resonator.
By applying mode selecting techniques, giant pulses that are
usually longitudinally multimode, can be obtained in a single
mode. For details the reader is referred to the literature [15].
A typical single mode giant pulse laser, using a Pockels-cell
electro-optical q-switch for additional synchronization, is given
in the upper part of Fig. 4. Mode selection is achieved by the
saturable absorber and the Fabry-Perot etalon output mirror.

In gain switched lasers, short pulse, high peak power capability is provided by fast pumping. High pumping rates are especially used in gas lasers to yield a rapid build-up of the
population inversion. Most transversely excited gas lasers
(see Fig. 4), such as excimer lasers (ArF: λ = 193 nm; KrF; λ =
248 nm; XeCl: λ = 308 nm; XeF: λ = 351 nm; Xe_2Cl: λ = 520 nm or
Kr_2F: λ = 436 nm, to mention only a few) or even CO_2 lasers, are
based upon this principle of operation. Often the transverse
discharges can be stabilized by resistive loading of the electrical circuit or by applying x-rays, uv-radiation from sparks or
electron beam injection to produce proionization of the lasers
medium.

CO_2 lasers emit in the infrared at λ = 10.6 μm. In spite of
the fact that no photographic film is available in this spectral
range, examples of the applicability of these lasers in infrared
optical imaging will be given in Section 8.

4.3 Mode Locked Pulses

The shortest pulses are achieved by the technique of mode-
locking. Strong phase coupling of all simultaneously oscillating
longitudinal modes can be achieved both by active and passive
modulation techniques [16]. Active modulation is applied to cw
lasers such as argon-ion lasers. These can then be used for
synchroneously pumping other lasers such as dye lasers. The
shortest pulses are generated using a colliding pulse technique.
Two ultrashort pulses are produced and propagated clockwise and
counter-clockwise in a laser ring cavity. The saturable absorber
cell, acting as a passive modulator, must then be aligned in such
a way that the two pulses overlap in the cell. The peak powers
achievable with these shortest pulses are relatively small. They
are of the order of several kW, so that these lasers are mainly
applied in spectroscopy. Higher intensities are achieved by
using pulsed solid-state or dye lasers. Active modulation, however, can no longer be used since the thermal drift during the
pumping process causes the frequencies and frequency spacings
between neighboring modes to be shifted. As first found by

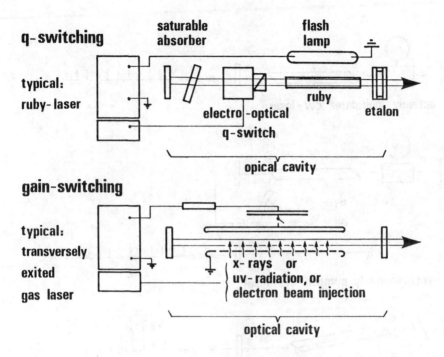

Fig. 4.   Experimental techniques for Q-switching and gain-switch-
          ing lasers.

De Maria [17] these matching conditions can also be fulfilled
automatically by using saturable absorbers which are similar to
those applied for q-switching. Trains of pulses with halfwidths
of a few ps, temporally delayed by the cavity round trip time,
(typically of the order of several ns) are thus obtained. Figure 5 schematically shows some of the methods used for generating
ps pulses. Their application in photography is mainly in strobo-
scopy. They are used in plasma physics for the investigation of
rapidly expanding plasmas or shock fronts.

4.4   Superradiant Pulses

       Most of the lasers usually require optical feedback which is
provided by an appropriate resonator. If the gain achieved in
the active medium is extremely high, photons that are spontaneously emitted in a small solid angle around the optical axis of
the lasing medium are strongly amplified by stimulated emission

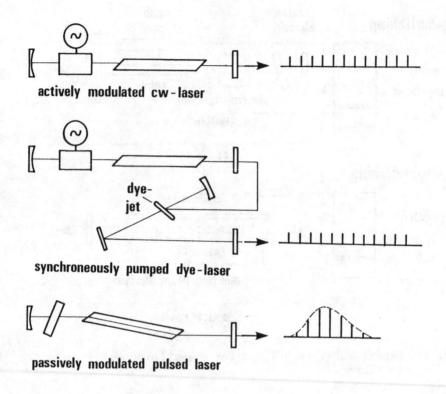

actively modulated cw-laser

synchroneously pumped dye-laser

passively modulated pulsed laser

Fig. 5. Picosecond pulse techniques.

processes. The emitted radiation is thus monochromatic but,
spatially and temporally, largely incoherent. Most nitrogen
lasers emitting in the UV, at $\lambda = 337.1$ nm (second positive
band system $B^3\pi_u - C^3\pi_g$), are operated in the supperradiant mode
(see Fig. 6) [18]. For fast photography these sources are
well suited as pulse durations of the order of several ns and
peak powers as high as megawatts are obtainable. If these lasers
are operated at higher pressures, pulse halfwidths of less than
1 ns (down to 60 ps), at several bars, have been reported [19].

These lasers provide convenient pumping sources for dye
lasers. They can attain high gain which allows operation in the
superradiant mode.

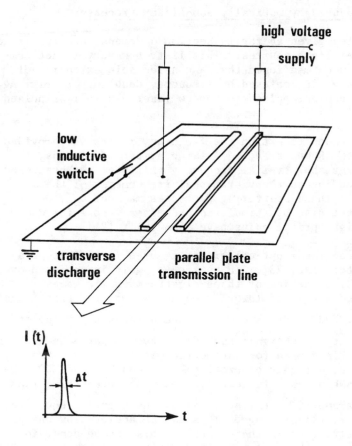

Fig. 6. Superradiant laser diagram.

For high speed photographic applications special types of solid state superradiant sources have been investigated such as ZnS, ZnO, CaSe, ZnTe or CdTe [20]. In this case, excitation is obtained by irradiating the samples with short pulse, high power, electron beams. Beams of some tens of ns halfwidth are obtainable with high voltage, vacuum field, emission discharges. The superradiant materials are deposited on foils which are positioned near the exit window of the electron beam gun. By using different materials, a large range of the visible spectrum can be covered. It should be remembered that the low inductance, high power sources, used for the excitation of the superradiant pulses, are from an established technology. These pulsers have been in use for many years in the field of x-ray photographic diagnostics.

4.5 Pulses Produced by Optically Non-Linear Processes

The examples given so far in Section 4 demonstrate the large variety of laser light sources. This list, however, is not complete. It has been mentioned that a considerable extension of the spectral range is achieved by frequency doubling or tripling. This technique can be applied to the cw lasers as well as pulsed lasers.

Further methods to extend the frequency ranges are provided by the use of the Raman effect. A large number of liquids, solids, and gases show typical frequency shifts of the scattered light, corresponding to the vibrational frequencies of the molecules involved. These scattering processes may not only occur spontaneously but also by stimulated processes. Powerful Raman lasers, with high optical gains have been built [21].

Another non-linear optical effect is used in optical parametric oscillators [22, 23]. In this case, gain is achieved by the non-linear interaction of three electro-magnetic waves: a pump wave v_p and a pair of lower frequency waves, referred to as signal wave v_s, and idler wave v_i. Above a threshold value of the pump intensity, the signal and idler waves experience net gain. The materials used for parametric oscillators are the same as those used for harmonic generation. Besides ADP and KDP, $LiNbO_3$ or $Ba_2NaNb_5O_{15}$ are frequently used. An interesting feature is that parametric oscillators allow the signal frequencies to be tuned over a large range. This can be achieved by changing the index of refraction in the crystal. This can be done, for example, by varying the angle of incidence between the three waves in the case of non-collinear interaction or by varying the refractive index by changing the crystal temperature. Tuning ranges from 0.684 µm to 2.36 µm have been realized in a double resonance oscillator, pumped by a frequency doubled neodymium laser. The tuning ranges thus obtained are considerably larger than those achievable with dye lasers.

5. FAST PHOTOGRAPHY USING LASERS IN CLASSICAL OPTICAL SYSTEMS

Sophisticated, short duration, pulsed light sources, such as flashes or sparks, have been developed and are still in use today. The shortest spark sources presently available, developed by H. Fischer [24] [25], have a steep rise time of a few tenths of a ns and a halfwidth of about 4 ns. With the use of lasers, however, the possibilities of conventional photography have been extended. It was shown in Section 4 that the spectral radiances exceed those of thermal sources and much shorter pulse durations can be achieved. Strongly self-luminous phenomena, subject to rapid motion, can thus be recorded both in reflected light and as silhouettes.

An example is given in Fig. 7. The rapidly expanding, strongly luminous, chemical reaction zone of an explosive has been recorded by means of laser photography. In this special case, a small bore of 2.5 mm in diameter, in a steel cylinder, was filled with nitropenta or with hexogen. Initiation was achieved by a small detonator, fixed at the rear sides, causing a nearly plane detonation wave to be propagated through the explosive. The free surface, which is covered by a glass plate, is intersected by this plane wave at the moment t_3. In order to stop the motion of this front at t_3, the illuminating giant pulse ruby laser ($\Delta t = 20$ ns) has to be properly synchronized. The left hand photograph of the eliptically shaped surface was recorded before the initiation of the detonator. The right hand pictures show records as taken at time t_3 giving some insight into the depth of the turbulent zone in the reacting explosives. The intense light of the explosion was suppressed by an interference filter in front of the camera. No additional high speed shutter has been required in this case.

5.1 Silhouette Recording

In flow visualization, such as aerodynamics or ballistics, silhouette recordings are commonly used for investigating transient rapid processes [26]. These techniques can also be used to study flames or high temperature plasmas. Lasers are much better, when compared to the thermal sources previously used for shadowgraphic or schlieren recordings. They allow nearly diffraction limited point light sources to be used.

A large number of optical systems have been used so far [27]. A few possibilities are given schematically in Fig. 8. The phase objects can be observed in converging, parallel or diverging illuminating beams. Lenses can also be replaced by spherical mirrors. In schlieren systems, condensor lenses or appropriate mirrors have to be used to recollimate the beam. The knife edge has to be placed in the image plane of the point light source. Similar approaches are also used in phase contrast measurements. The knife edge can be replaced by a half space, $\lambda/2$ phase shift, plate. Shadowgraphic systems can be made which do not use either edge or plate devices. As known, shadowgraphy yields refractive index induced intensity variations in the shadow plane or in the image plane of the shadow picture. These

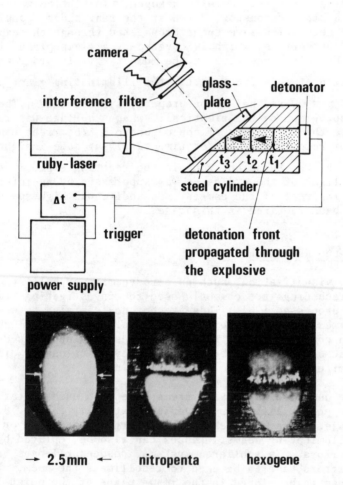

Fig. 7. Laser photography of a self-luminous explosive event.

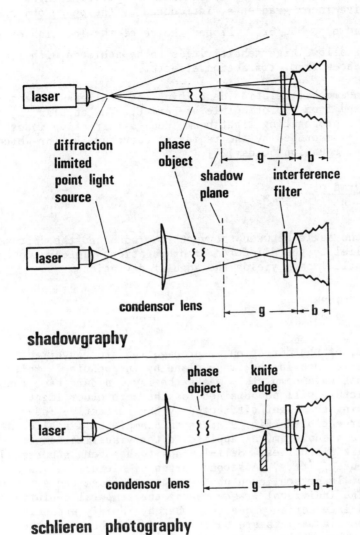

shadowgraphy

schlieren photography

Fig. 8.   Optical systems for shadowgraphy and schlieren photography.

intensity variations ΔI are proportional to spatial changes of the refractive index gradients, introduced by the phase objects, $\Delta I \sim \frac{\partial}{\partial r}$ (grad n) $= \partial^2 n/\partial r^2$. Proper choice of the location of the shadow plane allows high sensitivities to be achieved with minimum disturbances from diffraction effects.

The fundamental relation describing the schlieren effect can be deduced from the principle of Fermat. The angular rotation of a propagating light beam, due to refractive index gradients, is described by the radius of curvature R, for which the following equation holds [28],

$$\frac{1}{R} = \frac{|grad\ n|}{n}\ \sin \phi \ ,$$

where ϕ is the angle between the vectors grad n and the unit vector t, parallel to the direction of propagation. Integration along the optical path yields the angular deviations,

$$\varepsilon_x = \int_{s_1}^{s_2} \frac{1}{n} \frac{\partial h}{\partial x}\ ds \qquad \text{and} \qquad \varepsilon_y = \int_{s_1}^{s_2} \frac{1}{n} \frac{\partial n}{\partial y}\ ds.$$

These changes of the direction of propagation are converted to intensity variations in the image plane by introducing a knife edge. If this edge completely masks the rays, a dark background schlieren picture will be obtained, as the zero order light is stopped, making the light diffracted by the refractive index gradients perceptible. Fig. 9 gives an example of a schlieren picture and a shadowgram, as applied to the visualization of electric spark discharges. Self-luminosity has been suppressed by interference filters. A second series of pictures gives a comparison of laser schlieren photographs and image converter recordings. The individual frames reveal the temporal evolution of laser-plasma-interaction processes, whereby rapidly expanding laser induced plasma-jets are built up. As can be seen, the laser photographs yield considerably more information than the corresponding image converter recordings.

5.2 Interferometric Recordings

Interferometric measurements are most useful for quantitative evaluation. Fringe shifts are directly related to changes in the optical path lengths of the interfering beams. Classical interferometers are designed for non-monochromatic sources. All

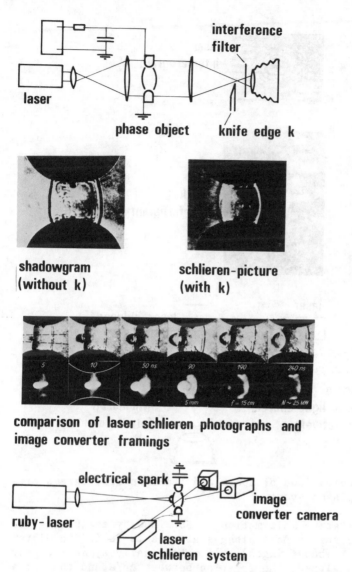

Fig. 9.   Comparison of shadowgraphy and schlieren photographs of
          a spark discharge.

the afore-mentioned advantages, however, hold if lasers are introduced as powerful monochromatic sources. The fringe contrast,
especially of the higher order fringes, is considerably improved
[29]. This is demonstrated in Fig. 10 where, for comparison, a
spark interferogram and a laser interferogram have been put together. The object under study is a laser induced gas breakdown.

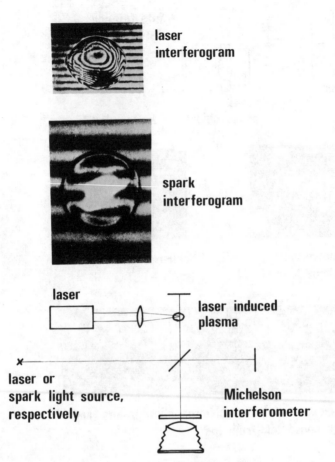

Fig. 10.   Comparison of spark and laser interferograms of a
           plasma event.

Quantitative information is obtained by evaluating the relative fringe shifts $\Delta a/a$ along a given axis (a is the distance between undisturbed fringes). If rotationally symmetric processes
are to be analyzed, the relation between $\Delta a/a$ and the refractive
index of the object under study is given by an Abelian integral
equation. Some of the most important formulas describing these
relationships are given in Fig. 11.

The refractive index finally allows densities to be calculated. If ionization can be neglected, the Gladstone-Dale equation can be applied to relate the index to the density, n [30].
In highly ionized plasmas, however, the refractive index profiles
are used to determine the electron density distribution. By recording several interferograms at various times both the temporal

rotational symmetric phase objects

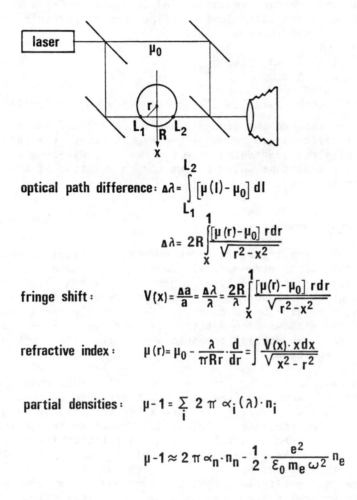

optical path difference: $\Delta\lambda = \int\limits_{L_1}^{L_2} [\mu(l) - \mu_0]\, dl$

$$\Delta\lambda = 2R\int\limits_{X}^{1} \frac{[\mu(r) - \mu_0]\, r\, dr}{\sqrt{r^2 - x^2}}$$

fringe shift: $\quad V(x) = \dfrac{\Delta a}{a} = \dfrac{\Delta\lambda}{\lambda} = \dfrac{2R}{\lambda}\int\limits_{X}^{1} \dfrac{[\mu(r) - \mu_0]\, r\, dr}{\sqrt{r^2 - x^2}}$

refractive index: $\quad \mu(r) = \mu_0 - \dfrac{\lambda}{\pi R r} \cdot \dfrac{d}{dr} = \int \dfrac{V(x) \cdot x\, dx}{\sqrt{x^2 - r^2}}$

partial densities: $\quad \mu - 1 = \sum\limits_{i} 2\pi\, \alpha_i(\lambda) \cdot n_i$

$$\mu - 1 \approx 2\pi\, \alpha_n \cdot n_n - \frac{1}{2} \cdot \frac{e^2}{\varepsilon_0\, m_e\, \omega^2}\, n_e$$

Fig. 11. Important equations for interferometric techniques.

evolution and the spatial distribution of electron densities can be determined [31].

Another advantage in applying lasers in interferometry is that two or even more interferograms, at different wavelengths, can be recorded simultaneously [32]. This has been shown by

using giant pulses of a ruby laser, transmitted through a frequency doubling crystal. These pulses thus contain two wavelengths, corresponding to the fundamental wave and their
harmonic. It is advantageous to apply multiple wavelengths
interferometry, if the phase objects are dispersive, as for example, in plasma physics (see Fig. 12). In this case, the wavelength dependent refractive index $\mu(\lambda)$ is described by a linear
superposition of the individual contributions of particle groups
i, with partial densities n_i [33],

$$\mu(\lambda) - 1 = \sum_i 2\pi \, \alpha_i(\lambda) \, n_i \, ,$$

where $\alpha_i(\lambda)$ are the polarizabilities of the atoms, molecules,

electrons, or ions in the ground state or in excited states. As
the polarizabilities are known or can be estimated, the inter-
ferometrically measured refractive indices at k wavelengths give
a system of k equations which allows the calculation of k unknown
partial densities n_i.

5.3 Cinematographic Applications

In laser cinematographic measurements, all types of mechanical, optical, or electro-optical cameras can be used. The
special characteristics of lasers, however, allow a large number
of cinematographic measurements to be performed without the
rather expensive high speed cameras. Some examples shall be
considered and discussed in detail. In some cases, lasers are
adapted to use only special components or parts of conventional
photographic equipment.

As an example, let us consider a multiple spark camera such
as developed by Cranz and Schardin, which has been improved later
on by Vollrath and Stenzel [34]. The schematic drawing is given
in Fig. 13. Electronically delayed, low inductive, electrical
sparks have been used as light sources. Repetition rates up to
10 MHz limited by the pulse halfwidths of the individual sparks,
have been achieved. With lasers, much shorter pulses, in the ns
or ps range, are available. Even shorter delay times can be
realized by optical delay lines such as schematically indicated
in the lower part of Fig. 13. These devices have, for example,
been successfully applied at ISL for investigating gas discharge
plasmas and laser induced plasmas. Some examples are given in
Fig. 14. A giant pulse from a ruby laser, after pulse shaping
to a halfwidth of 4 ns, has been used for the exposure of a four-
fold, schlieren, cinematographic system [35].

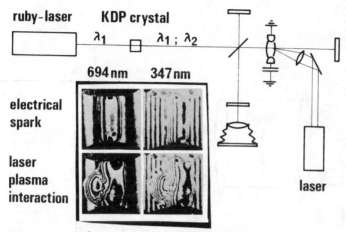

two - wavelengths interferometry

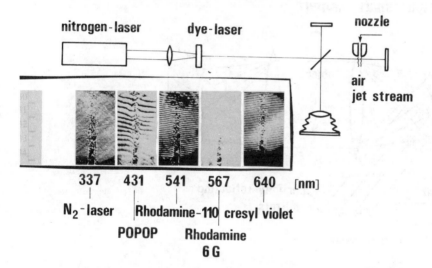

multiwavelengths interferometry

Fig. 12.   Interferometry of dispersive events.

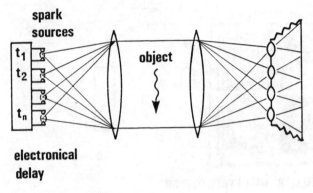

multiple spark camera

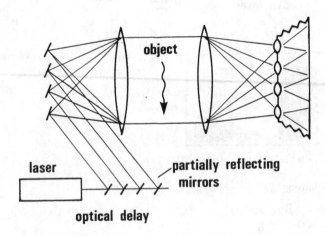

laser camera

Fig. 13.   Short multipulse techniques for illuminating transpar-
            ent objects.

multiple spark camera recordings

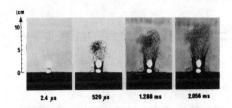

interaction of CO_2 laser pulses with water
$(7.8 \, J/cm^2)$

laser camera recordings

investigation of spark discharges

Fig. 14.   Examples of the use of short multipulse techniques to
study laser interactions with water and spark discharges.

Other lasers that are conveniently applied in fast photography and in cinematography are the gas lasers previously
mentioned (see Fig. 4). Nitrogen lasers are an example. At
atmospheric pressure, pulse halfwidths of less than 1 ns can be
attained. These lasers are also well suited for multichannel
systems. The principle of operation of such a device, as well as
the electrical equivalent circuit, are shown in Fig. 15 [36].

A segmented, parallel plate, transmission line, the upper
plates of which are charged to high voltages (up to several tens
of kV), is rapidly grounded on the left hand side by a low inductance pressurized spark gap. An electro-magnetic wave is thus
propagated along the line, initiating transverse discharges in

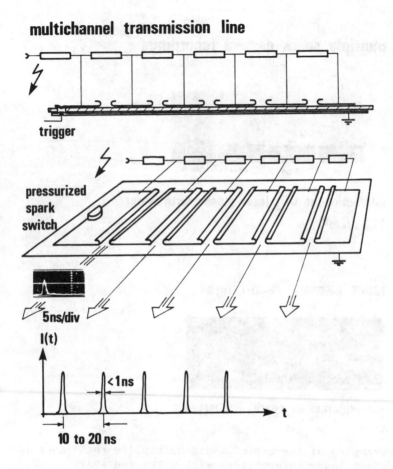

Fig. 15.   Schematic of a transmission line multilaser operating
            in the superradiant mode.

the gaps between the segmented upper plates. The time lag between successive impulses is given by the transient time of the
wave and the ignition delay times of the transverse discharges.
These rise times depend on E/p, where E is the electric field
strength across the gap and p is the pressure. The laser which
had been constructed and tested at ISL, allowed the delay times
to be varied between 10 to 20 ns, corresponding to repetition
rates of 50 to 100 MHz. Higher frequencies are achievable by introducing additional optical delays. As the multichannel laser
is operated in the supperradiant mode, both the pulses from the
front side and from the rear side can be used. The five channels
thus allow high speed cinematographic investigations with ten
pulses. In Fig. 16 a cinematographic measurement, using the

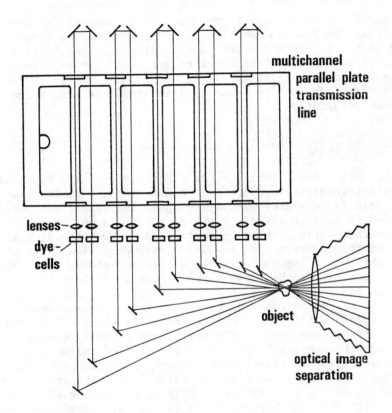

Fig. 16.   Ten-pulse laser using principle of optical image
           separation and dye cells for frequency control.

principle of optical image separation, is demonstrated.   Moreover, dye laser cuvettes can be arranged near the exit windows of
the laser, providing ten pulses, of subnanosecond duration, at
any wavelength throughout the range of dyes available.   Multiple
interferometric investigations have been performed in order to
demonstrate the multiwavelength capability [37].

        Further experiments, that have been carried out by Rückle,
et al [38], at University of Stuttgart, reported the use of ten
different nitrogen lasers, operated with 10 thyratrons that could
be triggered arbitrarily by an external trigger generator.   Such
systems allow the repetition rates to be adapted to larger velocity ranges of the objects under investigation.

6.   HOLOGRAPHIC RECORDING

Holography provides for three dimensional imaging of objects. It allows luminous wavefronts to be stored in a first step and to be reconstructed in a second step. In constrast to all other known photographic techniques, both the amplitudes and the phases of the light waves are recorded. The first experiments were performed by Gabor [39] in 1949. He also suggested the name holography.

6.1 General Remarks

Since photographic emulsions or other photographic materials store light intensities proportional to the product of the complex electric field vector with its complex conjugate, any phase information is lost in classical photography. In holography, this phase information can be reconstructed by superimposing a coherent background wave, the so-called reference wave, with the object light wave. The photographic plate thus stores the interference pattern of these two waves. Such methods of image information and subsequent wave front reconstruction have been discussed by many physicists since about 1920. It is most remarkable, however, that the first experimental results achieved by Gabor were obtained about 11 years before powerful coherent laser sources became available. In his so-called inline technique, he superimposed monochromatic light, scattered from rather simply shaped plane objects (such as slides) with the unscattered part, that is directly transmitted, thereby minimizing the optical path differences Δs between these two parts. These waves can only interfere if Δs is smaller than the coherence length $\Delta \ell$, which is inversely proportional to the spectral width. After the exposure, the photographic plate has to be processed and illuminated with a so-called reconstruction wave, some part of which is directly transmitted. Some other part, however, will be diffracted by the microscopically stored interference structures. These diffracted light waves form two images, a normal image and a conjugate image. A schematic representation is shown in Fig. 17. Technical applications of the in-line techniques are restricted, since the observation of the real image is strongly disturbed by the presence of the conjugate image.

These difficulties were widely eliminated by the use of lasers. As it was first demonstrated by Leith and Upatnieks [40], lasers allow an off-axis recording to be made. Due to the large coherence lengths of lasers, holograms of more complex objects, covering a considerable depth of field, can now be made both in reflected and transmitted light.

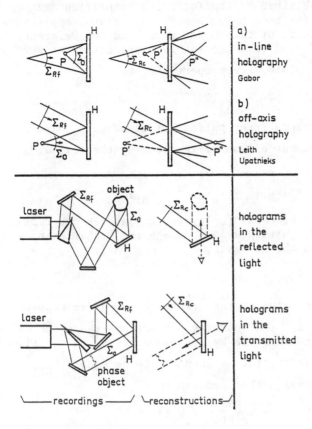

Fig. 17. Illustration of holographic techniques for a variety of
conditions.

6.2 Basic Principles

Referring to the off-axis technique, introduced by Leith and
Upatnieks [40], a schematic description of the experimental set-
up is given in Fig. 17. The interfering object beams and reference beams are, in this case, generated from a single laser by
using appropriate lenses, beam splitters, prisms or mirrors. For
reconstruction the developed plate again is illuminated by a reconstruction beam. Due to the off-axis condition, the real and
virtual images are then more easily observable, as they are geo-
metrically completely separated.

A basic understanding of the recording and reconstruction processes can be obtained by the following simplified mathematical description. If the complex analytical, signal amplitudes of the electric fields of the object waves and the reference waves are termed V_o and V_{Rf}, the intensities, acting at any point of the holographic plate, are proportional to

$$I \sim (V_o + V_{Rf}) \cdot (V_o + V_{Rf})^* \; .$$

If rectangularly shaped laser pulses, of duration t_E, are assumed, the total amount of energy is given by the following equation

$$E \sim t_E \; [\, |V_o|^2 + |V_{Rf}|^2 + V_o V_{Rf}^* + V_o^* V_{Rf} \,] \; .$$

The blackening of a processed holographic plate is described by its transmission T. Within a small range the transmission T depends linearly upon the energy E,

$$T = \bar{T} - \beta(E - \bar{E}) \; ,$$

where $\bar{T}$ is the mean value of the transmission, after exposure with a mean energy $\bar{E}$. The quantity β describes the slope of the T-versus-E curve and is given by the manufacturer of the film-plates. For simplicity, let us assume $\bar{E}$ to be equal $|V_{Rf}|^2 \cdot t_E$. The above given equation then reduces to

$$T = \bar{T} - \beta_i t_E [\, |V_o|^2 + V_o V_{Rf}^* + V_o^* V_{Rf} \,] \; .$$

The quantity β_1 replacing β to take into account the proportionality coefficient. The amplitude and phase information on the object wave V_o is thus stored by this process in the third and in the forth term of the above equation.

For reconstruction, the processed plate has to be illuminated with a reconstruction wave, which is described by the complex analytical signal amplitude V_{Rc}. This wave may differ from the reference wave both with respect to the wavelength and the angle of incidence. The transmitted reconstruction wave is then given by,

$$V_{Rc} T = V_{Rc} \cdot \bar{T} - \beta_1 t_E V_{Rc} \; [\, |V_o|^2 + V_o V_{Rf}^* + V_o^* V_{Rf} \,] \; .$$

The first two terms describe the directly transmitted parts of the incident beam that are attenuated but not deflected. The holographically stored information, concerning the object, is

contained in the third term directly and in the forth term with
inverted polarity of the phases.

It is clear that the simplified model given above, neglecting
volume effects due to the finite thickness of the photographic
emulsions, can help to provide some fundamental understanding.
More detailed mathematical descriptions are beyond the scope of
this paper. The interested reader is referred to the literature [41] [42] [43] [44].

6.3 Application of Holography in Recording Fast Phenomena

Due to short exposure times, even fast moving objects can be
visualized three- dimensionally by holographic techniques using
the reflected light. In the transmitted light, all optically
transparent material can be studied using the transmitted light.
Spatially, an isotropic refractive index distribution can be
studied holographically for solids (such as glasses, crystals or
plastics), liquids, gases, flames or high temperature plasmas.
From these measurements, pressure, density, or temperature fields
can be deduced with high spatial resolution.

Holograms of fast moving objects may be recorded both by
single shot, stopping the motion at any predetermined instant, or
by cinematographic techniques where series of holograms can be
recorded on a single rotating plate [45]. The pulse durations
must then be short enough that the optical path differences, introduced by the rotation during the exposure times, are much
smaller than fractions of a wavelength. One major advantage is
that holographically stored images can be subjected to different
optical processes. The holograms can, for example, be reconstructed as schlieren pictures or as shadowgrams. If classical
interferometers are applied, the reconstructions are recorded in-
terferometrically [45] [46].

In Fourier transform holography, use is made of the fact
that the complex amplitude distribution in an x-y-plane can be
deduced from the known distribution in the ξ-η-plane. Mathematically this process is described by the Fresnell-Kirchhoff integral relation [47]. If p denotes the distance between the two
planes, the far field is defined for $p \to \infty$. The amplitudes and
phases are then given by the Fraunhofer approximation. As can be
seen from Fig. 18 this equation is identical to a two-dimensional
Fourier transform, which is characterized by the spatial frequencies f_x and f_y, giving the lines per mm in x-direction and in
y-direction. Experimentally, the far fields are obtained in the
focal planes of objective lenses, systems of lenses, or mirrors.
Fourier transform holograms are thus obtained by superimposing

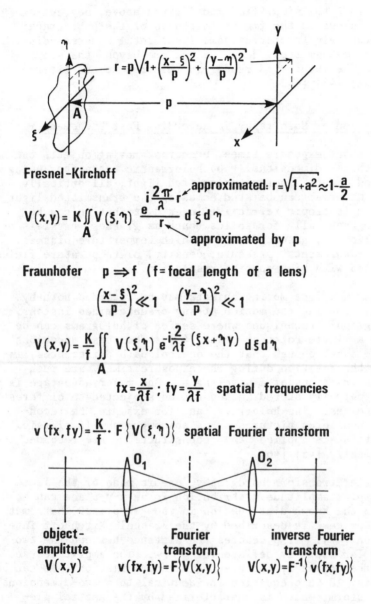

Fresnel-Kirchoff

approximated: $r = \sqrt{1+a^2} \approx 1 - \frac{a}{2}$

$$V(x,y) = K \iint_A V(\xi,\eta) \frac{e^{i\frac{2\pi}{\lambda}r}}{r} \, d\xi \, d\eta$$

approximated by p

Fraunhofer $p \Rightarrow f$ (f = focal length of a lens)

$$\left(\frac{x-\xi}{p}\right)^2 \ll 1 \qquad \left(\frac{y-\eta}{p}\right)^2 \ll 1$$

$$V(x,y) = \frac{K}{f} \iint_A V(\xi,\eta) \, e^{-i\frac{2}{\lambda f}(\xi x + \eta y)} \, d\xi \, d\eta$$

$$fx = \frac{x}{\lambda f} \; ; \; fy = \frac{y}{\lambda f} \quad \text{spatial frequencies}$$

$$v(fx,fy) = \frac{K}{f} \cdot F\left\{V(\xi,\eta)\right\} \quad \text{spatial Fourier transform}$$

object- amplitute $V(x,y)$	Fourier transform $v(fx,fy) = F\{V(x,y)\}$	inverse Fourier transform $V(x,y) = F^{-1}\{v(fx,fy)\}$

Fig. 18.   Fourier transform holography.

the object waves, in the focal plane of a suitably chosen lens system, with the off-axis incident reference waves. The off-axis reference waves may be nearly diffraction limited plane waves. However, they can be slightly convergent or divergent waves. The processed hologram thus represents the spatial frequency spectrum of the object. These techniques are applied in the field of data recognition, image processing, or improvements of blurred images by spatial filtering.

Among the many applications of holography, holographic interferometry has found the most widespread use for investigating fast processes. Due to the storage capability of photographic emulsions or other optical storage media such as thermoplastic materials, interferograms can simply be recorded holographically by superimposing two temporally delayed recordings on the same photographic plate. Delay times may vary from the μs-range or below to several minutes, if the experimental set-up is stable enough. If changes in the optical path lengths of the object wave are introduced between the two exposures, interference fringes will appear, allowing the optical path differences to be determined quantitatively. This technique is applied both in reflected and transmitted light. It provides valuable insight for rapidly changing phase objects such as those encountered in fluid dynamics (flow fields, boundary layers, acoustic waves or shock waves), high temperature physics or plasma physics (electrical discharges or laser induced processes). For specularly reflecting objects, applications of holographic interferometry are found in mechancial engineering, non-destructive testing, analyzing periodical vibrations, and deformations of shock loaded materials.

In contrast to classical interferometry, holographic interferometry does not require expensive optical components such as windows, lenses, beam splitters or mirrors. Fringe shifts are introduced by variations of optical pathlengths between the two exposures. All other phase disturbances are cancelled automatically.

Figure 19 shows the holographic interferometric recording of a spatially non-uniform electrical discharge. Spatial resolution is obtained because the object beam is transmitted through a ground glass plate, acting as a diffuse scattering plate. The giant pulse ruby laser and the transient discharges have to be properly synchronized. The distribution of the refractive index field can thus be observed directly in the diffracted light of the reconstruction wave. The location of the interference fringes in space can be resolved by focussing the objective lens to the various planes [48].

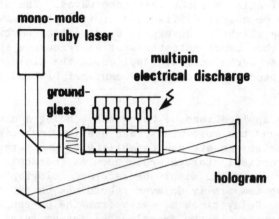

mono-mode ruby laser

multipin electrical discharge

ground-glass

hologram

1. exposure: without multipin discharge
2. exposure: with multipin discharge

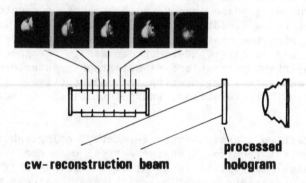

cw-reconstruction beam

processed hologram

Fig. 19. Holographic interferometric recording of a spark event.

Holographic interferometry can be applied to real time experiments. In this case, a real object is compared, interfero-
metrically, with a holographically recorded object. Temporal
changes can then be visualized directly. It should be mentioned
that in these experiments, synthetic holograms can be used which
have been calculated numerically. These techniques are used in
manufacturing processes such as in quality control of optical
components.

7. SPECKLE PHOTOGRAPHIC RECORDING

Speckles, considered in the previous sections, have provided some loss of information. It has been pointed out, however, that interesting new techniques have been developed by using speckles as two-dimensional spatial carrier frequencies, similar to the one-dimensional temporal modulation in high frequency or radio frequency techniques [49] [50] [51] [52].

7.1 Basic Considerations

Before lasers became available, similar methods were used by copying optical gratings, characterized by a given number of lines per mm (grid constant d), onto the photographs. The application of different gratings then allows a large number of photographs to be copied on one single slide. The information density is thus considerably increased. The superimposed photographs can be separated most easily by optical filtering. For this purpose, the slides have to be illuminated with a parallel, monochromatic beam. In the focal plane of the imaging lens, used in this process, the Fourier transforms of the superimposed object beams are formed (see Fig. 20). Each one of which is characterized by the zeroth and higher order diffraction maxima. Their distances are given by r_i - i λ · f/d, where (i = 0, ± 1, ± 2, ...), λ is the wavelength, f the focal length, and d the grating constant

(d = $1/f_x$, if the gratings are ruled parallel to the y-axis). Properly placed aperture stops in the focal plane are then used for separating the photographs. Spatial orientational modulation is obtainable with a single grating which has only to be slightly rotated for each photograph copied. Demodulation is achieved with rotating aperture stops, as all the diffraction maxima in the Fourier transform plane are then circularly arranged around the optical axis.

In laser photography, speckles automatically provide a spatial modulation. The modulation frequency, however, varies statistically around a mean frequency that is inversely proportional to the mean speckle diameter. According to Section 3.3, the speckle diameter can be estimated to be $\bar{d} = \lambda \cdot f/D_L$.

A special technique which uses two circular apertures in front of the objective lens shall be discussed briefly. Each one of these equally sized apertures of diameter D_L is then responsible for a speckle distribution with the mean diameter $\bar{d} = \lambda \cdot f/D_L$.

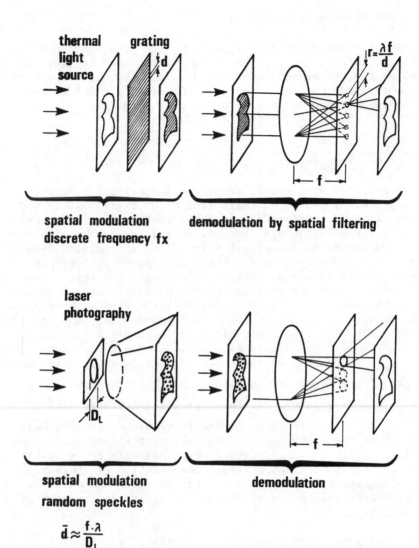

Fig. 20. Illustration of speckle photography.

Similar to Young's experiments, the two apertures, which are
separated by ℓ, are characterized by additional interference
fringes. The distances between adjacent maxima are given by
$g = \lambda \cdot f/\ell$. As ℓ is assumed to be larger than the diameters D_L,
each speckle grain thus contains several parallel interference
fringes which means that the speckles themselves are modulated
[53].

The spatial filtering process is the same as already discussed above. It has only to be considered that the diffraction maxima are considerably larger in diameter.

For proper recording the diameters d_H of adjacent halos are not allowed to overlap. These diameters are determined by the mean speckle size $\bar{d}$, $d_H \cong 2.4\ \lambda \cdot f/\bar{d}$, whereas the separation r, between the halos, is given by the interference fringe spacing, g, inside the speckles, $r = \lambda \cdot f/g$.

7.2 Applications of Speckle Photography

An example of the use of speckle photography for the investigation of surface deformations shall be considered in detail [54]. In this case two exposures are superimposed on high resolution film. A first picture is taken with the objects under study at rest and a second one after an arbitrary time interval Δt. It is assumed that any object point (ξ, η) has been displaced by $(\Delta\xi, \Delta\eta)$, during Δt. High resolution films are required since the information is provided by the small grains of the speckles. The intensities thus recorded by the photographic emulsion are $I(\xi, \eta)$ and $I(\xi + \Delta\xi, \eta + \Delta\eta)$. For simplicity let us consider in the following equations only one-dimensional notation $I(\xi)$ and $I(\xi + \Delta\xi)$. As already discussed in Section 6, we assume that the transmission of the processed plate linearly depends on the energy $E \cong I \cdot t_E$. The transmission is then given by

$$T \cong A - B\ [I(\xi) + I(\xi + \Delta\xi)].$$

The quantities A and B are constants that take into account the mean transmission $\bar{T}$, the mean exposure intensity $\bar{I}$, the slope of the T-versus-E curve, and the exposure time t_E.

For evaluating the local displacement $\Delta\xi$, the processed plate is illuminated with a parallel beam and the Fourier transform recorded in the focal plane of an objective lens.

The complex amplitudes are given by,

$$V(x) = F\{T(\xi)\} = F\{A\} - B\ [F\{I(\xi)\} + F\{I(\xi + \Delta\xi)\}].$$

Remembering that

$$F\{I(\xi + \Delta\xi)\} = \int_{-\infty}^{+\infty} \underbrace{I(\xi + \Delta\xi)}_{\cong\ I(\xi)}\ e^{-i2\pi f_x(\xi + \Delta\xi)}\ d\xi\ .$$

This approximation , $I(\xi) \cong I(\xi + \Delta\xi)$, is valid since, in practical experiments, the intensities are not strongly changing during Δt. The quantity f_x denotes the spatial frequency $f_x = \frac{x}{\lambda f}$. The Fourier transform of $I(\xi + \Delta\xi)$ is thus related to $I(\xi)$ by,

$$F\{I(\xi + \Delta\xi)\} \cong e^{-i\,2\pi f_x \Delta\xi} \underbrace{\int_{-\infty}^{+\infty} I(\xi)\, e^{-i\,2\pi f_x \cdot \xi}\, d\xi}_{F\{I(\xi)\}} \,.$$

The complex amplitudes in the Fourier plane are given by,

$$V(x) \cong F\{A\} - B \left[1 + e^{-i\,2\pi f_x \Delta\xi}\right] \cdot F\{I(\xi)\}$$

and the intensities are proportional to $V \cdot V^*$,

$$I(x) \sim (x) \cdot V^*(x) \underset{\sim}{\sim} 2 \underbrace{\left[1 + \cos 2\pi\, f_x \Delta\xi\right]}_{\cos^2 2\pi\, f_x \Delta\xi} \cdot F\{I(\xi)\} \cdot F^*\{I(\xi)\}$$

The term $F\{I(\xi)\} \cdot F^* \{I(\xi)\}$ describes the halo, the diameter d_H which depends only on the mean speckle size $\bar{d}$,

$$d_H \cong 2.44\, \frac{\lambda \cdot f}{\bar{d}} \,.$$

It can be seen that the light intensity inside the halo is not uniform. It is modulated by a cos-square function. The modulation period δ, corresponding to the distance between two maxima, can easily be found experimentally. The maxima occur at $\cos^2 2\pi f_x \Delta\xi = 1$, corresponding to,

$$2\pi f_x \cdot \Delta\xi = \frac{\pi}{2}, \frac{3\pi}{2} , \, \ldots\ldots$$

From the measured distance δ between two maxima, the displacement $\Delta\xi$ is obtained by

$$\Delta\xi = \frac{\lambda \cdot f}{\delta} \,.$$

If pulsed lasers are used, fast processes such as those observed in ballistics or detonics can conveniently be studied with speckle photographic techniques.

For example, assume a ruby laser, a camera with an objective lens of focal length f = 5 cm, and an aperture of 2 cm diameter, then the mean speckle size will be $\bar{d} \cong 1.75$ µm.

Arbitrary objects can be recorded with exposure times in the nanosecond range to the picosecond range. For quantitative evaluation, cw-lasers can be used, for example a low cost HeNe-laser. If the Fourier transform is considered with a 10 cm focal length lens or mirror, the diameter of the halos will be approximately 8.8 cm.

8. INFRARED PHOTOGRAPHIC RECORDING

In the infrared spectral range powerful lasers are available that can be effectively used for high-speed diagnostic techniques. Recordings of infrared images, however, present some difficulties since photographic emulsions are only available which are sensitive to about 1.2 µm.

At repetition rates of a few tens of Hz, highly sensitive electro-optical detectors, covering a large spectral range, can be used in vidicon systems. Higher rates are excluded by this technique, as the images have to be scanned.

Several techniques have been developed, however, to extend the spectral range to the middle infrared (for example CO_2-lasers at λ = 10.6 µm) or even to the far infrared up to several hundreds of µm (HCN-laser λ = 330 µm).

Most of these recording techniques make use of the temperature-dependent absorption or reflection, in the visible spectral range, of certain materials when irradiated with thermal radiation.

- Thin films of oil for example are partially evaporated [55] by the infrared radiation, thus yielding color changes if observed in the transmitted light (Czernys evaporograph).
- Liquid crystals yield temperature dependent changes of the reflectivity if they are illuminated with white light sources.

A large number of other systems have been developed for special purposes such as those that use the wavelength and temperature dependent characteristics of some semi-conducting materials such as Se-Cr. The infrared images are, in these cases, visualized by registrating the transmitted monochromatic light of a sodium lamp..

Previous experiments at ISL have been performed by using a CO_2 TEA laser as a short duration pulsed IR-source and a liquid crystal image converter which will be briefly discussed [56]. As already mentioned, the infrared images provide a color pattern that can directly be observed or that can be photographed by conventional flash photography. Mixtures of oleyl-carbonate and nonaocate have been applied to thin blackened plastic foils, permitting color changes of the reflected white light. These color changes go from red to blue for temperature variations of only a few tenths of degree C.

This technique has been used to investigate low inductance spark discharges (Fig. 21). All the optical procedures mentioned before, such as schlieren-techniques, shadowgraphy or interferometry have been applied [57]. IR holographic recordings have been excluded since the spatial resolution power of liquid crystal films is of the order of a few tens of lines per mm. At the expense of sensivity, considerably higher resolutions are achievable, with thin foils that are evaporated. This allows holography to be applied in the infrared spectral range.

Infrared photography is of great interest, especially in the field of plasma physics, since sensivities are considerably improved if long wavelength radiation is used.

As an example some results, obtained with a liquid crystal image converter, are shown in Fig. 21.

The infrared images provide, in this case, a color pattern that can directly be observed or that can be photographically recorded by conventional flash photography.

The improvement in sensitivity can be seen from the following equation

$$\mu \cong \frac{1}{2} \frac{\omega_p^2}{\omega^2} ,$$

which holds for highly ionized plasmas. The quantity μ is the refractive index, ω is the angular frequency of the laser, and ω_p the plasma frequency which is related to the electron density by,

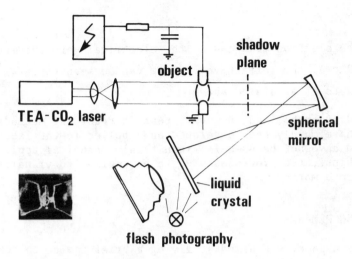

IR‑shadowgraphy

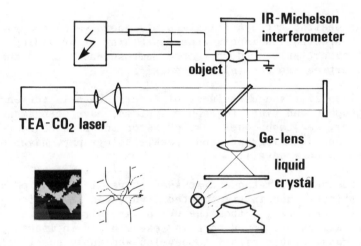

IR‑interferometry

Fig. 21.   Illustration of laser interferometry using IR techniques.

$$\omega_p^2 = \frac{n_e \, e^2}{m_e \, \varepsilon_o} \; ,$$

where m_e is the electron mass and ε_o the electrical permittivity. As $\omega = 2\pi\upsilon = 2\pi c/\lambda$, the refractive index μ can be seen to vary proportionally to λ^2 as stated above.

It should be emphasized, however, that IR recording methods using lasers have mainly been developed and applied in the laboratory and and have not be used for industrial, technical applications. Sophisticated processes have to be applied to visualize the IR image information.

9.   CONCLUDING REMARKS

The present paper was aimed at giving a brief review on some of the most important characteristics of laser radiation that make lasers most attractive as light sources for fast photography applications.

Among the desirable characteristics of lasers we have emphasized the high spectral radiances, the capability of providing short and ultrashort pulses (down to the femtosecond range), the coherence properties and the related speckles.

The applicability and usefulness of lasers, both in conventional photographic and cinematographic systems, have been emphasized. Moreover, new techniques making use of the high degree of coherence, such as holography and speckle holography, have been described in some detail.

Most of the optical methods discussed above, including photography in reflected and in transmitted light, can be applied over a large spectral range, from the UV to the infrared. Increasing applications are found both in research and in industry. There is also considerable effort to develop and apply numerical methods to image processing for quantitative evaluations.

Future trends are to extend the wavelength range towards shorter wavelengths. A considerable breakthrough can be expected if coherent x-ray sources become available.

REFERENCES

1. Sklizkov, G. V., Lasers in High-Speed Photography in Laser Handbook, vol. 2 edited by F. T. Arecchi, E.O. Schulz-Dubois, North Holland Publ. Company (1972).

2. Hecht, E. and Zajac, A., Optics, Addison-Wesley Publishing Company (1976).

3. Smith, H. M., "Holographic Recording Materials Topics," Appl. Physics, vol. 20, Springer-Verlag (1977).

4. Arecchi, F. T. and Schultz-Dubois, E. O., Laser Handbook, vol. 1, 2, North Holland Publ. Company (1972).

5. Stitch, M. L., Laser Handbook, vol. 3, North Holland Publ. Company (1979).

6. Beck, R., Englisch, W., and Gurs, K., Table of Laser Lines in Gases and Vapors, Springer Series in Optical Sciences, vol. 2 (1976) revised edition (1980).

7. Kogelnik, H. and Li, T., Appl. Optics vol. 5, p. 1550 (1966).

8. Grau, G., Optische Resonatoren und Ausbreitungsgesetze für Laserstrahlen, in Laser, edited by W. Kleen, R. Müller, Springer-Verlag Berlin (1969).

9. Hirth, A., Optics Comm., vol. 2, p. 139 (1970).

10. Shank, C. V., Science, vol. 219, p. 1027 (1983).

11. Born, M. and Wolf, E., Principles of Optics, Pergamon Press (1970).

12. Glauber, R. J., Optical Coherence and Photon Statistics in Quantum Optics and Electronics, edited by C. DeWitt, A. Blandin, C. Cohen-Tannoudji, Dunod, Paris (1964).

13. Dainty, J. C., Laser Speckle and Related Phenomena, Springer-Verlag Berlin (1975).

14. Könecke, A., and Hirth, A., Optics Communications, vol. 34, p. 245 (1980).

15. Kleen, W. and Müller R., Laser, Springer-Verlag Berlin (1969).

16. Ippen, E. P., Laser Focus, p. 80, January 1982.

17. DeMaria, A. J., Glenn, W. H., Brienza, M. J., Mack, M. E.,
    Proc. IEEE, vol 47, p. 2 (1969).

18. Leonard, D. A., Appl. Phys. Letters, vol. 7, p. 4 (1965).

19. Strohwald, H. and Salzmann, H., Appl. Phys. Letters, vol 28,
    p. 272 (1976).

20. Brewster, J. L., Barbour, J. P., Charbonnier, F. M.,
    Grundhauser, F. J., Proc. 9th Int. Congr. on High-Speed
    Photography and Photonics, Denver, p. 304, SMPTE (1970).

21. Kuhl, J. and Schmidt, W., Appl. Phys., vol. 3, p. 251 (1974)

22. Byer, R. L. and Herbst, R. L., Parametric Oscillation and
    Mixing in Non-linear Infrared Generation, edited by
    Y. R. Shen, Springer-Verlag, New York (1977).

23. Smith, R. G., Optical Parametric Oscillators in Laser Hand-
    book, edited by F. T. Arecchi, E. O. Schultz-Dubois, vol. 1,
    p. 837, North Holland Publ. Company (1977).

24. Fischer, H., J. Optical Society Am., vol. 51, p. 543 (1961).

25. Fischer, H., (Private Communication).

26. Lauterborn, W., and Vogel, A., Annual Review of Fluid Me-
    chanics, vol. 16, p. 223 (1984).

27. Schardin, H., Forschungsheft 367, VDI-Verlag Berlin (1934).

28. Ulmer, W., Dissertation, Univ. Stuttgart, (1967).

29. Hugenschmidt, M., and Vollrath, K., "Light Sources and Re-
    cording Methods in Fluid Dynamics," Methods of Experimental
    Physics, edited by R.J. Emrich, vol. 18-B, p. 687 (1981).

30. Merzkirch, W., "Density Sensitive Flow Visualization in
    Fluid Dynamics", Methods of Experimental Physics, vol. 18-A
    (1981).

31. Hugenschmidt, M., Angew, Z., Physik, vol. 30, S. 350 (1971).

32. Hugenschmidt, M. and Vollrath, K., Optics and Laser Tech-
    nology, vol. 3, p. 93 (1971).

33. Hugenschmidt, M., Proc. 15th Int. Congr. on High-Speed
    Photography and Photonics, San Diego SPIE, vol. 348, p. 59
    (1982).

34. Vollrath, K., and Thomer, G., Kurzzeitphysik, Springer-
     Verlag Berlin (1965).

35. Vollrath, K. and Hugenschmidt, M., Proc. 8th Int. Congr. on
     High-Speed Photography and Photonics Stockholm, John Wiley
     & Sons, New York, p. 284 (1968).

36. Hugenschmidt, M., and Wey, J., Optics Comm., vol. 29, p. 191
     (1979).

37. Hugenschmidt, M. and Vollrath, K., Proc. 13th Int. Congr.
     on High-Speed Photography and Photonics, Tokyo, Japan, Soc.
     of Precision Eng., p. 252 (1979) SPIE, vol. 189 (1979).

38. Rückle, B., Universität Stuttgart, Bericht IPF-81-1 (1981).

39. Gabor, D., Electron. & Power, vol. 12, p. 230 (1966).

40. Leith, E. N., and Upatnieks, J., J. Opt. Soc. Am., vol. 52,
     p. 1123 (1962).

41. Stroke, G. W., An Introduction to Coherent Optics and Holo-
     graphy, Acad. Press, New York (1966).

42. Kiemle, H., Röss, D., Einführung in die Technik der Holo-
     graphie Akad. Verlagsgesellschaft Frankfurt am Main (1969).

43. Vienot, J. C., Smigielski, P. and Royer, H., Holographie
     Optique, Dunod, Paris (1971).

44. Vienot, J. C., Holography in Laser Handbook, edited by
     F. T. Arecchi, E. O. Schulz-Dubois, vol. 2, p. 1487, North
     Holland Publishing Company (1972).

45. Merboldt, K. D. and Lauterborn, W., Optics Communications,
     vol. 41, p. 233 (1982).

46. Hirth, A., C.R. Acad. Sci., vol. 268-B, p. 961 (1969).

47. Menzel, E., Mirande, W., Weingärtner, I., Fourier-Optik
     and Holographie Springer-Verlay, Wien (1973).

48. Hugenschmidt, M., and Vollrath, K., Bericht des Deutsch-
     Französischen Forschungsinstituts Saint-Louis, ISL 21/71
     (1971).

49. Hopkins, H. H., and Tiziani, H., Applications de l'holo-
     graphie Compte rendu du Symposium Int. d'Holographie
     Besancon, édité par J. Ch. Vienot, J. Bulabois, J. Pasteur,
     p. 8 (1970).

50. Goodman, J. W., Proc. IEEE, vol. 53, p. 1688 (1965).

51. Lohmann, A. W., and Weigelt, G. P., Optics Communications, vol. 17, p. 47 (1976).

52. Hung, Yan. Y., Optics Communications, vol. 11, p. 132 (1974).

53. Köpf, U., Optics, vol. 35, p. 144 (1972).

54. Köpf, U., Siemens Forsch- und Entwicklungsbericht, vol. 2, Nr. 5, S. 277 (1973).

55. Därr, A., Decker, G., Röhr, H., Z. Physik, vol 248, S. 121 (1971).

56. Hugenschmidt, M., and Vollrath, K., Bericht des Deutsch-Französischen Forschungsinstituts Saint-Louis, ISL - 18/72 (1972).

57. Hugenschmidt, M., and Vollrath, K., C.R. Acad. Sci., vol. 274-B, p. 1221 (1972).

REFRACTIVE INDEX MEASUREMENTS

CONCEPTS AND ILLUSTRATIONS OF OPTICAL PROBING DIAGNOSTICS FOR
LASER-PRODUCED PLASMAS

J. A. Stamper, E. A. McLean, S. P. Obenschain
and B. H. Ripin

Naval Research Laboratory
Washington, DC
USA

1. INTRODUCTION

The laser-produced plasmas which occur in experimental
studies of inertial confinement fusion (ICF), with laser drivers,
present both challenges and opportunities for diagnostics. The
challenges arise because of the small size, short duration and
large inhomogeneity. Their small size is also an opportunity for
a particularly simple type of interferometry; folded-wavefront
interferometry. Actually, several rather simple diagnostics can
be realized by utilizing properties of a probing laser beam, such
as intensity, phase, polarization and angular scatter. Specific
diagnostic concepts based on these properties are described and
illustrated.

No attempt is made to review the field of optical probing
diagnostics, although a range of diagnostics is described.
Illustrations of the diagnostic techniques are taken from research at the Naval Research Laboratory (NRL). Many of the techniques were developed independently at NRL. However, some techniques, such as polarization interferometry, were first used as
a fast diagnostic at other laboratories.

The diagnostics described are fast diagnostic because of the
short duration probing laser pulses which were used. The electrical techniques, such as fast Pockels-cell triggering, for producing these pulses are not described. Instead, the emphasis is
on optical diagnostic techniques which use the short pulses. The
fast electrical techniques are discussed in other lectures of
this conference.

Since the illustrations refer to results obtained at NRL
over the past few years, a brief discussion is given of the typical experimental arrangement, as shown in Fig. 1. The main laser
beam (Nd-doped glass), with a 1.054 micron (presently) or a 1.064
micron (earlier) wavelength, is shown incident from the top.
The probing laser beam is shown incident from the side, parallel
to the target surface. The probing laser beam was produced by
splitting off a portion of the main laser beam and then frequency
doubling or tripling, and (sometimes) Raman-shifting the beam.
It was thus inherently timed with respect to the main beam. Most
recent studies were carried out by focusing 100 to 300 joules,
at 3 to 4 ns, to a large (e.g., 1 mm) focal spot on a thin-foil
target (typically 5 to 15 microns thick) to obtain moderate
irradances (around 10^{13} W/cm^2) and to study the ablative acceleration of ICF targets [1]. There was a diagnostic interest in both
the hot (few hundred eV), low density plasma on the front (laser)
side (e.g., interferometry and Faraday rotation) and in the cold
(few eV), dense target material accelerated to the rear (e.g.,
dark-field shadowgraphy). Earlier studies were done on plasmas
produced by focusing a short pulse (e.g., ps) laser, to a small
diameter (e.g., 25 microns), onto a solid slab so that a high
irradiance (10^{16} W/cm^2) was obtained. This regime is illustrated
later with a striking Faraday rotation light pattern. In the
photographic illustrations, one is typically looking parallel to
the target surface and into the probing laser beam. The last
illustration is an exception, as described later.

After a brief background discussion, various diagnostic concepts will be presented and most of them illustrated with photographic samples. Since the emphasis is on diagnostic concepts,
rather than diagnostic details or experimental results, much of
the diagnostic and experimental details will be omitted. However, this information can be obtained, either from the references or from the authors.

2.   BACKGROUND

A brief background discussion of interference fringes and
imaging is presented here which should allow a better understanding of some diagnostic concepts to be discussed later.

2.1 Interference Fringes

Interference fringes can be utilized in a variety of diagnostics, including the many forms of interferometry. Expressions
for the fringe spacing, with interfering planar wavefronts and
with interfering spherical wavefronts are reviewed. Formulae for

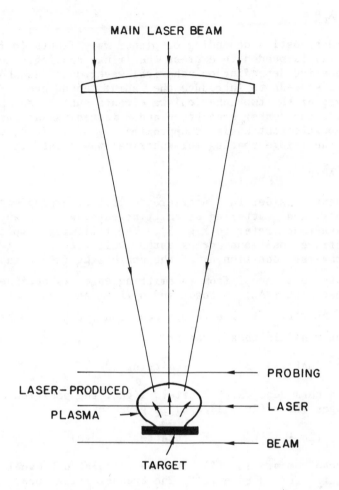

Fig. 1. Typical experimental arrangement for side-on laser
         probing.

fringes positions, resulting from the interference of an applied
planar wavefront with the diverging wavefront from an emitting
or scattering edge, are then derived. The formulae are used in
later sections.

First, let us consider the fringe spacings due to wavefronts
(with wave numbers $\underset{\sim}{k}$ and $\underset{\sim}{k}'$, where $\underset{\sim}{k} = \underset{\sim}{k}' = 2\pi/\lambda$) which are
at a small angle, θ, to each other. Then, $\underset{\sim}{\Delta k} = \sin(\theta/2) = \theta k$,
where $\underset{\sim}{\Delta k} = \underset{\sim}{k}' - \underset{\sim}{k}$. The fringe spacing is the distance along $\underset{\sim}{\Delta k}$
(i.e., nearly normal to $\underset{\sim}{k}$, $\underset{\sim}{k}'$) for which the phase difference
$\underset{\sim}{\Delta k} \cdot \underset{\sim}{x} = \underset{\sim}{\Delta k} \cdot s$ is 2π. Here, $\underset{\sim}{x}$ denotes a unit vector. Thus,

$$s = \lambda/\theta. \tag{1}$$

For diagnostics depending on planar wave fronts (θ fixed), only Eq. (1) is needed to express the fringe spacing. However, for diagnostics depending on spherical wavefronts (as shown in Fig. 2), one needs to introduce the radial dependence. Since the centers of the two spherical wavefronts subtend an angle d/R (where d is the center separation and R is radius) at the wavefront, the wavefront angle is approximately d/R. Thus, using Eq. (1), the fringe spacing for spherical wavefronts is

$$s \cong \lambda R/d . \tag{2}$$

We next consider the interference of a diverging wavefront, from a localized scattering or emitting edge, with an applied planar wavefront (refer to Fig. 3). The following simple derivation of fringe positions agrees rather well with Fresnel theory. The (transverse) position d_n of the nth bright fringe in a plane located at a distance L from an emitting edge, is obtained by noting that $L/(L + \Delta L)_n = \cos\alpha_n$ and $d_n/L = \tan\alpha_n$ so that (for $\alpha_n \ll 1$), $\Delta L \cong L\alpha_n^2/2$. The fringe position for the nth bright fringe ($\Delta L = n\lambda$) is thus given by,

$$d_n \cong 2n\lambda L . \qquad \text{(emitting edge)} \tag{3}$$

For a dark edge which scatters light, the position of the dark fringes ($\Delta L = (2n - 1) \lambda/2$) is given by

$$d_n \cong (2n - 1)\lambda L . \qquad \text{(scattering edge)} \tag{4}$$

One can compare Eq. (4) with the calculated Fresnel diffraction pattern from an edge [2]. The Fresnel frings positions can be expressed as $w_n^2\lambda L$, where the first four w_n^2 coefficients are 0.91, 3.0, 6.1 and 6.8. These compare favorably with the approximate coefficients (1,3,5,7) of Eq. (4).

These results (Eqs. (3) and (4)) can also be understood in terms of the wavefront angle $\alpha_n = \lambda/s_n$ (where s_n is fringe spacing) at the nth fringe. The spacing between adjacent fringes ($\Delta n = 1$) is $(\partial d_n/\partial n)\Delta n$ which, for the bright and dark fringe cases of Eqs. (3) and (4), gives s_n, respectively, to be $d_n/2n$ and $d_n/(2n - 1)$. Thus, the wavefront angle $\alpha_n = \lambda/s_n$ is given, respectively, for the bright and dark fringe cases, as $2n\lambda/L$ and $(2n - 1)\lambda/L$, as one could get directly from Fig. 2, when ΔL is $n\lambda$ or $(2n - 1)\lambda/2$.

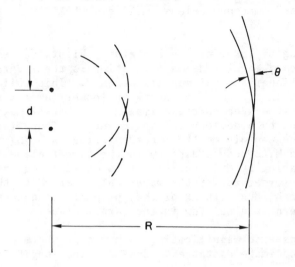

Fig. 2. The interference of spherical wavefronts.

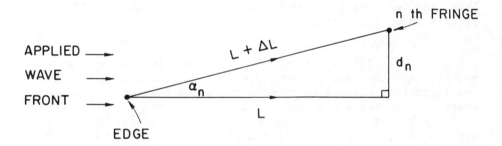

Fig. 3. Interference involving diverging wavefronts from an edge.

2.2 Imaging Considerations

One is interested primarily in interactions of the probing beam, with the plasma, in the region of the target. However, this information is usually deduced from data recorded on film in the image plane. A brief discussion is given here of how information is retained in the imaging process. This consists of a discussion of the lateral and longitudinal magnification relations which are necessary to understand the transformation of

parameters describing interference effects, and a discussion of
a commonly used, matched 2-lens imaging system.

2.2.1 Imaging of Interference Patterns. First, one should note
that the linear, angular de-magnification is the inverse of the
transverse, linear spatial magnification M. This follows from
the conservation of specific intensity [power/(area x solid
angle)] since the area increase implied by spatial magnification
is compensated by a decrease in solid angle. Thus, for interferometry, one can divide the recorded fringe spacing by the
magnification M to deduce density in the target plane. Division
of fringe spacing by M is equivalent to multiplying the wavefront
angle by M in order to get the wavefront referred to the target
plane. One can, thus, think of the, properly transformed, interference pattern as occurring in the target plane.

     The transverse magnification is $M = q/p$, where p and q are
the object and image distances. However, the longitudinal magnification dq/dp is M^2. This can be seen by taking the differential of the simple lens equation, $(1/p) + (1/q) = (1/f)$. The
result shows why the transverse fringe positions d_n of Eqs. (3,
4) scale as the square root of the longitudinal position L. One
should also note a distortion in the three dimensional images.
Thus, the image of an infinitesimal sphere is (when $M > 1$) a
prolate spheroid.

2.2.2 Matched 2-lens System. The light exiting the interaction
region, near the target plane, consists of both strongly scattered light that has passed through the steep gradient region and
nearly collimated light that has not passed through this region.
It is important for some imaging applications to faithfully relay, to the image plane, both the information contained in the
angular scatter and the phase information in the nearly collimated part. This can be done with a matched 2-lens system, as
shown in Fig. 4. It consists of two lenses, separated by the
sum of their focal lengths, with the object at the focus of the
first lens and the image at the focus of the second lens. The
scattered light is (assuming a small scattering region) collimated by the first lens and then focused by the second lens while
the collimated light is focused by the first lens and then recollimated by the second lens.

3.   FOLDED WAVEFRONT INTERFEROMETRY

     Because of the small size of the target plasma, the phase-
distributed part of the probe beam is small compared to the

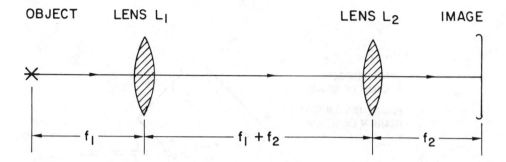

Fig. 4. Matched 2-lens system.

overall beam in the target region. One can, thus, by various
techniques, fold over the undisturbed part of the beam and use
this as a reference beam. One then has a particularly simple
type of interferometry, since the folding can be done outside of
the target chamber, after the, beam has passed through the target
region. Two types of folded-wavefront interferometers are described, illustrated and discussed. The shearing interferometer
uses translational folding (via reflections) and depends basically upon spherical wavefronts.

3.1 Shearing Interferometry

The principle of the shearing interferometer is shown in
Fig. 5. The probing beam is shown incident from the bottom at
45 degrees onto a shearing plate, and is large compared to the
plasma (shaded). The shearing plate is an uncoated flat plate
of high optical quality, with a small wedge (if background
fringes are desired). Rays of the reference beam and phase-
disturbed (by the plasma) beam are illustrated, respectively,
by solid and dashed lines. Due to the finite thickness of the
plate, the front and rear surface reflections undergo a relative
translation to the side, or are sheared. The phase-disturbed
region of the beam, for one surface reflection, thus falls on an
undisturbed region of the beam reflected from the other surface,
which serves as a reference beam. Note that two interferograms
are obtained: one with the front reflection as a reference and
one with the rear reflection as a reference. The interferogram
pairs (complementary interferograms) will be discussed later.

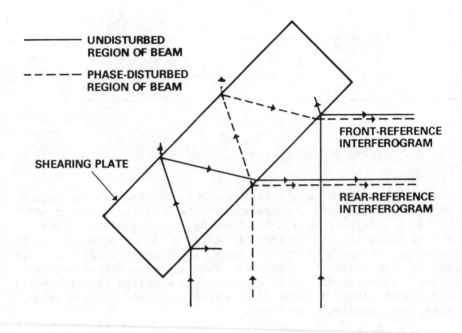

Fig. 5.   Principle of the shearing interferometer.

A shearing interferogram, taken with a $2\omega_o$ (5270 Å), 500-ps, probe pulse, 1 ns after the peak of the main laser pulse, is shown in Fig. 6.   The shearing plate was a 5-cm diameter by a 1-cm thick plate of fused quartz.   As is true in all of these illustrations, the probing beam is out of the sheet, toward the reader.   The 12-joule, 3-ns, main laser beam is incident from the left, onto a 6-μm thick by 300-μm diameter, plastic (CH) foil target.   A 400 μm size marker is shown.   Note the expected pair of interferograms, where the normal fringes (bent away from cen-ter) occur on the front (left-hand) side of the left-hand inter-ferogram and where the inverted fringes (bent toward the center) occur on the front side of the right-hand interferogram.   This behavior is examined quantitatively in a later section.

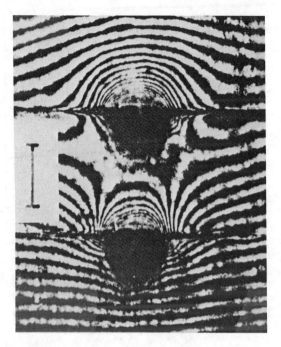

Fig. 6.   A shearing interferogram.

3.2   Polarization Interferometer

The principle of the polarization (Normarski) interferometer
[3] is shown in Fig. 7. The probing laser beam, polarized at
45 degrees to the vertical, is shown incident from the left and
is large compared to the plasma (shaded). Rays of the reference
beam and phase-disturbed beam are illustrated, respectively, by
solid and dashed lines. One should note that a single lens is
used so that the interferometer depends upon spherical wavefronts. After passing through the lens, the beam coverages to
a focus and then diverges in a spherical wavefront. As the diverging beam passes the interface of a polarizing (Wollaston)
prism, it is split (with angular separation θ_p) into two beams,
one with a vertical polarization (⇵) and one with a horizontal
polarization (↔). The diverging beams in the two polarizations
can be projected backwards to two apparent foci which are the
centers of the two spherical wavefronts for the interferometer.
A polarizer, oriented parallel or perpendicular to the 45 degree
incident polarization, is placed in front of the film plane so
that the transmission is the same for both polarizations and the

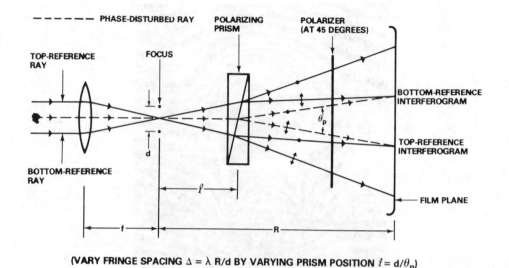

(VARY FRINGE SPACING $\Delta = \lambda R/d$ BY VARYING PRISM POSITION $\ell = d/\theta_p$)

Fig. 7.   Principle of the polarization interferometer.

two beams will interface at the film plane. The angular separation produces, at the film plane, a spatial separation such that
the phase-disturbed region of one beam falls on the undisturbed
region of the other beam, which serves as a reference. From
Eq. (2), the fringe spacing is $\lambda R/d$, where R is the (radial) distance from the focus to the film plane and the wavefront center
separation d is $\ell\theta_p$, where ℓ is the distance of the prism interface from focus. Thus, this interferometer has the advantage
that one can easily vary the background fringe spacing by moving
the prism.

An interferogram taken with a polarization interferometer,
using a Wollaston prism ($\theta_p = 1$ degree), is shown in Fig. 8.
This was taken with a 300-ps, third harmonic (3500 Å), probing
laser pulse, and was timed at the peak of the main laser pulse.
The 6 joule, 4 ns main laser beam is incident from the left, onto
a 1 mg/cm^2 carbon foil target. The main laser focal spot diameter was 100 microns. One sees, as expected, two interferograms.
The qualitative appearance of this interferogram pair is similar
to that of the shearing interferogram pair. These interferogram
pairs are discussed next.

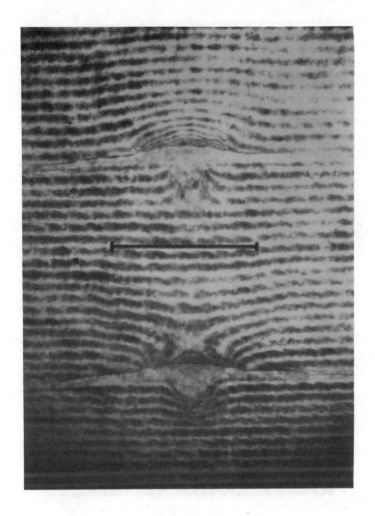

Fig. 8. A polarization interferogram.

3.3 Complementary Interferograms

     A typical interferogram pair, obtained with folded-wavefront
interferometry, is shown in Fig. 9. These interferograms were
taken with a shearing interferometer and only the top-front quad-
rant of each is shown. The axis of symmetry is at the bottom
in both cases and a 200-µm marker is shown. The five fringes,
coming in from the top from a large radial distance, are seen to
curve away from the plasma in the normal interferogram (more
easily Abel inverted), shown on the left, while they curve into
the plasma in the inverted interferogram, shown on the right.

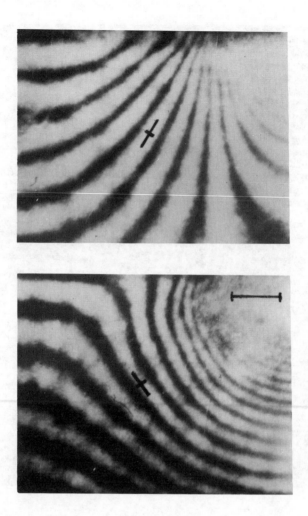

Fig. 9.   Complementary interferograms.

These two interferograms are complementary in that one can be obtained from the other by an inversion of the background wavefront
angle. This is done physically by a left-right reversal, by in-
terchanging the reference and phase-disturbed beam or by reflecting both beams off of a flat surface.

     In order to facilitate the Abel inversion, it is desirable
to arrange the diagnostic components such that the region of
interest (e.g., front-side plasma) occurs on the outside (not in

the overlap region of the interferograms) of the normal interferogram. This arrangement requires that one understand the detailed qualitative nature of the interferograms. Three things
should be kept in mind: (1) The phase contributions from the
plasma and background dominate in different regions. (2) There
are background phase contributions from both beam collimation and
the effect of the folding element. The total background sign is
reversed by a left-right reversal or by reflecting off of a
mirror or splitter. For a shearing interferometer, the contribution due to a wedged shearing plate can only be reversed by
rotating the plate 180 degrees about its (approximate) axis of
symmetry. (3) The merging region of the background and plasma
dominant regions has a normal or inverted signature, depending on
the net sign of the total background contribution.

In order to analyze the problem quantitatively, we consider
a cylindrically symmetric medium (as in Fig. 9) and take a cylindrical (ρ,z) coordinate system [4]. The refractive index $\bar{n}$ is
then a function of the radial (ρ) and axial (z) coordinates.
A laser beam passing through the medium, perpendicular to the
z-axis, is combined with the reference beam at a small wavefront
angle $\theta_w = \lambda/d$, where d is the fringe spacing and λ is the wave-
lenght. The wavefront intersection is taken, for simplicity, to
be perpendicular to the z-axis. In this plane, each fringe is an
equiphase curve that is a function of z and the projected radial
distance r (ray impact parameter) of the ray from the z-axis.
The phase of a ray, along a path $s = \rho^2 - r^2$, can then be expressed as,

$$\delta(r,z) = \pm \frac{2\pi}{d} z + \frac{2\pi}{\lambda} \int_{-\infty}^{\infty} \bar{n}\ (r,z,s)ds, \tag{5}$$

where, in the first term, two signs are allowed for the phase of
the applied wavefront. The fringes define a set of curves in the
r-z plane, each having a constant phase. The phase change along
a fringe $(\partial\delta/\partial r)dr + (\partial\delta/\partial z)dz$ thus vanishes. This shows that
the fringe slope dz/dr can be expressed as $(-\partial\delta/\partial r)/(\partial\delta/\partial z)$.
Using Eq. (5), this gives,

$$\frac{dz}{dr} = -\frac{\theta_r}{\theta_z \pm \theta_w}, \tag{6}$$

where $\theta_r = \int (\partial\bar{n}/\partial r)ds$ and $\theta_z = \int (\partial\bar{n}/\partial z)ds$ are the refraction
angles for gradients along the r and z axis, respectively, and
θ_w is the wavefront angle. The two signs of the applied wavefront phase, or angle, in Eqs. (5) and (6) correspond to the two
complementary interferograms obtained with a folded-wavefront

interferometer. In fact, one can determine the refraction angles
(without Abel inversion) at a given r-z position by using both
interferograms and applying Eq. (6) to the two slopes [4].

3.4 Discussion of Shearing and Polarization Interferometers

The shearing and polarization interferometers each have some
advantages such as the simplicity of the shearing interferometer
and the easily varied fringe spacing of the polarization interferometer. There are also some limitations that should be noted.
Shearing interferometry may not be applicable to probing pulses
of very short duration since there is a time delay between the
front and rear reflections. For an angle of incidence θ (with

respect to the normal), onto a plate of thickness t and index $\bar{n}$,
this delay is

$$\tau = (2t/c) \sqrt{ke(\bar{n})^2 - \sin^2\theta} . \tag{7}$$

Thus, the delay in picoseconds, for a fused quartz plate ($\bar{n}$ =
1.45), at 45-degrees incidence, is 84.6 times the plate thickness
in centimeters.

When focusing through a Wollaston prism, as in polarization
interferometry, there is a polarization dependent focal shift and
astigmatism. For example, when focusing through a plate of index
$\bar{n}$ and thickness t, there is an increase Δf_o in the paraxial focal
length of magnitude $(\bar{n} - 1)t/\bar{n}$. The difference $\Delta\bar{n}$, in the indices of the component prisms of a Wollaston prism, is small com-
pared to the average index $\bar{n}$. Hence, one might think that the
difference Δf in the paraxial focal lengths, for the two polarizations, would be small compared to Δf_o. However, this is generally not true, due to a subtle asymmetry that can be seen by
referring to Fig. 10. The separation of the two polarizations,
at the exit face of the Wollaston prisms, is greater for the
bottom rays than for the top rays, since the bottom rays travel
a distance greater (by $\ell_{eff} \cong d \cot \alpha$) than that travelled by the
top rays, after encountering the interface. The convergence
angles are very nearly the same for the two triangles in the two
polarizations so that their different bases (widths at the exit
face) imply a different height, i.e., a focal shift. This focal
shift is evaluated, by a paraxial ray-trace, to be $f_o \theta_w \cot\alpha$,
where $\theta_w = 2\Delta\bar{n} \cot\alpha$ is the external Wollaston separation angle
and f_o is the average focal distance from the Wollaston prism.

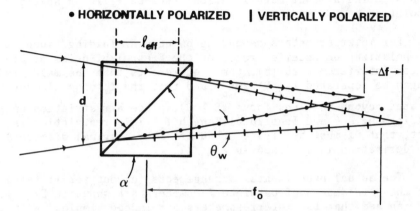

$F \equiv f_0/d$ (f/No.); F, θ_w, ℓ_{eff}/f_0 SEPARATELY ASSUMED SMALL

$\Delta f \cong F\ \theta_w\ \ell_{eff}$ (SEPARATION ANGLE $\theta_w \cong 2\Delta\bar{n}\ \cot\alpha$)

Fig. 10.   Focusing through a Wollaston prism.

The polarization dependent focal shift also implies a polarization dependent astigmatism since the polarization separation only
applies (in the case illustrated) in the vertical direction. The
focal shift and astigmatism will not be bothersome when the focal
shift is comparable to or less than the focal depth.

For subnanosecond resolution and a large phase-disturbed
region (in mm), one must (for shearing interferometry) choose a
compromise between a thin plate (favoring temporal resolution)
and a thick plate (favoring spatial separation). The lateral
spatial separation of the exiting front and rear surface ray
reflections is, for an incident angle θ onto a plate of thickness
t and index $\bar{n}$,

$$s = \frac{\sin 2\theta}{\sqrt{\bar{n}^2 - \sin^2\theta}}\ t\ . \tag{8}$$

Thus, the separation, at 45-degrees incidence onto a quartz
plate is 0.79 times the plate thickness. Since this separation
applies to the magnified image, one may obtain a greater separation by using a lower magnification; thus sacrificing spatial
resolution.

For interferometers depending on angular folding, such as
the polarization interferometer or the Fresnel bi-prism or bi-
mirror interferometers [5], the requirements that the magnified
images be separated ($z_2 \theta > 2Mr_o$) and that the folding element be
located beyond the lens focus ($\ell > 0$), place a restriction on the
folding angle θ and lens focal length f. This restriction is, in
fact, that θf must be greater than $2r_o$ (the original diameter).
The derivation is outlined in Fig. 11.

The actual experimental arrangements (except for filters)
for these folded-wavefront interferometers are shown in Fig. 12.
One can see that the interferometers are indeed simple. With the
exception of the matched 2-lens system, which images the target
region onto the film, the only component of a shearing interferometer is the shearing plate. Similarly, only the components
(lens, prism, polarizer) used in discussing the principle of the

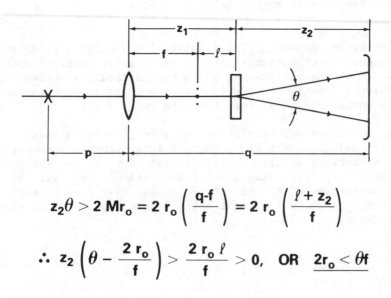

$$z_2\theta > 2\,Mr_o = 2\,r_o\left(\frac{q\text{-}f}{f}\right) = 2\,r_o\left(\frac{\ell + z_2}{f}\right)$$

$$\therefore\ z_2\left(\theta - \frac{2\,r_o}{f}\right) > \frac{2\,r_o\,\ell}{f} > 0,\ \ \text{OR}\ \ \underline{2r_o < \theta f}$$

Fig. 11.   Restriction on folding angle and focal length.

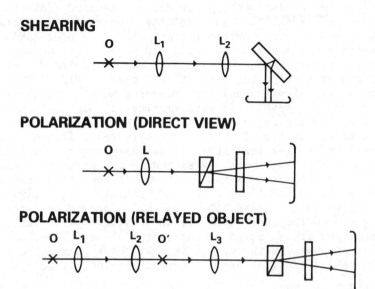

Fig. 12. Arrangements for interferometry.

polarization interferometer are needed for its realization. However, for the polarization interferometer (where the fringe-
spacing and image separation requirements are interrelated), it
may be desirable to enhance the flexibility of applying the basic
interferometer to a relayed object. This arrangement, shown at
the bottom, uses a matched 2-lens system to produce the real
image for the relayed object.

4.   OTHER "FRINGE" BENEFITS

Even without an interferometer, interference fringes can
occur due to interfacing wavefronts produced (perhaps accident-
ally) by optical components or by structure in the target region.
One can also benefit from the information contained in these
fringe patterns. Three examples are discussed here.

Whenever the target is large, so that some edge is not in
focus, one is likely to see the Fresnel diffraction pattern of
this edge. One could, for example, use Eq. (4) to estimate how
far the edge is from the focus.

Fringes were also seen in our dark-field study (discussed later) and were identified as resulting from reflections off of a slightly-wedged interference filter. These fringes could be seen in the regions which refracted the probe light strongly enough to get past the dark-field mask and be recorded. Such fringes can be seen in the front (left) side in Fig. 13A, where the spacing varies with position according to the local refraction. The fringe spacing is much smaller (11 microns) in the strongly refracting region of the rear-side plasma, as shown in the enlargement in Fig. 13B. These interference fringes, due to the filter, differ from those in ordinary interferometry in that both of the wavefronts correspond to the phase-disturbed region. Nevertheless, one can use these fringes to obtain relative phase shifts and local refraction angles. For a given location in the target plane, the phase shift due to refraction is superimposed on the locally linear phase shift due to the filter. All quantities are referred to the target plane. Thus, the rate of change of phase, with respect to distance z normal to the fringes, is $2\pi[(1/d_o) + (\theta/\lambda)]$, where d_o is the fringe spacing (110 microns) without refraction and θ is the angle of refraction. The discussions in Sections 2.2 and 3.3 may be helpful

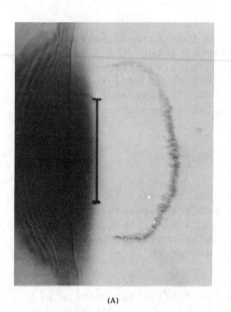

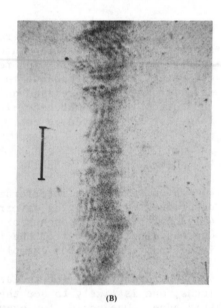

(A)   (B)

Fig. 13.   Dark field study fringes.

here. This rate of change of phase can also be expressed as $2\pi/d$, where d is the fringe spacing in the refracting region. Thus, the refraction angle is λ $(d_o - d)/dd_o$, or 3 degrees. This is light which just clears the mask.

In an interesting application of interference fringes, associated with refraction in the target region, Michaelis and Willi have indicated how to use these fringes to measure density [6]. For a spherical target, rays at various impact parameters and scattering angles will tend to converge to form a ring focus at some axial distance from the target center. The diverging wavefront from this ring focus interferes with the still-collimated part of the beam to form a fringe pattern described by Eq. (3) for a bright edge. Bennatar and Popovics [7] have analyzed the problem and developed a method (spherical quantitative refractometry) for making quantitative density measurements.

5. POLARIZATION EFFECTS AND USED

5.1 Faraday Rotation

Large (100T) magnetic fields are produced as a result of the laser-plasma interactions [8]. These fields have been measured in the focal region, using Faraday rotation of a probing laser beam [9-11]. Faraday rotation is the rotation of the plane of polarization of a plane-polarized, electro-magnetic wave which propagates along a magnetic field. A plane-polarized wave, propagating along the axis of the optically active medium, can be resolved into positive helicity (wave number k_+) and negative helicity (wave number k_-) waves. The rate of change (for the plane-polarized wave) of the right-handed rotation angle θ, with respect to distance z along the axis, is $d\theta/dz = (k_- - k_+)/2$.

For waves propagating along a magnetic field, the dispersion relation is $(ck_{\pm}/\omega)^2 = 1 - \omega_p^2/\omega(\omega \mp \omega_{cz})$, where ω is the wave frequency and the plasma frequency, ω_p, and electron cyclotron frequency, ω_{cz}, are defined by $\omega_p^2 = 4\pi ne^2/m$ and $\omega_{cz} = eB_z/mc$, respectively. If $n\omega_{cz} \ll (n_{cr} - n)\omega$, where the critical density, n_{cr}, is the electron density, n, at which $\omega_p = \omega$, then $d\theta/dz \cong (\omega_p/\omega)^2 (\omega_{cz}/2nc)$. Thus, the Faraday rotation angle is proportional to the path integral of density times the magnetic field component along the path. The rotation angle can be expressed in convenient units as,

$$\theta(\text{deg}) = 1.51 \ \lambda_p^2 (\mu m) \int \frac{n(cm^{-3})B_z(MG)dz(\mu m)}{10^{21} \left[1 - (n/n_{cp}) \right]} \ , \tag{9}$$

where $\bar{n} = 1 - (n/n_{cp})$ is the refractive index and λ_p and n_{cp} are the wavelength and critical density, respectively, for the probing beam.

With an independent determination of density, via interferometry, one can unfold the integral to get the magnitude of the magnetic field. The rotation is right-handed when $d\theta/dz > 0$, i.e., $k_- > k_+$. For Faraday rotation, this means that the rotation is right-handed or left-handed, respectively, according to whether the wave propagates parallel or antiparallel to the magnetic field.

Faraday rotation probing of a laser-produced plasma is shown schematically in Fig. 14. The magnetic field is in an azimuthal direction, about the axis of the main laser beam. The (initially vertically polarized) probing laser beam is incident from the right, parallel to the target surface, and acutally is much larger than the laser-produced plasma at focus. A small portion of the probe beam, directed antiparallel to the magnetic field, is pictured undergoing a left-handed rotation.

The light patterns recorded with the Faraday rotation study can be understood by referring to Fig. 15. The (initially vertically polarized) probing laser beam is directed parallel to the target surface and comes out of the page, while the main laser beam is incident from the right. The orientation of the thermally generated magnetic field is depicted on the left-hand side.

The magnetic field is in the direction of $\overline{\nabla T} \times \overline{\nabla n}$, with the temperature gradient cylindrically in toward the laser axis and the density gradient into the target. Thus, the magnetic field is azimuthal about the main laser axis, into the page at the top and out at the bottom.

Now consider the light pattern which is expected when looking, through an almost-crossed polarization analyzer (right-hand side of Fig. 15), at the oncoming probing laser beam which has passed out of the page through the magnetic field structure shown on the left. In regions (below the laser axis) where the beam propagates parallel to the magnetic field, the wave E-vector is rotated to the left so that it is more nearly cross-polarized than the incident beam to the analyzer. However, in regions where the beam propagates antiparallel to the magnetic field

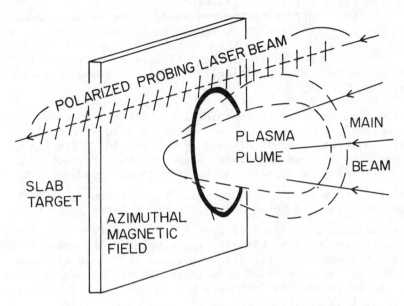

Fig. 14.   Faraday rotation probing of a laser-produced plasma.

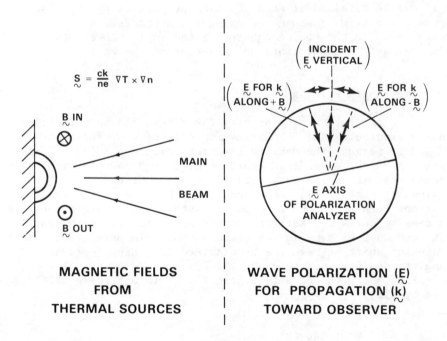

Fig. 15.   Background for understanding Faraday rotation light
           pattern.

+(above the laser axis), the E-vector is rotated to the right and
makes a smaller angle with the analyzer. This gives an enhanced
transmission. Thus, one would expect an up-down asymmetry, with
a lighter region at the top than bottom. Further, if the analyzer is rotated clockwise so that its axis is slightly clockwise
of the cross-polarized position, the up-down asymmetry should re-
verse.

The expected Faraday rotation light pattern and its dependence on analyzer orientation are actually seen in Fig. 16.
Two of the three channels of data are shown [10]. The plasma was
produced with a 75-ps main laser beam, tightly focused ($\sim$ 25 µm)
onto a slab of polystyrene (CH) to produce an irradiance of
around 10^{16} W/cm^2. A small pre-pulse was incident 2 ns before
the main pulse. The 50-ps duration probing laser pulse had a
(Raman-shifted) wavelength of 6329 $\overset{o}{A}$. These data were taken
50 ps after the peak of the main laser pulse. The analyzer was
rotated 15 degrees counter-clockwise from the cross-polarized
position for the right-hand photograph and was rotated 7.5 degrees clockwise for the left-hand photograph. In addition to the
asymmetric light pattern and its reversal with analyzer orientation, one can see the expected darkening (below background) at a
position opposite to the bright region. The Faraday rotation
angle can be obtained at each position by correcting the observed
exposure for film response and using the dependence of polarizer
transmission on the cosine squared of the angle between the
polarizer axis and the (rotated) E-vector of the wave.

5.2 Steep-Angle Probing

The polarization of a probing laser beam can be affected by
density gradients [12]. Polarization effects can, therefore,
be used to measure the density gradient scalelength [13]. This
density gradient dependence could be neglected for the side-on
probing used in the Faraday rotation studies discussed in the
previous section [9-11]. However, the effect is particularly
important when the probe beam makes a steep angle with respect to
the density surface. Refraction determines the ray path and
this path must be used in the calculations. In particular, there
is a phase shift between the S-polarized component (polarized
along the density surfaces) and the P-polarized component
(polarized along the density gradient). This S-P phase shift is
given by [12]

$$\delta = \frac{1}{2k} \int \nabla^2 \; (\frac{1}{n}) ds, \qquad\qquad (11)$$

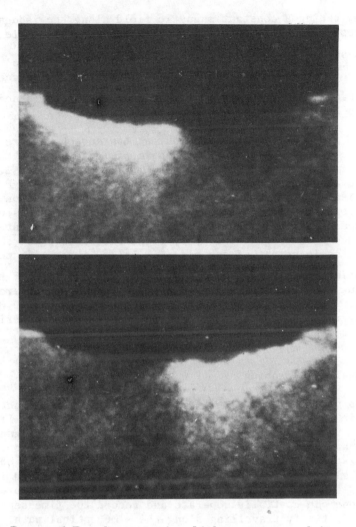

Fig. 16.   Expected Faraday rotation light pattern and its
           dependence on analyzer orientation.

where k is the wave-number of the probing radiation, $\bar{n}$ is the
refractive index, and the integral is taken along the ray path.
For a laminar density, $n/n_c = \exp(-x/L)$, one can integrate along
the path (determined by a ray trace) and express δ in the form
$C \lambda/L$ where the coefficient C is 3.0 and 0.4, respectively, for
angles of incidence of 22.5 degrees and 45 degrees. Assuming the
magnetic field varies slowly compared to the density, one can estimate the Faraday rotation angle, θ, along the same steep ray

path by taking a constant magnetic field along the isodensity surfaces. The ratio, θ/δ, of the Faraday rotation angle to the S-P phase shift, can then be expressed as $C'B(MG)L^2/\lambda\lambda_o$, where λ is the probe wavelength and λ_o is 1.054 microns. The coefficient C' is 0.02 and 0.2, respectively, for angles of incidence of 22.5 degrees and 45 degrees.

An experiment was carried out at Lawrence Livermore National Laboratory in which the average angle of incidence (of the 1 micron radiation) was 22.5 degrees [13]. For incident light polarized at 45 degrees to the plane of incidence, the S-P phase shift is determined by the degree of linear polarization of the scattered light [13]. A phase shift of 2.5 radians was observed. By assuming a linear density profile and numerically integrating the S and P wave equations, they found that this phase shift implied a scale height of one micron. It is interesting that the above simple analysis (where $\delta \cong 3\lambda/L$), when applied to their experiment, also implies a gradient scale length, L, of around one micron. This analysis also shows (because C' is only 0.02) that Faraday rotation could be ignored in the Livermore experiment - even for magnetic fields in the megagauss range.

5.3 Dual-Time Probing

In order to record the velocity of an ablatively-accelerated foil, it was necessary to record its position at two different times. Since, for most of the probing studies, the magnetic fields and density gradients have a negligible effect on the polarization [12], one can utilize the polarization to obtain this dual-time data [14]. The technique is illustrated in Fig. 17. By utilizing the linear polarization of the probing beam, one can spatially separate and record two time-separated pulses initially travelling along the same optical path.

As shown at the top of Fig. 17, the horizontally polarized incident laser beam (30 ps, 5270 Å) is first rotated 90 degrees so that it is vertically polarized as it enters the beam-splitter. There is a variable optical delay in one leg and a 90-degree polarization rotator in the other leg. The two beams are then re-combined in a polarization-dependent beam combiner. The combined beam consists of two, time-separated probe pulses which are orthogonally polarized (vertical beam is delayed) and travelling along the same optical path. As shown at the bottom of Fig. 17, the combined beam is directed into the target parallel to the target surface. Since both pulses travel along the same optical path, a single, fast collector can be used.

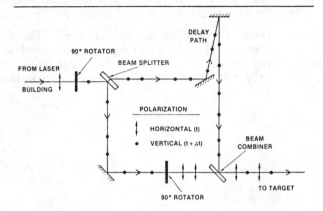

DUAL-TIME PROBING VIA POLARIZATION - ENTRANCE OPTICS

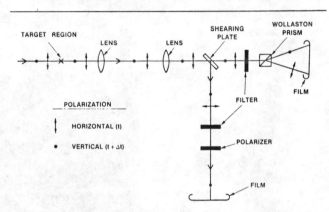

DUAL-TIME PROBING VIA POLARIZATION - COLLECTION OPTICS

Fig. 17. Principle of a dual-time probing, utilizing polarization.

As shown at the bottom of Fig. 17, a matched 2-lens system (described earlier) is used to image the probe pulses onto the film planes. In the case illustrated here, a shearing plate was used so that one simultaneously records a shearing interferogram (reflected) and dual-time shadowgrams (transmitted). For the dual-time shadowgrams, a Wollaston prism (7-degree separation angle) was used to angularly separate the two polarizations (times) so that they could be recorded side-by-side on the same photograph.

A dual-time shadowgram is shown in Fig. 18. The target was a 300-micron diameter by 6-micron thick foil of polystyrene (CH). One can see how the plasma has evolved between one nanosecond before the peak of the main laser pulse and one nanosecond after the peak. The plasma shadow has moved about 200 microns in

2 nanoseconds, giving a velocity of 10^7 cm/sec. The shearing interferogram, shown earlier in Fig. 6, was obtained on the same shot. A polarizing prism of sufficiently large aperture was not on hand so that only the late-time (t = +1 ns) interferogram was recorded.

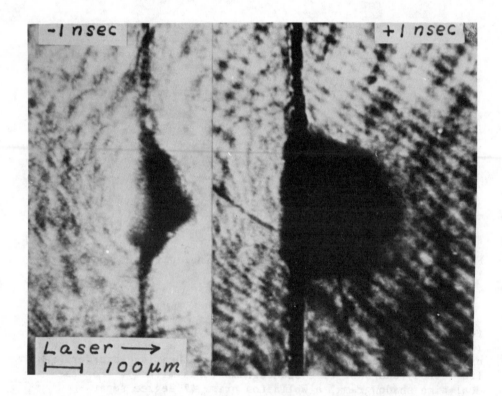

Fig. 18. Dual time shadowgraph.

6.   DIAGNOSTICS USING THE FOURIER-TRANSFORM PLANE

The two planes, oriented perpendicular to the optical axis of a lens and symmetrically located one focal length from the lens, are called the front and back focal planes. The light field at the back-focal plane of a lens is the spatial Fourier transform of the light field at the front focal plane [15]. This back-focal plane is thus referred to as the Fourier-transform plane, or FTP. Since the smaller wavenumber components are located near the axis, or center of the FTP, one can use an on-axis opaque mask in the FTP to filter out the longer wavelengths and thus observe the interesting steep-gradient (large wave-number) structure. This is used in the dark-field method. The FTP light pattern can also be characterized as an angular mapping of light rays in the front-focal plane (or target) region. This is seen, intuitively, since all of the rays leaving the target region in a particular direction are focused to the same point in the back-focal plane. Thus, the angular distribution of scattered light can be obtained by photographically recording the FTP light pattern.

A single-time, dark-field, shadowgram of the cold, dense target material, ablatively accelerated with 10^{13} W/cm^2 in a 1 mm spot, is shown in Fig. 19. The probing pulse was a Raman-shifted, second-harmonic (6258 Å), pulse of 100 ps duration and timed 2.6 ns after the peak of the main pulse. In order to accentuate information in the very steep gradient regions of the accelerated target (11 micron, CH foil) material, a large (5 mm diameter) mask was used. A matched 2-lens system was used to record the shadowgram. Note the turbulent appearance.

When there is an ambient gas, photoionized by the laser-produced plasma, one can use dark-field shadowgraphy to study the momentum coupling region between the ablated target material and the ambient. A dual-time, dark-field shadowgram is shown in Fig. 20 of the coupling region between a laser-ablated Al plasma and an 1.5 Torr ambient of 0.9 N_1, 0.1 H_2. Second-harmonic (5270 Å) probe pulses of 300 ps duration were used. The times are 52 ns after the peak of the main laser pulse. The probe pulses at these two times were filtered with separate masks of about 1 mm dimension. The low density of the scattering region ($\sim 10^{17}$ cm^{-3}) placed a rather severe restriction on the FTP filtering. The probe foci in the FTP had to be less than 100 micron from the mask (razor blade) edge. A single lens was used to record the shadowgram.

Fig. 19.   Single-time, dark-field shadowgram of ablatively-
           accelerated target.

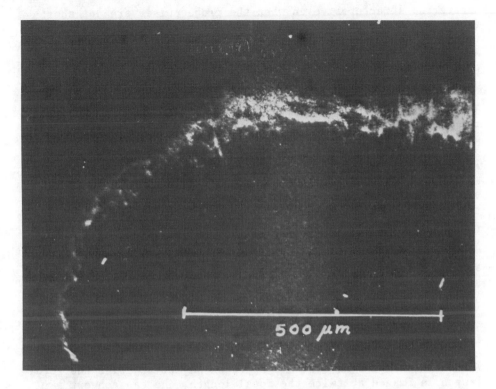

Fig. 20.   Dual-time, dark-field shadowgram of coupling region
           between ablated target material and ambient gas.

6.2 Angular Scatter

Measuring the angular distribution of scattered light is a
general and powerful technique for obtaining information about
the scattering medium. For example, in earlier work [17], we
were able to utilize the angular distribution of scattered light
from the main or heating pulse to measure the amplitude and
period of critical surface perturbations. We consider here only
two, rather simple, cases: gross-effects and periodic surface
structure. In these studies, the angular mapping of scattered
probe light was achieved by recording the image of the back-focal
plane (Fourier Transform plane, or FTP) of a lens having the
scattering medium at its front focal plane.

6.2.1 Gross shape and size effects. The gross features (orientation and extent) of the angular distribution of scattered light
can give information on the general shape and size of fast-moving

scattering structures, even when the probe pulses are not short
enough to freeze the motion and thus directly record this information. These features are related through simple, heuristic
arguments to the FTP pattern and are then illustrated with a recorded pattern.

The orientation of the FTP light pattern gives information
on the shape of the scattering structure. For example, horizontal filaments would scatter light vertically. The angular extent
of the FTP light pattern gives information on the size of the
scattering structure since smaller structure scatters or diffracts light through a larger range. For a quantitative example,
we consider filaments with axes in the x-direction and Gaussian
variation of spatial width, α, in the y-direction. Thus, the
transverse wave vector k_y is $2\pi/\alpha$ and the angular spread $\theta = k_y/$
k_z is λ/α. This would produce an FTP spatial spreds s of $\theta_y f$,
where f is the focal length of the lens. One would thus observe
a light pattern smeared in the y-direction and the width s of the
smear would give the size of the scattering structure since α is
$\lambda f/s$.

Just before taking the angular scatter data of the next
illustration (Fig. 21), we had been studying the ablative acceleration of dense target material, using dark-field shadowgraphy,
with a matched 2-lens system. Since an object at twice the focal
length is imaged at twice the focal length, one can convert a
matched 2-lens imaging system (discussed earlier) to a FTP
imaging system by replacing the second lens with one of half its
local length. For a relatively weak lens, this can be accomplished, without disturbing the original 2-lens system, by
placing an identical lens to the second lens in contact with that
lens. This was done to the matched 2-lens system in our dark-
field study in order to record the FTP light pattern shown in
Fig. 21. We left the dark-field mask in.

The FTP light pattern shown in Fig. 21 is for a 7-micron,
CH foil target irradiated with 125 J. One should ignore the central dark circular region (due to the dark-field mask) and the
narrow streaks (due to the un-irradiated target). A 100 mrad
size marker is shown and a 10 cm focal length lens was used for
the collector. We had seen, in both the dark-field and bright-
field shadowgrams, taken with rather long probe pulses, fine-
scale, filamentary-appearing structure with the axes generally
along the direction of motion. If the structures were actually
filamentary in nature, rather than being due to time smearing,
then the FTP light pattern would have been noticeably elongated
in the direction normal to the direction of motion. However,
the basic pattern in Fig. 21 is fairly symmetric, indicating

Fig. 21. Fourier-transform-plane light pattern for an ablative-
          ly-accelerated simple foil. Probe beam is parallel to
          target.

that the filamentary appearance was probably due to time smearing. In fact, subsequent data with shorter probing pulses (such
as Fig. 19) did not show this filamentary appearance. Since the
first lens had a focal length of 10 cm, the 1 cm width of the FTP
light pattern is consistent with structure of a few microns
($\lambda f/s$) size.

6.2.2 Scattering from a periodic surface. Targets with initial
periodic perturbations are useful in studying growth of the
Rayleigh-Taylor instability. The regions of interest in such a
study are at a high density. These regions can be better sampled
with a probing laser by directing the probe beam onto the target
at some angle θ, with respect to the target normal, rather than
by using side-on probing (see Fig. 22). However, such steep-
angle probing does not allow direct imaging of the density structure. Thus, information in the angular scatter or FTP pattern
is of particular value. For the special case of scattering by
periodic structure, the scattered light is concentrated in those
directions which are consistent with diffraction (e.g., grating
equation). Hence, a localization of the scattered light can be a

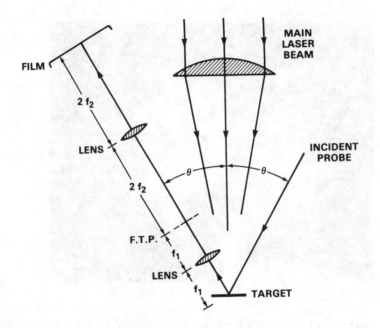

Fig. 22.   Experimental arrangement for recording the FTP light
           pattern when probing at a steep angle of incidence.

signature of a periodic structure. However, specular reflection
from a rather flat surface also provides a localized pattern and
must be considered.

For the example considered here [18], the angle of incidence
θ in Fig. 22 was 30 degrees and a second-harmonic probe was
used. Thus, densities up to 3×10^{21} cm^{-3} was sampled. By
photographing the FTP light pattern, a record is obtained of the
entire angular distribution of scattered light collected by the
first lens. It was inconvenient to directly record the light
pattern in the FTP of the first lens since the lens was rather
fast (an f/2 lens with a 10 cm focal length). Thus, a second
lens, with a 50 cm focal length, was used to relay the FTP
light pattern, one-to-one, onto the film. The conversion of the
matched 2-lens system, discussed with the first illustration of
scattered light, ends up with an equivalent optical system.

The FTP light pattern for an ablative accelerated, 6-micron-
thick, carbon target, with initial perturbations having a 50-mi-
cron period, is shown in Fig. 23. The thick-to-thin ratio was
1.26. A simple periodic structure in the accelerated target

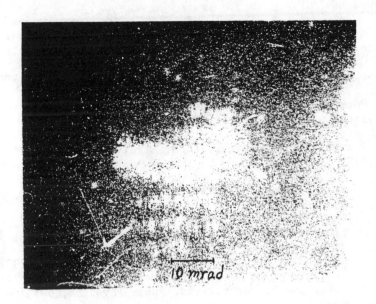

Fig. 23.   Fourier-transform plane light pattern showing angular
           scatter of light from a periodically-perturbed target.

plasma would produce the observed intense pattern and angular
width. The expected width is the ratio of the .5 micron probe
pulse wavelength to the 50 micron perturbation wavelength, i.e.,
10 mrad, as observed. The narrower angular spread in the other
direction (vertical) is probably specular since, at early times,
the accelerated target is still rather parallel to the grooves.
There is also a set of rather faint, vertically-repeated patterns. These are probably caused by interference due to reflections off a filter in front of the recording camera. Thus,
although complicated in detail, the observed angular scatter, off
of the accelerated target, is consistent with a simple, periodic
plasma structure.

Acknowledgements: The authors wish to acknowledge contributions
from S. H. Gold, R. H. Lehmberg and L. Sica. The technical
assistance of N. Nocerino, E. Turbyfill, M. Fink, L. Seymour, and
B. Sands is greatly appreciated. The work was supported by the
U.S. Department of Energy and the Defense Nuclear Agency.

REFERENCES

1. B. H. Ripin, R. Decoste, S. P. Obenschain, S. E. Bodner, E. A. McLean, F. C. Young, R. R. Whitlock, C. M. Armstrong, J. Grun, J. A. Stamper, S. H. Gold, D. J. Nagel, R. H. Lehmberg, and J. M. McMahon, Phys. Fluids 23, 1012 (1980).

2. A. Sommerfeld, Optics (Volume IV, Lectures on Theoretical Physics), (Academic Press, New York, 1950).

3. R. Benattar, C. Popovics, and R. Sigel, Rev. Sci. Instrum. 50, 1583 (1979).

4. J. A. Stamper, Applied Optics 19, 3053 (1980).

5. R. Illingworth, R. , . Thareja, A. Raven, P. T. Rumsby, J. A. Stamper, and O. Willi, J. Appl. Phys. 51, 1435 (1980).

6. M. M. Michaelis and O. Willi, Optics Comm. 36, 153 (1981).

7. R. Benattar and C. Popovics, J. Appl. Phys. 54, 603 and 609 (1983).

8. J. A. Stamper, NRL Memo Report 3872, October 27, 1978.

9. J. A. Stamper and B. H. Ripin, Phys. Rev. Lett. 34, 138 (1975).

10. J. A. Stamper, E. A. McLean, and B. H. Ripin, Phys. Rev. Lett. 40, 1177 (1978).

11. A. Raven, O. Willi, and P. T. Rumsby, Phys. Rev. Lett. 41, 554 (1978).

12. R. H. Lehmberg and J. A. Stamper, Phys. Fluids 21, 814 (1978).

13. D. W. Phillion, R. A. Lerche, V. C. Rupert, R. A. Haas, and M. J. Boyle, Phys. Fluids 20, 1892 (1977).

14. J. A. Stamper, S. E. Bodner, D. G. Colombant, R. Decoste, S. H. Gold, J. Grun, R. H. Lehmberg, W. M. Manheimer, E. A. McLean, J. M. McMahon, D. J. Nagel, S. P. Obenschain, B. H. Ripin, R. R. Whitlock, and F. C. Young in Laser Interaction and Related Plasma Phenomena, edited by H. Schwarz, H. Hora, M. Lubin, and B. Yaakobi (Plenum, New York 1981), Vol. 5, p. 638.

15. J. W. Goodman, Introduction to Fourier Optics (McGraw-Hill, New York, 1968).

16. J. A. Stamper, S. H. Gold, S. P. Obenschain, and E. A.
     McLean, J. Appl. Phys. 52, 6562 (1981).

17. B. H. Ripin in NRL Memo Report 3591 (1977), pg. 128-177,
     unpublished.

18. J. A. Stamper, S. P. Obenschain, B. H. Ripin, E. A. McLean,
     J. Grun, and M. J. Herbst, NRL·Memo Report 5093 (1983).

19. C. M. Vest, Holographic Interferometry (Wiley, New York,
     1979).

20. J. Sheffied, Plasma Scattering of Electromagnetic Radiation
     (Academic, New York, 1975).

21. G. R. Allen et al., Proceedings of the Fifth International
     Conference on High-Power Particle Beams, San Francisco
     (Sept. 1983), pp. 362-5, and references therein.

22. U. Kogelschatz and W. R. Schneider, Appl. Opt. 11, 1822,
     (1972).

23. A. G. M. Maaswinkel, R. Sigel, H. Baumhacker and
     G. Brederlow, Max Planck Institut Für Quantenoptik Report
     MPQ 66, (1982) unpublished.

24. D. T. Atwood, D. W. Sweeney, J. M. Auerback and P. H. Y.
     Lee, Phys. Rev. Lett. 40, 184, (1978).

25. A. Raven and O. Willi, Phys. Rev. Lett. 43 278, (1979).

26. R. Benattar and C. Popovics, Phys. Rev. Lett. 45, 1108,
     (1980).

27. C. L. Shepard, G. E. Busch, R. R. Johnson, R. J. Shroeder
     and J. A. Tarvin, Bull. Am. Phys. Soc. 28, (1983).

28. K. Lee, D. W. Forslund, J. M. Kindel and E. L. Lindman,
     Phys. Fluids 20, 51, (1977).

29. C. E. Max and C. F. McKee, Phys. Rev. Lett. 39, 1336,
     (1977).

30. J. Virmont, R. Pellat and P. Mora, Phys. Fluids 21, 567,
     (1978).

31. C. Popovics, Thesis of Paris XI University no 2427, (1981).

32. C. M. Vest, Appl. Opt. 14, 1601, (1975).

33. G. Gillman, Opt. Comm. 35, 127, (1980).

34. D. T. Atwood, IEEE J. Quantum Electron. QE14, 909, (1978).

35. R. Benattar, C. Popovics and R. Sigel, Rev. Sci. Instrum.
     50, 1583, (1979).

36. Yung-Lu Teng, R. Fedosejevs, R. Sigel, K. Eidmann, R. Petsch
     and G. Spindler, Max Planck Institut für Quantenoptik Report
     PLF 41 (1980) unpublished.

37. B. Grek, F. Martin, H. Pepin, G. Mitchel, T. W. Johnston
     and F. Rheault, Phys. Rev. Lett. 41, 1811, (1978).

38. O. Willi, P. T. Rumsby, A. Raven and Z. Q. Lin, Opt. Commun.
     41, 110, (1982).

39. K. Eidmann, A. Masswinkel, R. Sigel, S. Witkowski,
     F. Amiranoff, R. Fabbro, J. D. Hares and J. D. Kilkenny,
     Appl. Phys. Lett. 43 440, (1983).

40. J. Grun, S. P. Obenschain, B. H. Ripin, R. R. Whitlock,
     E. A. McLean, J. Gardner, M. J. Herbst and J. A. Stamper,
     Phys. Fluids 26, 588, (1983).

41. M. M. Michaelis and O. Willi, Opt. Commun. 36, 153, (1981).

42. R. Benattar and C. Popovics, J. Appl. Phys. 54, 603, (1983).

43. R. Benattar and C. Popovics, J. Appl. Phys. 54, 609, (1983).

44. J. A. Stamper, NATO ASI proceedings (Italy 1983). First
     paper of present chapter.

45. D. T. Atwood, SPIE 97, p. 413, High Speed Photography
     (Toronto, 1976).

46. R. Benattar, C. J. Walsh and H. A. Baldis, Opt. Commun.
     47, 324, (1983).

47. N. Niyanaga, Y, Kato and C. Yamanaka, Opt. Commun. 44, 48,
     (1982).

48. R. Benattar and J. Godart, On the use of X-ray radiation to
     probe laser created plasmas by refractometry, to be pub-
     lished in Opt. Commun.

49. F. C. Jahoda, LA-3963-MS (1968).

50. J. E. Hammel et al Nuclear Instruments and Methods, 207, 161, (1983).

51. W. T. Armstrong and P. R. Forman Applied Optics 5 320 (1979).

52. A. E. Siegman, Optics Communications 31 257 (1979).

53. L. K. Lam et. al. Optics Letters 6 475 (1981).

54. Jack Feinberg Optics Letters 7 486 (1982).

55. G. Dodel and W. Kunz Infrared Physics 18 773 (1978).

56. M. D. Bausman et. al. Fourth International Conference on Infrared and Millimeter Waves and Their Applications, Miami Beach Florida (1979).

57. R. M. Erickson et. al. to be published IEEE Transactions on Plasma Science, Dec. 1984.

58. J. E. Murray, "Temporal Compression of Mode-Locked Pulses for Laser Fusion Diagnostics", IEEE Jour. Quantum Elect., 17, 1713, Sept. 1981.

59. Gar. E. Busch, "Multiframe Holographic Shadowgraphy and Interferometry of Laser Target Plasmas", SPIE vol. 348, High Speed Photography (1982).

60. D. W. Sweeney, J. Opt. Soc. Am. 64, 559 (1974).

61. J. W. Mather, Phys. Fluids 8 (1965) 366.

62. R. Haas, H. Krompholtz, L. Michel, R. Fuehl, K. Schoenbach, G. Herziger, Phys. Lett 88A (1982) 403.

63. H. Krompholz, G. Herziger in: Chaos and order in nature, ed. H. Haken, Springer series in Synergetics, vol. 11 (Springer, Berlin, 1981) pp. 131-141.

64. H. Krompholz, E. Grimm, F. Ruehl, K. Schoenback, G. Herziger, Phys. Lett. 76A (1980) 255.

65. K. Schmitt, H. Krompholz, F. Huehl, G. Herziger, Phys. Lett. 95A (1983).

66. H. Krompholz, W. Neff, F. Ruehl, K. Schoenbach, G. Herziger,
    Phys. Lett 77A (1980). 246.

67. G. Herziger, H. Krompholz, L. Michel, K. Schoenbach, Phys.
    Lett. 64A (1978) 390.

SUB-NANOSECOND, FOUR-FRAME, HOLOGRAPHIC INTERFEROMETRY DIAGNOSTICS

G. R. Allen, H. P. Davis, L. P. Mix, and J. Chang

Sandia National Laboratories
Albuquerque, NM
USA

1.   INTRODUCTION

A holographic interferometry diagnostic technique has been developed and implemented on pulsed power experiments at Sandia National Laboratories. It incorporates a frequency-doubled Nd:YAG laser, with an extra-cavity pulse slicer, which produces a sub-nanosecond ($\tau \sim 350$ ps), green ($\lambda = 0.532$ µm) laser pulse. Using beamsplitters and varied optical path lengths, a train of 4 scene and 4 reference beam pulses is generated with variable inter-pulse time delays of 4 to 28 ns. The laser is command triggered with $< \pm 5$ ns jitter, and is rep-ratable at 1 Hz for easy alignment in the field. The diagnostic is portable, with all optics mounted on a 4' x 8' optical table. All vulnerable electronics are shielded for EMP protection. These capabilities make it a powerful diagnostic for plasma (with $\int n_e \cdot d\ell = 4 \times 10^{16}$ cm^{-2} yielding 1/10 fringe shift) and surface motion studies in our pulsed power accelerator experiments, where typically $\lesssim 1$ ns resolution is required during an event lasting 10-100 ns. It is particularly suited to the study of surface motion of imploding ICF targets where implosion velocities of ~ 20 cm/µs require $\lesssim 500$ ps resolution to measure the velocity with $\sim \pm 10\%$ accuracy. This paper describes the operating principles of holographic interferometry, the apparatus of this diagnostic, and results from its application to pulsed-power experiments.

The new four-frame holographic interferometry diagnostic technique has been developed using frequency-doubled (green) Nd:YAG, laser light. The green system is a similar, but

improved, version of an older, four-frame holography system which
uses a pulsed ruby laser. Both the green system and ruby system
are now operational.

Holographic interferometry [1] has a wide variety of applications in the particle beam fusion and related pulsed programs
at Sandia. The original ruby system has been used extensively
to diagnose the plasma in the A-K gap of electron and ion beam
diodes, the interaction of particle beams and projectiles with
matter, and the ablation and implosion dynamics of spherical and
cylindrical targets irradiated by electron and ion beams. The
new green system has the same capabilities with an improved
alignment technique and a shorter laser pulse, 350 ps (versus
4 ns for ruby). The faster time response afforded by the shorter
laser pulse will provide excellent (~ 100 µm) spatial resolution
of imploding targets with velocities as great as 30 cm/µs.

The four-frame holographic interferometry technique is
demonstrated schematically in Fig. 1. A laser, whose properties
will be discussed later, generates a single pulse which is split
into a reference beam and scene beam. Each beam is further split
to produce 4 scene and 4 corresponding reference beams. Each
scene beam traverses a different path length before recombining
into a colinear train of four laser pulses which exit the aluminum enclosure of the diagnostic setup. The scene beams are expanded in diameter, traverse through the experimental medium to
be diagnosed, and return to the interior or the diagnostic enclosure. A lens in the scene beam path images the plane of the
experimental medium onto the film plane, where the four reference
beams arrive coincident with their corresponding scene beams.
Each reference beam is directed onto the film at a 5° angle, from
the scene beams, which are normally incident. A scene beam and
its corresponding reference beam create a pattern of interference
fringes on the film with a fringe spacing of $\lambda/\sin \theta$ (= 6.1 µm in
this case). (High-resolution holographic film must be used. The
infrared light of the Nd:YAG laser has been frequency-doubled, to
green light, for which the film sensitivity is high. Any refractive medium that distorts the phase fronts of the scene beam,
relative to its reference beam, will result in a bending of the
fine scale interference fringes. Two exposures are made on the
same film plate, one exposure before and one during the presence
of the refractive medium (e.g., plasma) to be diagnosed, so that
two fine-scale (6.1 µm) interference patterns are super-imposed
on the film. In the reconstruction phase, where light passes
through the hologram, wave fronts are diffracted by each set of
fine-scale fringes, and the two wave fronts interfere to create a
macroscopic interference pattern (which is photographed with
Polaroid film) with fringes due to the difference in refraction
of the scene beams between the two exposures. Any refraction
encountered by the scene beam during both exposures (such as

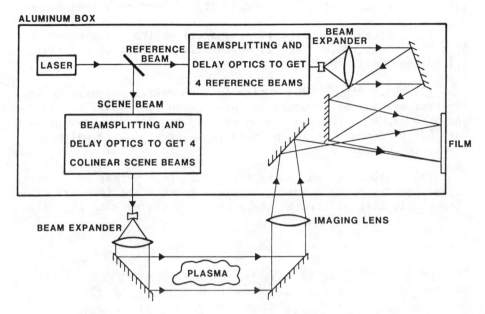

Fig. 1. Schematic of the four-frame holographic interferometer.

imperfections in the optics) is canceled out and does not appear
in the reconstruction macroscopic interferogram. A set of background fringes is sometimes produced on the macroscopic inter-
fer by slightly tilting a mirror in the path of the scene beams
between the first and second exposures. This enables the detection of fractional fringe shifts (e.g., due to a low density
plasma) as a slight bending of the background fringe pattern.

In this diagnostic arrangement, four frames of a dynamic
event are captured by virtue of the four-pulse laser train. The
framing time is the width of a single laser pulse (350 ps, FWHM),
and the inter-frame time, determined by the difference in optical
path lengths of the beams, is variable from 4 to 27 ns. The
macroscopic interferogram displays a full fringe shift for a
plasma electron like density (n_e) of

$$n_e L = 4.2 \times 10^{17} \text{ cm}^{-2}, \tag{1}$$

or a neutral gas density (n_o) of

$$n_o L \sim 7.6 \times 10^{18} \text{ cm}^{-2}, \tag{2}$$

where L (in cm) is the chord length of the laser beam through
the plasma or gas, assuming spatial uniformity along the chord.

Additional features which make this a particularly well-suited diagnostic method for accelerator experiments include:

(1) Portability. The laser and all optics are mounted on a 4' x 8' optical table (Fig. 2) which allows access to essentailly all pulsed-power experiments.

(2) Ruggedness. The entire optical system is enclosed in a light-tight aluminum housing for protection from dust, debris, and stray light. The power supplies and control electronics have been EMP-hardened and proven reliable in proximity to pulsed power accelerators.

(3) Easy alignment. The four scene beam pulses are colinear as they exit the aluminum housing, so that all four are simultaneously aligned through the experiment and back to the film

Fig. 2.   Photograph of the holographic interferometry diagnostic system.

plate. The Nd:YAG laser is rep-rated at 1 Hz so that the path of
all the scene and reference beams is easily and quickly aligned
to the film, without the use of an additional alignment laser.
Realignment of the diagnostic equipment between accelerator shots
is usually trivial.

(4) Schlieren photography is also done, coincident with
holographic interferometry, simply by splitting off part of the
return scene beams and passing that light through a schlieren
apparatus to a separate film plate (Polaroid film). With green
light, the sensitivity of the schlieren technique, which re-
sponds to density gradients, as opposed to densities for interferometry, exceeds the sensitivity of interferometry for density
gradient scale lengths less than 1 mm. In plasmas or gases with
strong density gradients, but relatively low densities, where
less than 1/10 fringe shift (roughly the minimum detectable on
the interferogram) is obtained, a detectable image of the density
gradients can be recorded with schlieren techniques.

2.   DESCRIPTION OF THE APPARATUS

The apparatus is assembled on two levels (Fig. 2): the
laser and pulse-tailoring components are rail-mounted on the
upper level; and the beam-splitting and imaging components are
mounted on the 4' x 8' optical table, below. The Nd:YAG oscil-
lator/amplifier/pulse-tailoring components are shown schematically in Fig. 3. The laser optical train begins at the laser
oscillator, consisting of two roof prisms (RP), a Pockels cell
(PC_0), rod and linear flashlamp, iris aperture for TEM_{00} mode
selection, and a beamsplitter (BS_1) which extracts half of the
oscillator energy as a usuable laser pulse of 10 mJ, 8 ns (FWHM),
at λ = 1.064 μm.

The HeNe laser and mirrors M_9 and M_{10} (removable) are used
to align the oscillator components. One-fourth of the oscillator
energy is split out (BS_1) and focused into a laser-triggered
spark gap (LTSG) which breaks down and transmits the high voltage (HV) pulse to the laser pulse slicer. The four-pass path
length extender utilizes prisms to delay the laser pulse so that
it arrives at the first Pockels cell (PC_1), in the laser pulse
slicer, simultaneously with the high voltage pulse. The linear
polarizer (LP_1) improves the polarization of the beam to increase
the effectiveness of the pulse slicer. The timing is such that

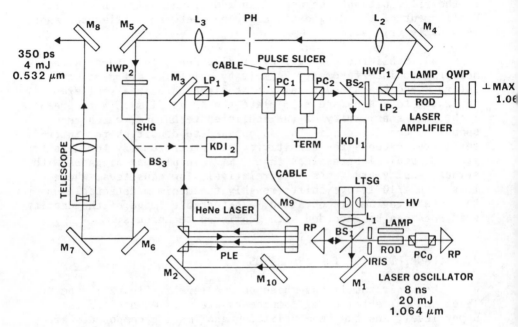

Fig. 3. Schematic of the laser and pulse-tailoring components of
         the diagnostic system.

PC_1 becomes transmitting just prior to the peak of the 8 ns
pulse, and PC_2 becomes opaque just after the peak. A shorter
pulse is thus sliced out of the 8 ns pulse. A beamsplitter (BS_2)
and photodiode ($KD1_1$) monitor the pulse shape.

The pulse proceeds to a double-pass amplifier consisting of
a half-wave plate (HWP), linear polarizer (LP_2), YAG rod, linear
flashlamp, quarter-wave plate (QWP), and rear mirror. A telescope (lenses, L_2 and L_3) with spatial filtering (pinhole, PH)
improves the intensity profile of the beam and reduces the beam
diameter by 2.6X. The increased beam energy density improves the
conversion efficiency of the second harmonic generator (SHG). A
photodiode ($KD1_2$) monitors the 4 mJ green beam, providing a signal to which the experimental event is synchronized. When sub-
nanosecond laser pulses are generated, a streak camera is used to

monitor the green beam. The next telescope expands the beam by 4X to a 6 mm diameter and a mirror (M_8) directs the beam vertically down, through a mechanical shutter, to the optical table surface inside the camera box.

Here the single laser beam is split into four scene beams and four reference beams, and the interframe time delays are introduced (Fig. 4). Before leaving the camera enclosure, en route to the experimental event, the polarization of scene beams 2 and 4 are rotated by 90°, and all 4 scene beams are recombined into a colinear pulse train. After passing through the experimental area, the scene beams then re-enter the camera box, where beams 1 and 3 are separated from 2 and 4 with a polarizing beam-splitter, and are directed to the film plane. The four reference beams remain inside the camera box, traveling the same path distance as, and recombining with, their respective scene beams, at the film plates, to create the holographic interferograms.

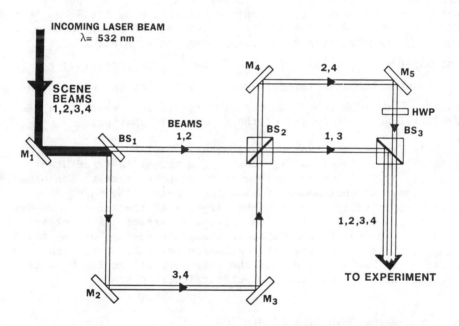

Fig. 4.   Schematic showing some of the components in the beam-splitting and interframe time delay optics.

Only two film plates are needed to record four holograms.
Frames 1 and 3 co-exist on one film plate. Both scene beams 1
and 3 illuminate a single film plate at normal incidence, while
the reference beams of frames 1 and 3 are incident at 5° to the
vertical and horizontal, respectively. In the reconstruction
phase, scene 1 is diffracted vertically from the optical axis of
the hologram, and scene 3 is displaced horizontally. Frames 2
and 4 on the second film plate are treated likewise.

3.   EXAMPLES OF DATA

Two characteristic examples of the type of data obtained on
pulsed power experiments at Sandia are shown in Figs. 5 and 6.
Figure 5 shows the development of the plasma in the A-K gap of an
electron beam diode, with 8 ns inter-frame time and 4 ns time
resolution (ruby laser diagnostic). The last 4 frames were obtained on a single accelerator shot; the first frame is from a
separate shot. From this interferogram, a profile of n_e across
the gap can be plotted at any lateral position along the gap,
with a symmetry assumption along the direction of the laser beam.
On the final frame, when the gap closes and the electron beam is
expelled, several fringe shifts are seen on both electrodes,
corresponding to n_e of a few x 10^{17} cm^{-3}. The rapid increase in
n_e, between the last three frames, demonstrates the need for a
laser pulse width (equal to the time resolution) which is a
smaller fraction than 1/2 of the inter-frame spacing.

In Fig. 6, the implosion of a hollow cylindrical target
foil (50 μm of aluminum) is shown. The target is irradiated
radially by an azimuthally symmetric, focused, intense ion beam
in an inertial confinement fusion experiment. The holography
laser passes axially through the target. As the target implodes,
the fringes are bent at the target inner surface due to plasma
ablation, and the fringes stop at the inner surface of the solid-
density target material. The implosion velocity of the inner
surface is thus measured, and the intensity of the ion beam is
inferred from hydrodynamic computer calculations.

4.   SUB-NANOSECOND TIME RESOLUTION

The implosion velocity of 2 cm/μs measured in Fig. 7 is slow
compared with the 20-30 cm/μs obtainable with thinner foils or
higher ion beam intensities. The time resolution (laser pulse
width) required to yield ~ 10% accuracy, in the measurement of
the implosion velocity (requires 100 μm spatial resolution), of a
3 mm diameter target, is

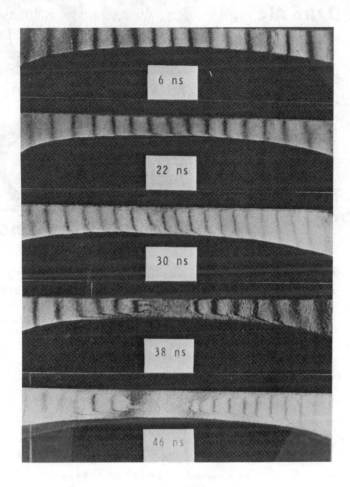

Fig. 5. Holographic interferograms of the dynamic plasma in the
A-K gap of an electron beam diode.

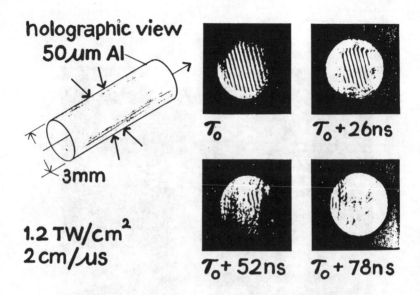

Fig. 6.   Holographic interferograms of the implosion of a hollow
          cylindrical target irradiated by an intense ion beam.

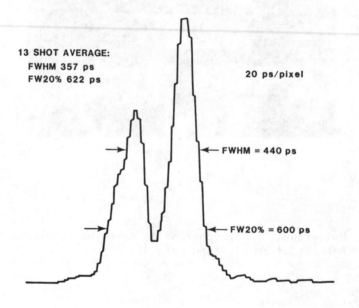

Fig. 7.   Streak camera record of a typical sub-nanosecond laser
          pulse.

$$\tau = \frac{10 \text{ ns}}{v(\text{cm}/\mu s)} \ . \tag{3}$$

The 4 ns ruby laser pulse used here, to measure 2 cm/µs, was sufficient. The 350 ps laser pulse of the Nd:YAG laser diagnostics system extends the range of accurately measurable velocities to 30 cm/µs.

The technique for slicing the laser pulse was described in Fig. 3. To obtain the 350 ps FWHM pulse, the critical components include a Lasermetrics Model SG-201 laser-triggered spark gap, operated at 15 kV DC voltage, 0.52 mm gap, and 100 psia nitrogen fill, and two Lasermetrics Model 1080 Pockels cells with 150 ps, 10-90% optical risetime. All components are matched to the terminated, 50Ω coaxial-cable transmission line.

A typical laser pulse detected by a streak camera (20 ps/ pixel), just following the second-harmonic generator crystal in the laser system, is shown in Fig. 7. The double spike is due to the spiky, multimode, output of the oscillator cavity which has no longitudinal mode selection. Good quality holograms are obtained in spite of this temporal structure. The FWHM of this pulse is 440 ps, and the full width at 20% is 600 ps. The average FWHM of 13 pulses is 357 ps, and the average full width at 20% is 622 ps.

The four-frame, sub-nanosecond capability is effective only if the inter-frame time can be made short enough to span only the duration of interest in the experiment. The faster events, requiring sub-nanosecond resolution, typically last 10-20 ns. By careful placement of the dozens of 2 inch diameter mirrors, needed to manipulate 8 different laser beams, the inter-frame time delay has been made as short as 4 ns, so the span of the 4-pulse train is as short as 12 ns. In addition, the laser is command triggered with approximately $\pm$ 3 ns (1σ) jitter for accurate synchronization with fast experiments.

To demonstrate the effectiveness of the faster laser pulse, we have diagnosed a high-velocity ionization front resulting from the breakdown of air by a ruby laser. This front is shown in Fig. 8, propagating from left to right in each frame. The top two were exposed with slow temporal resolution (7-ns FWHM laser pulse); the bottom two frames, with fast resolution (subnanosecond laser). The holographic interferograms displayed in the figure show sharp fringes with the subnanosecond exposure, which are noticeably blurred with the 7-ns exposure. The purpose of this test is to demonstrate the extent to which fringes become blurred and velocity measurements become inaccurate, with a longer laser pulse.

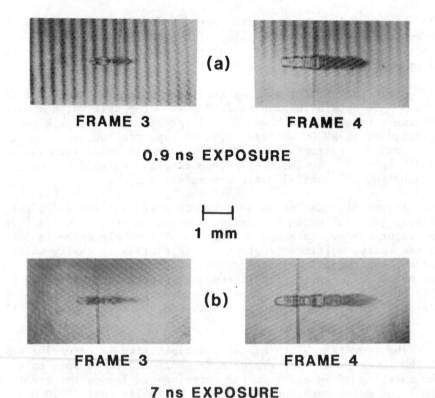

Fig. 8.   Holographic interferograms of a ruby-laser-produced
          air spark with time resolution (YAG laser FWHM) of
          (a) 0.9 ns and (b) 7 ns.

     The shortcoming of the longer laser pulse is made visually
obvious in this test. However, in an imploding target experiment, the position of the inner surface of the cylindrical foil
target is not visually blurred because the background fringe pattern merely terminates at the optically thick target material.
The uncertainty in the position of the imploding material versus
time (and, therefore, the uncertainty in velocity) is not apparent. But is must be calculated from the known pulse duration
of the diagnostic laser.

5.   CONCLUSION

A sub-nanosecond, four-frame holographic interferometry diagnostic has been developed and field tested on pulsed-power accelerator experiments. It has many desirable features which makes it a very useful and powerful diagnostic for plasma or target experiments of 10-100 ns duration. An experiment is being planned for the near term with an intense ion beam driven target where inner surface velocity and outward ablation velocity, both in excess of 10 cm/µs, will be measured.

6.   ACKNOWLEDGEMENT

We gratefully acknowledge the loan of Fig. 6 from Frank Perry of Sandia National Laboratories.

REFERENCE

1. C. M. Vest, Holographic Interferometry (Wiley, New York, 1979).

LASER INTERFEROMETRY: STREAKED SHADOWGRAPHY AND SCHLIEREN IMAGING

C. Popovics and R. Benattar

Laboratoire de Physique des Milieux Ionises,
Ecole Polytechnique, 91128 Palaiseau
France

1. INTRODUCTION

Optical diagnostics are extensively used to probe plasmas because they do not perturb the measured medium. Among them, the measurements of the refractive index and its variations give direct access to the electronic density, its gradients and the hydrodynamics of the plasma. Schlieren and shadowgraphy show qualitatively the density gradients and can even give quantitative information if certain conditions are fulfilled, in particular symmetry. Laser interferometry gives a very precise density measurement in a cylindrical plasma. Time resolution is achieved either using ultra-short laser pulses or using a streak camera. For characteristic times longer than a few nanoseconds, framing cameras are also of great interest.

We illustrate such methods, describing some optical diagnostics of a laser produced plasma, i.e.:

(i) the interferometric measurement of the sharpening of a density gradient by the light pressure of the laser creating the plasma, in the case of local deposition of the laser energy;

(ii) the velocity and acceleration measurements of thin targets by streaked shadowgraphy;

(iii.) and the measurement of the time evolution of a density gradient by streaked schlieren imaging.

In a transient, completely ionized plasma, such as a laser induced plasma, the main access to the electronic density is refractive index measurements, using a laser probing beam. An important advantage of such a technique is that is does not perturb the plasma. If the refraction of the probe beam is not too large , interferometry can be used and provides the electronic density profile, when the plasma is axisymmetric. Shadowgraphy and schlieren imaging give a direct access to density gradients in cylindrical or spherical plasmas. They are simplier to set up than interferometry, but the information obtained about a transient plasma is less quantitative than for interferometry: indeed, one has to use several repetitive shots, or a more sophisticated method, such as schlieren with grids [1], in order to obtain the density profile from a transient shadowgraphic or schlieren experiment. However, the usual arrangement provides the density profile in one shot, in the special case when "refracting fringes" are observed (see section 5).

Time resolution in this type of experiment is determined either by the probing laser pulse duration or by the camera used to record the images. When the plasma is one dimensional, either plane or spherical, a streak camera can be used to achieve the time resolution. The spatial resolution is fixed by the imaging optics used in the experiment. One micron resolution is reached when the configuration of the plasma is such that a microscope objective can be set up close enough to the plasma.

In the following, we shall discuss first the general concepts used in such optical diagnostics. The second section will describe the measurement of a density profile steepening by the ponderomotove force in a laser-target interaction. This will point out how the experimental characteristics (probe beam wavelength in particular) have to be chosen to reach a particular range of electronic density and gradient scale lengths. The next section will describe briefly the interest of streak shadowgraphy. In the last part we describe a time resolved, density profile, measurement using quantitative refractometry; that is a schlieren technique in the presence of refractive fringes. We also discuss, in this last section, the conditions of observation of refractive fringes.

2.   OPTICAL PROBING OF A LASER-TARGET INTERACTION PLASMA

As a laser beam does not propagate in a plasma in which the density is higher than its cut-off density and is strongly refracted near this critical density, its wavelength must be carefully chosen, depending on what is to be measured (range of density, refraction angle, or interference fringe shift in particular). The way to achieve temporal resolution is also an important feature of an experiment. The time scale, the geometry of the plasma and the required spatial resolution are among the main parameters to consider before choosing the way to time resolve the measurements.

2.1 Laser Driven Plasma Characteristics

The diagnostics shown further in this paper are set up for laser-target interaction experiments. Let us briefly consider the density range, time and spatial scales of such an experiment.

A high power, pulsed laser beam is focused on a planar or spherical target inside a 10 to 300 microns range focal spot. Pulse durations lie between 30 ps to a few nanoseconds, mainly in the 100 ps - 1 ns range. The wavelength of the driving laser gives the order of magnitude of the density range to be probed. Indeed, in these experiments, optical probing is often used to measure the electronic density gradient scale length between 0.1 to 10 times the driving laser cut-off density n_c ($n_c = \frac{\omega_o^2 \, m_e}{4\pi^2}$),

where ω_o is the laser frequency m_e and e the mass and charge of the electron. The absorption and transport of the laser energy to the target occur in this region, as shown on Fig. 1. Absorption of the laser is in the 0.1 - 1.0 n_c region. Parametric instabilities occur mainly at n_c, $n_c/4$, and between 0.1 n_c and n_c. Hydrodynamic instabilities and transport of the laser energy to the inside of the target occur at higher densities (a few tens of n_c). This transport region is difficult to reach with probing beams since only x rays propagate in such a dense plasma, except in the CO_2 laser interaction, for which the cut-off density is much lower.

The driving lasers extensively used are the neodymium-glass laser at its first, second, third, and fourth harmonic, the CO_2 laser and the iodine laser (1st and 3rd harmonic). Table 1 gives their wavelengths and cut-off densities.

Measuring a density profile in this type of experiment requires the ability to probe a high plasma density with extremely good spatial and temporal resolution.

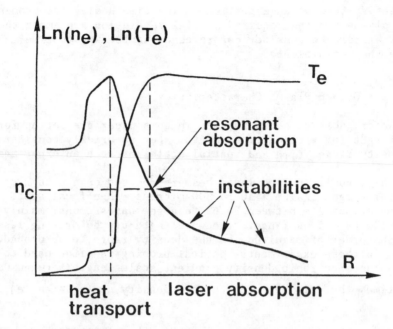

Fig. 1.   Schematic diagram of the electron density and tempera-
          ture profiles, along the laser axis for a laser-target
          interaction and localization of the main physical pro-
          cesses.

2.2   Probe Beam Intensity and Wavelength

A laser beam can be used as a probe beam if its wavelength
is sufficiently small for the beam to propagate at the density
which is to be probed and high enough for the refractive index
modification be measurable. Moreover, its intensity must be
higher than the self emission of the plasma and low enough to not
perturb the plasma.

Plasma intensity is generally not a problem. It is usually
not too high, since the probing beam power is about 10^{-4} to 10^{-5}
times the main laser power and the self emission of the plasma
can be made negligible by raising the intensity of the probing
beam. Special care must be taken when the frequency is a harmonic of the driving laser since non-linear effects in the plasma
give rise to these harmonics. Fortunately, in this case, the
self emission is localized in a very small plasma volume.

Wavelength considerations present much more severe limitations. Indeed, the probing beam wavelength has to be much lower

Table 1. Driving lasers wavelengths and corresponding cut off
          densities.

Laser	Harmonic Frequency	Wavelength (µm)	Cut off density (cm^{-3})
Nd	1w	1.05	$1.0 \ 10^{21}$
Nd	2w	0.53	$4.0 \ 10^{21}$
Nd	3w	0.35	$9.0 \ 10^{21}$
Nd	4w	0.26	$1.6 \ 10^{22}$
CO_2	1w	10.6	$1.0 \ 10^{19}$
I	1w	1.32	$6.6 \ 10^{20}$
I	3w	0.44	$5.9 \ 10^{21}$

than that of the driving laser. Therefore, the tendency to go to
short wavelength driving lasers makes the optical probing more
difficult. For CO_2 experiments (λ = 10.6 µm), a ruby laser
(λ = 0.69 µm) or the second harmonic of a neodymium-glass laser
(λ = 0.53 µm), is very convenient as a probing laser. For
neodymium experiments with the first harmonic (λ = 1.06 µm), its
fourth harmonic has been extensively used as a probing beam. For
UV experiments done with the third (λ = 0.35 µm) or the fourth
(λ = 0.26 µm) harmonic of a neodymium-glass laser, the probing
beam can no longer be from a laser since its wavelength has to be
in the x-ray range. Shadowgraphy has been done with a laser produced plasma used as an x-ray source. A good range of wavelengths would be 10 to 50 nm, but a monochromatic intense source
in this range, synchonised with the driving laser, does not
exist.

2.3 Spatial and Temporal Resolution

A first estimate of the temporal resolution needed in an experiment is linked to the driving laser pulse duration. Having a
probing resolution of about a tenth of the driving laser pulse
duration is a good order of magnitude estimate. However, in a
laser created plasma, time and space resolution are not independent since the plasma expands very quickly. A characteristic
scale for the speed of an isodensity contour is 100 km/s, i.e.,
0.1 µm/ps. So, if a 1 µm spatial resolution is needed, 10 ps
time resolution must be achieved. Such a resolution is not possible for every experiment since it is the best performance which

can be achieved at present. A usual technique is to probe the
plasma with an ultra short laser pulse. This can be obtained by
selecting one of the pulses of a mode locked train, or by cutting
a long pulse with an electronic shutter, or shortening a pulse by
non linear optical effects. A resolution of 100 ps can be obtained with electronic shutters (Pockels cell). The best resolution which has been achieved is 3 ps [2] with an a pumped dye
laser ($\lambda = 0.58$ μm).

Synchronization of the probing beam with the driving laser
is not a trivial problem. In several experiments [3-6], this has
been solved using a part of the driving laser pulse and quadrupling its frequency. The wavelength of the probing beam is
then much shorter than the driving laser one, and the pulse
duration is shortened by a factor of two in the frequency quadrupling. For the probe beam to be short enough so as not to degrade spatial resolution, very short driving laser pulses have to
be used (30 to 50 ps). However, this shortening is not always
sufficient. Even in experiments with a 30 ps duration driving
pulse, spatial resolution is limited by the probing pulse drua-
tion.

Another very convenient apparatus is the streak camera,
whose resolution is about 5 to 10 ps. The image of the plasma
is recorded through an entrance slit. This camera, therefore,
only records one dimensional images (planar or spherical). Usually, the geometry of the experiment cannot be controlled and,
as a result, a noticeable uncertainty affects the measurement.
If a diagnostic method which is able to check the geometry is
set up, the streak camera becomes the simplest way to achieve
high temporal resolution, as shall be seen in Section 5. In some
cases an electronic aperture framing camera can be used. The
performance achieved is about a 1 ns exposure time, which is
scarcely short enough for many experiments. However, when this
resolution is sufficient, this is the easiest way to obtain two
dimensional images.

We shall discuss now a few experiments which use these techniques.

3.   INTERFEROMETRIC MEASUREMENT OF A DENSITY PROFILE STEEPENING

This first example has been chosen to show how the parameters of an experiment have to be checked carefully in order to
measure a calculated physical process.

3.1 Steepening of a Density Profile

During the interaction of an neodymium-glass laser beam
(λ = 1.06 µm), with a solid target, the absorption of the laser
energy is primarily due to resonant absorption. In this process, the absorbed energy is deposited in plasma waves excited
near the critical surface. An electric field is raised locally.
The associated ponderomotive force steepens the density scale
length and of the height of the step. These have been calculated
analytically and in simulations [7-9]. What is expected, for the
interaction of a neodymium-glass laser of 3.10^{14} and $1.5 \ 10^{15}$
W/cm^2 intensity, normally incident on a flat target, is shown in
Fig. 2 [10]. The electronic density in the step is expected to
be between $0.5 \cdot 10^{21}$ and $1.4 \cdot 10^{21}$ cm^{-3} and its gradient scale
length is around 1 µm.

3.2 Requirements for the Experimental Arrangement in a Neodymium Glass Laser Interaction with a Target

3.2.1 Wavelength and pulse duration. Since the density step is
expected to be in a high density region, the wavelength of the
probing beam, which has to penetrate through this region, has to
be very short. The shortest wavelength probing beam available is
the fourth harmonic of the neodymium-glass laser (λ = 0.26 µm),
which has a cut off density of $1.6 \ 10^{22}$ cm^{-3}. The way to obtain
this probing beam, synchronized with the driving laser, is to
sample a part of the main pulse, in the amplification chain, and
pass it through two KDP crystals, in order to quadruple its frequency. This non-linear quadrupling also shortens the pulse by a
factor of two.

We have seen in Section 2.3 that 10 ps time resolution would
be required, to have 1 µm spatial resolution, in such an expanding plasma. Therefore, the experiments have been done with the
shortest pulses available, which are main pulses of 30 ps or
50 ps and probing pulses of 15 ps or 25 ps.

3.2.2 Size of the plasma. The measured interference pattern is
an image of the optical path length, integrated along the probing
beam. The radial dimension of the plasma, therefore, has to be
small enough for the gradient steepening not to be smoothed by
the outer regions of the plasma, and for the interference order

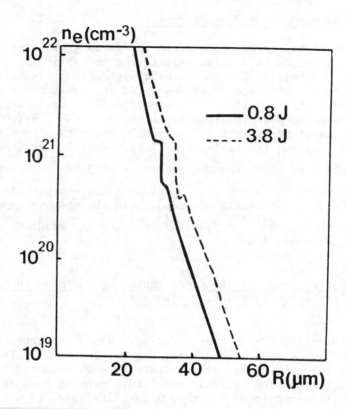

Fig. 2. Hydrodynamic simulation of the electron density gradient
steepening at the critical density, due to ponderomotive
effects, in laser-target interaction. A neodymium-glass
laser beam is incident on a glass target. The laser in-
tensity is $3 \cdot 10^{14}$ W/cm^2 (full line) and $1 \cdot 5 \ 10^{15}$ W/cm^2
(dotted line) [10].

to be of a few units for the probing rays going through the
region of interest. The following calculation shows that the
critical density ($n_c = 10^{21}$ cm^{-3}) is not reached with a UV probe
beam unless the plasma is no more than a few tens of microns
thick.

The order of interference, k, is linked to the electronic
density, n_e, by the relation

$$k = (2 \ n_{pr} \ \lambda_{pr})^{-1} \int_c n_e(r) \ dr ,$$

where λ_{pr} is the wavelength of the probing beam and n_{pr} its density cut off.

If we assume that the plasma is homogeneous over a length ℓ and negligible elsewhere, this relation becomes

$$n_e \ell = 2 n_{pr} \lambda_{pr} k,$$

and for $\lambda_{pr} = 0.26$ µm,

$$n_e \ell \ (cm^{-2}) = 8.5 \times 10^{17} k \ .$$

Therefore, $\ell = 24$ µm for $k = 3$ and $n_e = n_c = 10^{21}$ cm^{-3}. The 24 µm length is indicative of the fact that the plasma has to be very small for the critical layer to be probed.

3.2.3 Diffraction and refraction of the probe beam. Diffraction of the probing beam on the target and refraction in the plasma are likely to modify the phase shifts measured on a recorded interferogram. Their effects can be eliminated if the target is spherical and the imaging optics carefully focused.

Diffraction occurs when a sharp edge is out of focus. This is the case when the plasma is created on a large target. The imaging optics have a large aperture, in order to reach high spatial resolution and, as a consequence, their focal tolerances are very small.

Refraction cannot be neglected in such a plasma in which the density is very high and the density gradient is very steep. An order of magnitude estimate of the refraction angle α can be made assuming that the probing beam is incident on a 100 µm thick plasma, normal to a 1 µm scale length density gradient. The angle α is then

$$\alpha = (n_e/2n_{pr}) \cdot (1/L),$$

where n_e/n_{pr} is the ratio between the electronic density and the probing beam cut off density, l is the size of the plasma, and L the density gradient scale length. For a UV probing beam ($\lambda = 0.26$ µm), propagating in a 10^{21} cm^{-3} electron density, this estimate leads to a refraction angle of 3 degrees. However, for interferometric deconvolution, refraction is not taken into account, since it has been shown [11-12] that the modification of the phase shifts, due to refraction can be neglected if the imaging optics are accurately focused. This assumption has been verified for such a plasma [13] by varying the focusing for a holographic interferometer.

Thus, focusing of the imaging optics is a crucial point. This is why several laboratories use holographic interferometry [3-6] in which focusing is made during the reconstruction of the image.

3.3 Reliable Measurements

The occurence of the steepening of the density gradient, by ponderomotive effects, at the critical layer in the interaction of a neodymium laser with a target has been demonstrated experimentally in 1978 [3] and studied in more detail during the following years [4-5].

There experiments have been performed with short, neodymium-glass laser pulses (30 ps or 50 ps) on very small microspheres (40 μm diameter). This provided a short UV probe beam (15 ps or 25 ps) and a small size for the plasma, with no diffraction effects. The probing pulse duration was short enough to avoid smearing of the fringes by the hydrodynamic movement. Figure 3a shows an inteferogram, from which a density gradient steepening has been measured, with a holographic interferometer [3]. A polarization interferometer [14] has also been used to study the density step and its evolution with time and laser intensity. An arrangement using this interferometer is shown in Fig. 5 (see Section 3.2). Figure 3b shows the density profile for various laser irradiances. For the lower energy, the step is not completely shaped. For the higher energy, a blurring of the fringes does not allow measurement of the whole density profile. The 3.10^{14} W/cm^2 shot is a confirmation of the steepening obtained previously from Fig. 3a.

4. SHADOWGRAPHY

Shadowgraphy is a diagnostic method which is easy to set up and which is often used to obtain a qualitative description of the shape and the size of the plasma. It gives a map of the dense regions of the plasma and of the strong gradients. It is of special interest in the unstable plasmas, where strong inhomogeneities occur.

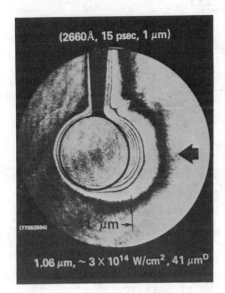

(a)

(b)

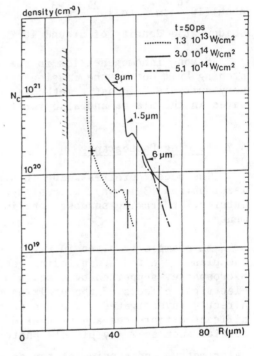

Fig. 3a. Interferogram at peak irradiation of a neodymium-glass
         laser, incident on a 41-µm diameter microballoon. Inci-
         dent laser intensity is $3 \cdot 10^{14}$ W/cm^2 [3].

     3b. Electron density measured by interferometry in Nd-glass
         laser target interaction. Density profile steepening
         is observed for $3 \cdot 10^{14}$ W/cm^2. For higher energy
         ($5 \cdot 10^{14}$ W/cm^2), the step could not be measured in this
         experiment [5].

4.1 Principles of the Diagnostic Method

When a probing beam is incident on a plasma, the rays which are absorbed, reflected, or refracted out of the imaging optics do not reach the detector. Therefore, dense regions which reflect or absorb the probing beam, and strong gradients which refract it, are represented by shadows in the image plane. In laser-target interactions, the target is surrounded by a shadow. Its edge represents an iso-optical path in the plasma.

This shadow follows a restricted range of densities, even if it is not exactly an isodensity contour, since the depth of the plasma crossed by the probe beam varies from one ray to the other. With an optical probing beam, this range varies between about 10^{19} cm^{-3} to 10^{21} cm^{-3}. An x ray probing beam gives a shadow at a density of around 10^{22} cm^{-3}.

Strong inhomogeneities in the plasma strongly refract the probing beam out of the imaging optics. Shadowgraphy is then a very suitable diagnostic method for the propagation of shock waves in the plasma and also for filamentary structures [15-17].

4.2 Framing Photography

The recent progress made in synchronization of very short laser pulses (3-15 ps), with high power lasers, has raised the interest of framing shadowgraphy in laser interaction experiments.

An example is given in Fig. 4 of a multiframe shadowgraph of the plasma created by an iodine laser ($\lambda = 1.3$ μm) focused on a microsphere, supported by a stalk or in free fall [18]. The electrical effects of the stalk on the interaction had never been directly investigated. The two sets of shadowgrams show no significant differences and neither do the other simultaneous diagnostics such as x ray pinhole photography, k_α spectrometry, ion calorimetry, and studies of the angular distribution of the fast ions accelerated during the interaction.

The arrangement of this multiframe imaging system [2] is shown in Fig. 5. The 3 ps pulse of a dye laser is temporally split by a set of 6 optical delays made of partially reflecting mirrors. Each beam is sent to the plasma by a part of a multiprism and collected by a second multiprism. An objective is inserted, to image the plasma on the film plane, giving 6 images,

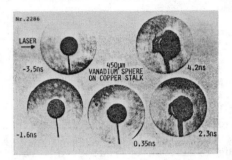

(a)

(b)

Fig. 4.   Multiframe shadowgrams of iodine laser irradiated tar-
          gets [18].
          a)   Sphere on a stalk.
          b)   Free falling sphere.

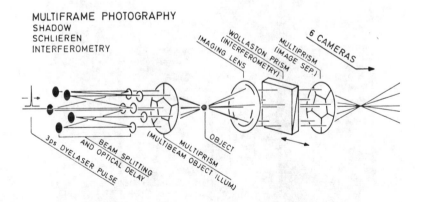

Fig. 5.   Mounting of a picosecond multiframe shadowgraphic imag-
          ing system.   Interferograms can be obtained by inserting
          a Wollaston prism and two crossed polarizers [2].

registered at 6 different times. A Wollaston prism and polar-
izers can be inserted to transform the shadowgraph system into a
multiframe inteferometer.

4.3 Streak Shadowgraphy

Shadowgrams are often recorded on a streak camera in order
to get time resolution. As the image is recorded on a slit,
streak shadowgraphy has to be applied to one-dimensional problems. The study of acceleration of thin foils, irradiated with a
large laser focal spot, is one of the main applications of streak
shadowgraphy, the probe beam being either an optical or a x-ray
source.

An an example, Figure 6 shows the streak record of the interaction between a neodymium-glass laser (λ = 1.06 µm) and a
double foil composed of two thin foils (a target and an impact
foil) separated by a variable distance [19]. The probing beam
is the second harmonic of the main laser (λ = 0.53 µm). The aim
of the experiment is to measure the velocity of the dense plasma,
created by the laser, on the first foil. This is done by measuring the time of impact of the target, on to the impact foil, as
a function of the distance between them. This velocity is linked

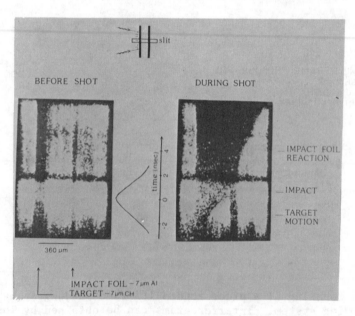

Fig. 6.   Streaked shadowgrams of a target colliding with an
          impact foil [19].

to the hydrodynamic efficiency of the interaction, i.e., the efficiency of conversion of laser energy into acceleration of the dense plasma. Figure 6 shows the shadow starting from the plastic target which is as irradiated by the laser (target motion). This shadow reaches the second foil (impact) and, a few nanoseconds later, a plasma expands behind the impact foil. This plasma is created when the shock, induced by the impact of the dense part of the target, reaches the rear of the impact foil. Figure 7 shows the velocity deduced from the shadowgrams, recorded for several foil separation distances and for several thicknesses of the foils. It shows that the measurement is independent of the impact foil thickness and that the velocity decreases when the target thickness increases. It can be noticed that the velocity, which can be deduced from the shadow of the low density plasma expanding behind the target, is twice the velocity deduced from the impact. This indicates that the low density plasma expands more quickly than the dense plasma. However, this measurement is very imprecise since an isodensity does not follow exactly the shadow. This is because the size of the plasma crossed by the probe beam increases (see Section 4.1).

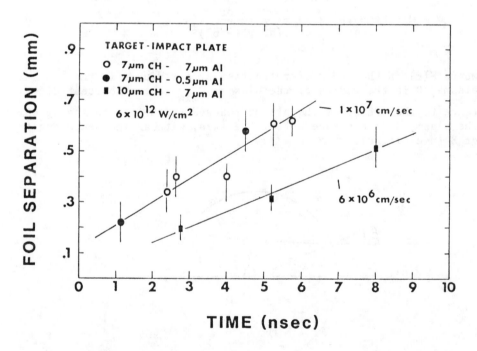

Fig. 7.   Graph of target impact foil separation versus time
           (arbitrary time origin), deduced from Fig. 6. The
           slopes of the lines give the velocity of the dense
           target [19].

X-ray shadowgraphy is now widely used [19] since it gives a
more direct measurement; the shadow lying in the dense part of
the plasma.

5.   SCHLIEREN IMAGING

A shadowgraphic mounting can easily be transformed in to a
schlieren arrangement by inserting a mask in the focal plane of
the imaging optics. The rays which are refracted by the plasma,
or diffracted by the edge of the mask, are collected in the image
plane. The image of the plasma is formed only by the refracted
rays. Such a technique has been widely used to probe the density
gradients of a plasma [1, 11]. This is because the angle of refraction, of a ray passing through the plasma, is directly linked
to the density gradient.

If the plasma is spherical, Bouguer's law leads to an integral relation between the index of refraction, the impact parameter p of a ray, and its refraction angle α:

$$\alpha = \Pi - 2B - 2p \int_{R_{min}}^{R} \frac{dr}{r(N^2 r^2 - p^2)^{1/2}} , \qquad (5.1)$$

where $N(r)$ is the index of refraction at a radius r in the
plasma, R is the radius of the plasma, R_{min} is the shortest distance to the center of the plasma reached by the ray inpinging on
the plasma at a distance p from the axis, with an incidence angle
of B (see Fig. 8).

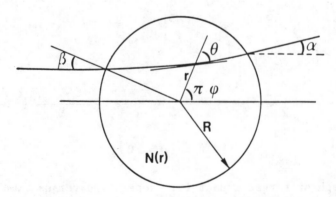

Fig. 8.   Schematic diagram of an optical ray refracted in a
          spherical plasma.

5.1 Classical Technique

For spherical (or cylindrical), plasmas the schlieren technique has been used for several years. In these cases the relation (5.1) can be inverted to obtain the index of refraction N, at a given distance r, in the plasma [1],

$$N(s) = N_o \exp(-\Phi(s)),$$

$$r(s) = s \exp(\Phi(s)),$$

with $s = r\,N(r)$ (5.2)

$$\Phi(s) = \frac{1}{\pi} \int_s^R \frac{\alpha(p)\, d\,p}{(p^2 - s^2)^{1/2}} \ .$$

In a completely ionized plasma, the electron density n_e is then deduced from

$$N^2 = 1 - n_e / n_{pr} \ ,$$

where n_{pr} is the cut off density of the probing beam.

The usual way to measure the $\alpha(p)$ relation is to move the schlieren mask, measure its position and the position of the associated changes in the image. However, for laser target interaction, this technique is difficult due to the non reproducibility of the plasma from shot to shot, and the delay between two shots. An alternative has been proposed, namely to replace the mask by a grid in order to measure several angles of refraction at the same time [1]. Two main difficulties have limited the use of such grids: 1) the range of the measured refraction angle has to be known to construct the grids; 2) how to relate each part of the image to a given selected angle of refraction is not obvious.

5.2 Quantitative Refractometry with Refractive Fringes

An interesting alternative has been proposed for laser target interaction and, more generally, for spherical plasmas, presenting a decreasing density profile, such that the $\alpha(p)$ relation can be approximated by a linearly decreasing function [20-22].

In such a plasma, the refracted rays can focus on a ring and the $\alpha(p)$ relation can be measured from the interference pattern formed in the image plane (refractive fringes).

5.2.1 Principle of the method. Let us assume that the refracted
rays, collected in the imaging optics, focus on a ring R
(Fig. 9). Then, the refracted rays reach the image coming from
a virtual or going to a real image R' of R. R' is real or virtual
depending on the position of R versus the focus of the imaging
optics. The schlieren stop diffracts the light. Therefore, an
interference pattern is formed in the image plane, with the light
coming from the two sources R' and S. The interference pattern
consists of annular fringes, separated by a distance d, proportional to the square roots of integers [23],

$$d_k{}^2 = 2 k \lambda_{pr} / (1 / L_1 \pm 1/L_2), \qquad\qquad (5.3)$$

where k = 1, 2, 4, . . . , λ_{pr} is the probing beam wavelength,
and L_1 and L_2 are the distances from the schlieren stop and
the source R' to the image plane. The sign depends on the nature
of R' (+ for R' real, - for R' virtual).

    It should be noticed that such fringes appear in shadow-
graphy when the image plane is out of focus. This can also
be used for electronic density measurements.

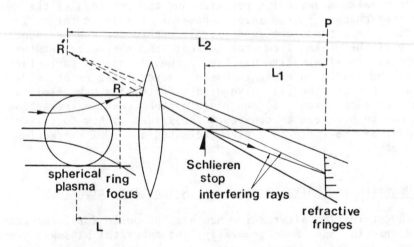

Fig. 9.   Schematic representation showing the formation of re-
          fractive fringes, in the schlieren image, of a spherical
          plasma. Refracted rays focus on a ring R. R' is the
          image of R, through the imaging optics. The image plane
          P is represented titled by 90 degrees for better illus-
          tration [21].

5.2.2 Existence of the Ring Focus. The ring focus does not exist in all experimental situations. We shall consider the theoretical conditions for its existence. But, first, let us look at a simulation which shows that, for a selected range of angles of refraction, the refracted rays focus in a ring [25]. Figure 10 is the ray tracing, of probing beams at 0.26 μm and 0.53 μm, crossing a spherical plasma where the density profile is exponential. Is is clear that the rays, shown as the dashed lines, focus in a ring and that selecting these rays by inserting a mask, which eliminates the non refracted rays, would lead to refractive fringes. Experiments have also pointed out the occurence of such a focus [20-22].

Theoretically, the occurence of this ring focus has been demonstrated to be linked to the shape of the function $\alpha(p)$, where α is the angle of refraction and p the impact parameter of a probing ray [21].

The conditions of observation are the following:

(1) The plasma is spherical (an extension to cylindrical plasmas could be found).

(2) α and p satisfy the relation,

$$p = R_o \cos \alpha - L \sin \alpha ,\qquad\qquad (5.4)$$

where R_o is the radius of the plasma and L is the distance between the ring plane and the center of the plasma (see Fig. 11).

When the ring is far from the plasma ($L / R_o \gg 1$) and the angle α small, this relation can be linearly approximated. For a restricted number of rays focusing on a ring of radius R, it becomes,

$$p = R - L \alpha .\qquad\qquad (5.5)$$

The condition for the linear assumption is then,

$$L / R > 3 .\qquad\qquad (5.6)$$

The schlieren technique is a way of selecting a range of impact parameters where the refraction angles α are small (limited by the aperture of the imaging optics) and the function $\alpha(p)$ is approximately linearly decreasing.

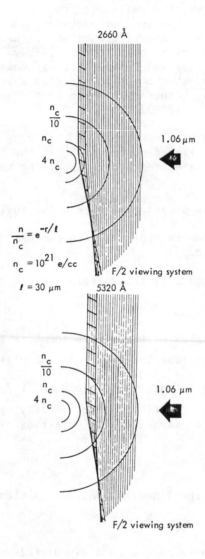

Fig. 10. Ray tracing of the refraction, of a probe beam, in a
          spherical plasma with an exponential density profile.
          For a small range of the impact parameter (inside the
          dashed pencil of rays), the rays focus on a ring [24].

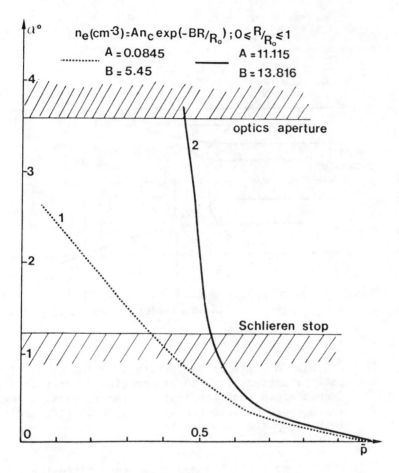

Fig. 11.   Schematic ray tracing in a ring-focusing plasma [24].

Figure 12 shows typical $\alpha(p)$ functions, calculated from two given exponential density profiles. The dashed parts of the plane represent the selection of the angle of refraction, due to the optics aperture and the position of the schlieren stop, which would be appropriate to probe such a density profile.

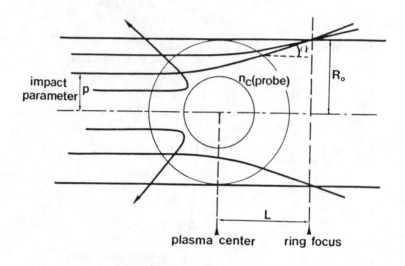

Fig. 12.   Calculated angle of refraction as a function, of im-
           pact parameter, for two exponential density profiles.
           Dashed areas show the limit of the refraction angle
           which could be selected by a schlieren stop and the
           imaging optics aperture [24].

5.2.3 Time resolved schlieren measurements. Since these measurements have to be done on spherical plasmas, streak camera recording can be combined with such a diagnostic method, giving
time resolved measurements of the density profiles. It can be
noted that the occurence of refractive fringes, separated by distances proportional to $(k)^{1/2}$, is a test of the quality of the
diagnostic. From the experiments done [21, 22, 25], it can be
seen that the fringes have been observed mostly for long laser
pulses and moderate intensities, where spherical symmetry is more
likely to occur. It can be noticed also that, for a given shot,
the fringes are not seen during the entire lifetime of the
plasma.

Figures 13 and 14 show an example of a streaked shadowgram and the density profiles deduced at different times [21]. The plasma is created by a long, neodymium-glass laser pulse having a 2.5 ns pulse duration, incident on a 10 μm diameter glass microsphere. The probing beam is the third harmonic of the driving laser, with longer duration (20 ns).

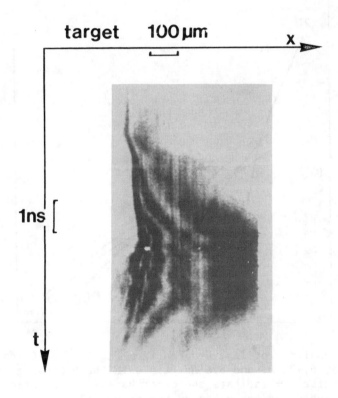

Fig. 13.   Streaked schlieren image of a plasma created by a 70 μm diameter glass microsphere, irradiated by a neodymium-glass laser [24].

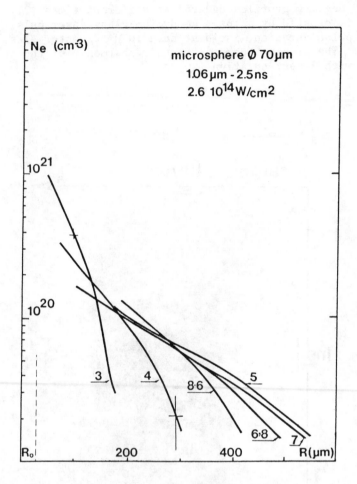

Fig. 14. Density profiles, deduced from Fig. 13, at times 3, 4,
          5, 6, 7, 8, and 8.6 ns (t = 0 is the beginning of the
          pulse). + indicate the error bars. The interaction
          conditions are given on the figure [24].

6.   CONCLUSION - PROSPECTS

     Among laser interaction diagnostics, optical probing, in-
cluding laser interferometry, shadowgraphy, and schlieren
imaging, is the diagnostic method with the best time and space
resolution.   One to ten microns and 3 to a few tens of picoseconds are presently the performance values of these diagnostic
methods.   These methods provide the electron density profile and
determine the location of discontinuities, such as shock waves.

They can be considered as basic measurements of the electron density, but the experiment has to be carefully adapted to probe the desired range of densities, depending on the size of the plasma. These diagnostics are also very sensitive to instabilities, turbulence or simply asymmetry. This can prevent density provile measurements, but can also be used to diagnose such processes.

Presently two main development directions seem to be adequate, for these diagnostics, in laser target interaction experiments. First, higher densities have to be probed in order to study transport processes and hydrodynamic instabilities. X-ray sources must be improved. X-ray shadowgraphy is already used in the 1-10 keV range [19]. After source development, quantitative schlieren or shadowgraphy could then be extended to the x-ray domain. Studies are beginning in this field [26-27]. Secondly picosecond synchronized dye lasers open new areas of underdense plasmas diagnostics with framing photography. The knowledge of the density profiles, in large underdense plasmas, would be of great interest for parametric instability studies. However, the inhomogeneities which occur, in such large plasmas, could be an important obstacle to reach quantitative measurements with optical probing.

REFERENCES

1. U. Kogelschatz and W. R. Schneider, Appl. Opt. 11 1822, (1972).

2. A. G. M. Maaswinkel, R. Sigel, H. Baumhacker and G. Brederlow, Max Planck Institut für Quantenoptik Report MPQ 66, (1982) unpublished.

3. D. T. Atwood, D. W. Sweeney, J. M. Auerback and P. H. Y. Lee, Phys. Rev. Lett. 40, 184, (1978)

4. A. Raven and O. Willi, Phys. Rev. Lett. 43, 278, (1979).

5. R. Benattar and C. Popovics, Phys. Rev. Lett. 45, 1108, (1980).

6. C. L. Shepard, G. E. Busch, R. R. Johnson, R. J. Shroeder and J. A. Tarvin, Bull. Am. Phys. Soc. 28, 1058, (1983).

7. K. Lee, D. W. Forslund, J. M. Kindel and E. L. Lindman, Phys. Fludis 20, 51, (1977).

8. C. E. Max and C. F. McKee, Phys. Rev. Lett. 39, 1336, (1977).

9. J. Virmont, R. Pellat and P. Mora, Phys. Fluids 21, 567, (1978).

10. C. Popovics, Thesis of Paris XI University no 2427, (1981).

11. C. M. Vest, Appl. Opt. 14, 1601, (1975).

12. G. Gillman, Opt. Comm. 35, 127, (1980).

13. D. T. Atwood, IEEE J. Quantum Electron. QE14, 909, (1978).

14. R. Benattar, C. Popovics and R. Sigel, Rev. Sci. Instrum. 50, 1583, (1979).

15. Yung-Lu Teng, R. Fedosejevs, R. Sigel, K. Eidmann, R. Petsch and G. Spindler, Max Planck Institut für Quantenoptik Report PLF 41 (1980) unpublished.

16. B. Grek, F. Martin, H. Pepin, G. Mitchel, T. W. Johnston and F. Rheault, Phys. Rev. Lett. 41, 1811, (1978).

17. O. Willi, P. T. Rumsby, A. Raven and Z. Q. Lin, Opt. Commun. 41, 110, (1982).

18. K. Eidmann, A. Maaswinkel, R. Sigel, S. Witkowski, F. Amiranoff, R. Fabbro, J. D. Hares and J. D. Kilkenny, Appl. Phys. Lett. 43, 440, (1983).

19. J. Grun, S. P. Obenschain, B. H. Ripin, R. R. Whitlock, E. A. McLean, J. Gardner, M. J. Herbst and J. A. Stamper, Phys. Fluids 26, 588, (1983).

20. M. M. Michaelis and O. Willi, Opt. Commun. 36, 153, (1981).

21. R. Benattar and C. Popovics, J. Appl. Phys. 54, 603, (1983).

22. R. Benattar and C. Popovics, J. Appl. Phys. 54, 609, (1983).

23. J. A. Stamper, this publication, Chapter 6.

24. D. T. Atwood, SPIE 97, p. 413, High Speed Photography (Toronto, 1976).

25. R. Benattar, C. J. Walsh and H. A. Baldis, Opt. Commun. 47 324, (1983).

26. N. Niyanaga, Y. Kato and C. Yamanaka, Opt. Commun. $\underline{44}$, 48, (1982).

27. R. Benattar and J. Godart, On the use of X-ray radiation to probe laser created plasmas by refractometry, to be published in Opt. Commun.

MOIRE-SCHLIEREN, TIME-DIFFERENTIAL INTERFEROMETRY, AND ENHANCED SENSITIVITY OF FARADAY ROTATION MEASUREMENTS.

P. R. Forman

Los Alamos National Laboratory
Los Alamos, NM
USA

1.   MOIRE-SCHLIEREN

There are experiments where the density or the density gradients are too large to allow conventional interferometric measurements. In such cases a schlieren measurement can yield useful information. Moiré-schlieren has the distinct advantage over conventional schlieren in that a quantitative result is obtained without having to resort to intensity calibration [1]. The method is illustrated schematically in Fig. 1. Two coarsely ruled transmission gratings are mounted after the plasma. They are separated by a distance, D, and their rulings are slightly missaligned so that Moiré-fringes are produced. The beam passes through a lens and is incident upon a masking screen, which is located a focal distance away from the lens. The coarsely ruled gratings produce many orders in this image plane. One of these orders is selected by an aperture and allowed to be incident upon the recording film. Rays from the collimated laser beam which are deflected by the density gradients of the plasma, which are perpendicular to the direction of the coarse gratings, distort the Moiré-pattern, giving a large effective amplification to the deflection. The results appear to be the usual interferograms with the usual background fringes. The interpretation of such recordings are naturally different. The fringes are Moiré fringes and a deflection of a "fringe" is the result of the plasma density gradients displacing the shadow projection of the rulings of the first grating by one ruling period of the second grating. The sensitivity is then determined by the grating period δ and the separation of the two gratings D. The angular deflection θ, which gives one Moiré-fringe, is given by

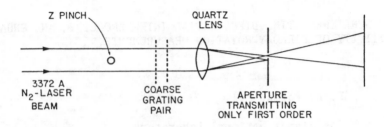

Fig. 1.   Schematic drawing of Moiré-schlieren arrangement.
          Nitrogen laser beams incident from the left.

$$\text{Tan}(\theta) = \delta/D.$$

Such measurements have been performed on a laser initiated
z pinch [2]. This z pinch produces a 100 micrometer diameter
plasma with a peak density of 10^{19} to 10^{20} cm^{-3}. The total duration of the plasma is about 100 ns so the laser pulse must be
short in order to avoid blurring of the Moiré-fringes. A pulsed
nitrogen laser, with a time duration of 2 to 3 ns, was chosen as
a cheap convenient light source. A sample of the Moiré-fringes
measured on the z pinch are shown in Fig. 2. It is such fringe
displacements, along with the assumption of cylindrical symmetry,
which allows Abel inversion of the data, that the densities of
the z pinch have been determined.

The holographic analog of the scheme has previously been
demonstrated [1]. Holographic recording allows the experimenter
to set up the Moiré-gratings at his leisure, after the fact, and
adjust the sensitivity or the direction of the density gradients
that are being measured. Unfortunately, the coherence quality of
the pulsed nitrogen laser, that was used, did not allow this possibility.

The limiting sensitivity is determined either by how well
one can collimate the laser light or by diffraction smearing of
the shadow, of the first Moiré plate, at the second plate. This
then sets a practical upper limit to how far apart one may

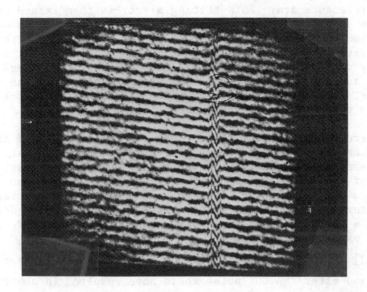

Fig. 2. Moiré-schlieren fringes of the laser initiated z-pinch.
         Density distribution is deduced by Abel inversion of
         these data.

separate the two Moiré-screens. Other experiments [1] have shown
that the limit of sensitivity is at least

$$\int \frac{\partial n}{\partial x} \, d\ell < 10^{15} \, cm^{-3}.$$

2. TIME DIFFERENTIAL HOLOGRAPHIC INTERFEROMETRY

Ordinary, double exposure, holographic interferometry is
time differential interferometry. Before the plasma is produced
the first exposure of the film is recorded, thus producing a
record of the phases produced by the vacuum windows, etc. Typically about two minutes elapses while the laser is recharged and
it is triggered again, this time while the plasma is present. On
this second exposure the same phase distortions produced by the
windows are recorded in addition to those that are a result of
the plasma being present. Upon reconstruction it is the interference of these two holograms which result in the appearance
of fringes. In this two minute interval an additional very useful step is often included. Usually something is done to the
reference so that it is incident upon the film in some slightly
different way. This is done so that background fringes are produced, which appear even if no plasma is present. One of the
most practical methods of doing this is to slightly tilt a mirror

in the reference arm. This tilting mirror is then imaged onto
the film so that there is no motion of the reference beam on the
film plane when this tilting is done. The net result is that,
on the second exposure, the reference beam is incident at a very
slightly different angle. Because of the long time between exposures and the fact that no plasma was present during the first
exposure, double exposure holographic interferometry is seldom
referred to as time differential.

There are situations however when the measurement of interest is the change in the plasma over a short time interval. In
these cases it would be very advantageous to make the first exposure of the film microseconds before the second exposure, thus
recording only the change in the plasma that occurred during that
time interval. Such an occasion arose when we wanted to determine if a theta pinch suffered any measurable change in density
when it was subjected to a 100 ns, 150 J pulse from a CO_2
laser [3]. Taking the difference of two ordinary holograms taken
before and after the CO_2 pulse would have resulted in subtracting
very large numbers from each other, with the resultant large uncertainty. Time differential interferometry is then a natural
solution. The problem arises as to how to perform something
equivalent to the tilting of the mirror in the microsecond available between exposures.

The solution is illustrated in Fig. 3. The "tilting mirror"
is a Rochon prism which is imaged onto the film. This Rochon
prism deflects the reference beam for both of the exposures but
it is polarization sensitive. If it is contrived that the two
laser pulses have orthogonal polarizations the prism will produce
slightly different deflections to the reference beams. The
spacing of the background fringes can be easily adjusted by varying the magnification of the imaging of the Rochon prism onto the
film. The method of superimposing two orthogonally polarized
beams was by the use of a calcite beam displacer. In fact, this
calcite prism was included inside the laser cavity of a Q
switched ruby laser, thus enabling the single laser, with the
addition of a second Pockels' cell, to provide two separate
pulses which are exactly orthogonal in polarization and which ex-
hibit excellent colinearity [3]. This scheme is possible because
the aperture that is actually used in a holographic ruby laser
varies from between 1.5 to 2 mm while the rods are often ~12 mm
in diameter.

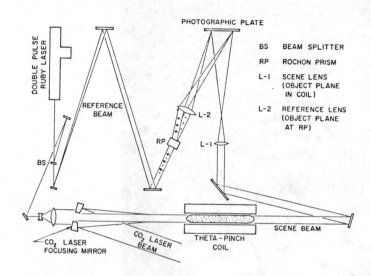

Fig. 3.   Layout of the time differential holographic interfero-
          meter.   The Rochon prism introduces a slightly different
          angular deviation for the two different polarizations.

        Figure 4 shows some typical data which were obtained with
this holographic setup. Figure 4a shows the density profile of
the theta pinch taken at a time of 3.6 microseconds after the
initiation of the current. For this interferogram the first
cavity of the laser was triggered 100 μs before the current
initiation so that it represents a conventional double exposure
interferogram. This illustrates a second advantage of the scheme
which was used here. If one is working in a hostile environment,
where waiting the 1 to 2 minutes to take the second exposure
might subject the equipment to vibrations which would introduce
unwanted fringe shifts; these can be avoided by the short time
interval between exposures. Figure 4b shows a time differential
interferogram which was obtained on an equivalent discharge at
the same time after current initiation. In this case the time
interval between exposures was 600 ns. Notice that the fringes
are deflected in the opposite direction in this case. This results from the fact that the plasma is decaying in time. By comparing such interferograms, made with and without the CO_2 laser,
it was possible to put an upper bound on any change in density
that the laser produced.

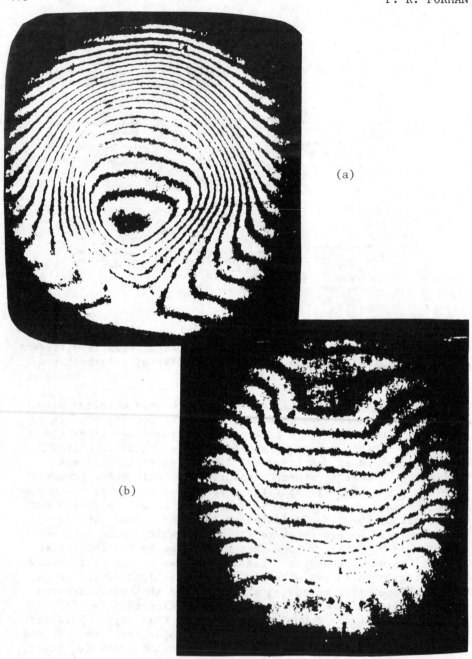

Fig. 4. a) Conventional interferogram of theta pinch taken at
             3.6 μs.
         b) Time differential interferogram (.6μs) taken at
             same time.

A new and better way of doing time differential interferometry may be possible for short time differences [4]. It was noticed that when very dense plasmas were examined, using the scheme just described, shadowgraph effects obscured the interferograms. The fringes appeared to be straight but the intensity modulation confused the interpretation of the interferograms. It may be possible to avoid this problem by the technique shown in Fig. 5. This is a conventional Michelson interferometer, with the single exception that the mirror in the plasma leg is not an ordinary mirror but rather a phase conjugation mirror. It will have to be an optical medium, which is actively pumped by two oppositely directed colinear laser beams, which are derived from the same laser beam that is the scene beam. The scene beam of the interferometer interacts with one of the pump beams, in a non-linear fashion, within the medium, which is the phase conjugation mirror, to form a volume diffraction grating. This diffraction gating is perfect if the incoming scene beam is a perfect plane wave. If this is not the case, but the scene beam has been distorted by the presence of the plasma, an imperfect grating will be formed. The imperfections of this volume diffraction grating will be such that the other pump beam, when it is diffracted by this grating, will be traveling in the opposite direction to the scene beam, hence it is a mirror, and it will have exactly the negative phase that the entering scene beam had, hence conjugation. This beam, upon traveling back through the plasma, will follow the exact path that the scene beam took on the way through, on its way to the phase conjugation mirror. If no changes have occurred in the time it took for the scene beam to exit the plasma and return, any phase change introduced by the plasma in the first traversal will be exactly cancelled by the return trip. If, however, there are changes, exact cancellation will not occur and a time differential interferogram will result. The time difference is determined by the time it takes for the scene beam to exit the plasma and the conjugated beam to return.

This time can easily be made several ns, say 5 ns, which is ideal for studying the development of the laser initiated z pinch which was described in the section on Moiré-schlieren. Phase conjugation has been investigated at reasonable power levels with nanosecond response time [5]. We are currently investigating this possibility at LANL. We are starting with a barium titanate, self conjugating crystal to learn some of the problems involved [6].

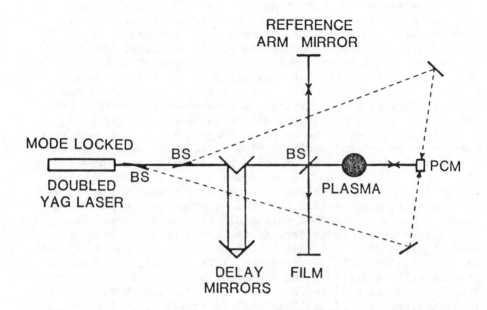

Fig. 5.   Michelson interferometer with a phase conjugating mirror. Beams for the phase conjugating mirror (dashed
lines) are obtained from the laser by beam splitting.

3.   FARADAY ROTATION

     One of the least known quantities of magnetically confined
plasmas is the internal magnetic field. This is particularly
true of high beta plasmas, where beta is defined as the plasma

pressure, $nk(T_e+T_i)$, divided by the magnetic pressures, $B^2/8\pi$.

The reversed field pinch, or RFP, is particularly subject to this
lack of knowledge because it depends upon the details of the
field distribution for its stability. No physical probe is allowed to be put into the plasma, as a physical probe would so
perturb the plasma that the results of such a measurement would
be seriously questioned and also the plasma often destroys such
probes, resulting in a severe vacuum leak.

     Because of this need to know, we are presently making
Faraday rotation measurements on ZT-40M. The ZT-40M experiment
is a toroidal plasma experiment with a 0.4 m diameter and a
2.28 m major diameter. The location of the beam on which we are
making this measurement is shown in Fig. 6. The laser beam
enters off-axis, vertically, and passes through the plasma and

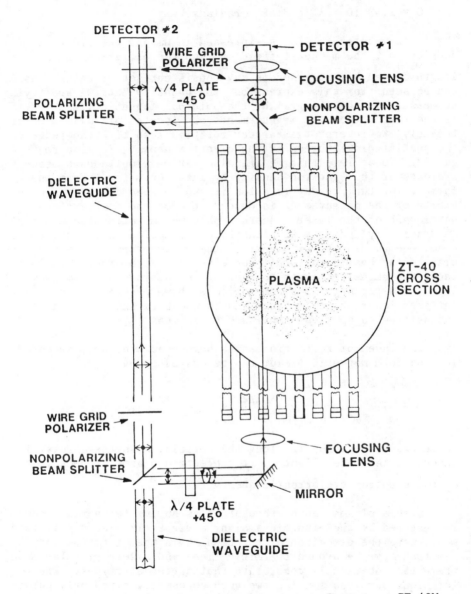

Fig. 6. Faraday rotation and interferometer layout on ZT-40M.

experiences a very slight change in the direction of the polarization. It is this change in polarization that is to be measured. The amount of rotation that the beam experiences is given
by

$$\theta = 2.63 \ 10^{-17} \lambda^2 \int n_e B \cdot d\ell \quad \text{radians.}$$

The laser wavelength, λ, is expressed in μm and the electron
density, n_e, the magnetic field, B, and the differential unit of
length $d\ell$ can be expressed in any consistent set of units. As
can be seen from this expression, the quantity that we are trying
to measure is the integral of the poloidal magnetic field. This
is the reason that the measurements are located off axis. Unfortunately, to interpret a Faraday rotation result, a knowledge of
the spatial distribution of the electron density is also required. To achieve this we are also making simultaneous measurements of the electron density by interferometry. From the
formula for the Faraday rotation it can be seen that the sensitivity of the measurement is dependent upon the square of the
wavelength of the laser. Another phenomenon which also depends
quadratically upon the wavelength is refraction of the probe
beam, due to density gradients in the plasma which are perpendicular to the beam path (schlieren effect). Such effects could
cause the signal strength to vary by refracting the beam away
from the detector. As a compromise we have elected to use a far
infrared laser at 185 microns and to use a technique which is
insensitive to spurious amplitude variations [7,8].

The index of refraction, of a homogeneous plasma, embedded
in a uniform magnetic field, can be expressed by

$$n = (1 - \frac{\omega_p^2}{\omega^2 (1 \pm \omega_b/\omega)})^{1/2} \, ,$$

where the propagation is along the B field, ω is the angular frequency of the laser light, ω_p the electron plasma frequency, and
ω_b the electron cyclotron frequency.

Such a plasma can be thought of as having two normal modes
represented by the plus/minus sign; a right circularly polarized
mode, were the direction of rotation is in the direction that
electrons gyrate around the field lines and a left circular mode,
where the rotation is counter to that of the electrons. The
technique involves sending two such oppositely circularly polarized beams along the same path through the plasma and measuring
the difference in phase shift that the two beams experience in

traversing the plasma. If these two beams have slightly different frequencies, such that $w_1 - w_2 = \Delta w$, the resultant beam is
equivalent to a single, linearly polarized beam whose direction
of polarization rotates at half the difference frequency, Δw. If
such a beam is passed through a linear polarizer, the signal observed by a square law detector will be

$$V = A \sin(\Delta w t).$$

When the additional plasma phase shift, due to Faraday rota-
tion is present, $\phi = (\phi_\ell - \phi_r)/2$, the ac signal will be modified
to

$$V = A \sin(\Delta w t + \phi(t)).$$

The modulation frequency is chosen to be very large, with respect
to the rate of change of ϕ, so that the time dependent phase
shift, due to the Faraday effect can be determined from the zero
crossings of the oscillating signal, thus making the measurement
insensitive to amplitude modulation.

To achieve this situation experimentally the output of
the FIR laser is divided into two co-propagating, frequency offset, beams by means of the beam splitters and rotating grating,
as shown in Fig. 7. The half-wave plate, which is located
immediately outside the FIR laser, allows the polarization of the
laser to be rotated with respect to the wire grid polarizer, thus
allowing a balance to be achieved between the two recombined
beams, so that approximately equal amplitudes of the counter and
co-rotating polarization beams are achieved. The rotating
grating is machined on the periphery of a 30 cm diameter aluminum
disk. By rotating this disc at reasonable rates, a Doopler shift
of up to about 1 MHz is imparted to the beam that reflects off
it. The quarter wave plate converts the orthogonal linearly
polarized beams into right and left circularly polarized beams.
Figure 6 shows the set-up that is used, which allows simultaneous
measurements of Faraday rotation and density. After the nonpolarizing beam splitter, one of the two beams within the reference arm is eliminated by the polarizer. The interferometer
portion of the scene beam is converted back to linear polarization by the quarter wave plate, after a portion of the beams have
passed through the nonpolarizing beam splitter. The unshifted
beam is not reflected by the polarizing beam splitter so the
light incident on the interferometer detector is a reference beam
at the unshifted frequency and an orthogonally polarized, frequency shifted, scene beam. The last polarizer is used to take
components of the two beams so that interference will occur.
Figure 8 shows a typical density measurement. The top trace

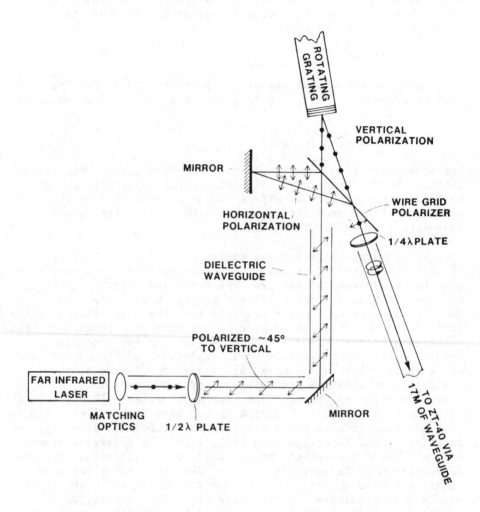

Fig. 7.   Schematic of method used to obtain two orthogonally
          polarized frequency offset beams by use of a rotating
          grating.

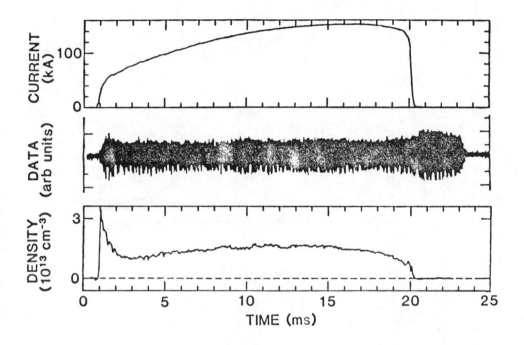

Fig. 8.   Typical density measurement on ZT-40M. Upper trace
          shows the time history of the toroidal current. Middle
          trace shows the raw data. Lower trace shows the density
          as a function of time.

shows the toroidal current in ZT-40M as a function of time. Below it is the raw signal of the detector, as recorded on a digital recorder which samples the signal every 2 microseconds. On
this set of data the wheel only produced a 40 kHz offset so this
sampling interval was sufficient. There is a pronounced increase
in signal when the plasma terminates. We attribute this to the
fact that there is significant refraction occurring when the
plasma is present. This fact would preclude any measurement
technique which was sensitive to amplitude variations. The last
trace is the integral of the density, expressed in fringes.
Figure 9 shows a recent Faraday rotation trace along with the
current trace.

     In order to assess the possible errors, which would neces-
sarily arise due to imperfections in the experimental set up, a
computer code was written to evaluate these errors [9]. First,

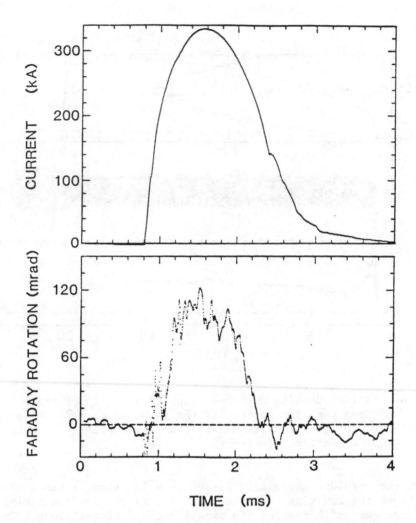

Fig. 9.   Faraday rotation signal on a short duration 330 kA dis-
           charge.

the quarter wave plate thickness would not be exact so a thickness error, δ, is assumed.   Second, the transmission is not expected to be exactly the same along both axes so a transmission
ratio (r) is assumed.   Third, there will undoubtedly be slight
misalignments of the axes of the quarter wave plate with respect
to the laser polarizations, leading to the introduction of an
error angle, Ψ.   ZT-40M also has a toroidal magnetic field in
addition to the polodial field which we wish to measure.   Such a

field makes the plasma slightly birefringent, due to the Cotton-Mouton effect. The effect of this transverse field is introduced by a birefringence angle BR.

It has been possible to obtain analytic expressions in limiting cases. One of these is the case where θ, the orientation angle of the polarizer, is set to zero degrees. In this case

$$\text{Phase shift} = \phi \cdot (\frac{\sin BR}{BR} \cdot \cos \delta \cdot \frac{1}{\cos(2\Psi)} \cdot r) \quad \theta = 0 \ ,$$

where each effect appears as a multiplicative factor, times the desired Faraday rotation term. The effect of birefringence is negligible, as expected in out case, because sin(BR)/BR is almost 1 for small angles. In other experiments this may not be true. Errors in the thickness of the quarter wave plate are not too serious if they are not large, as $\cos(\delta)$ is a weak function, for small δ.

The last two terms represent the possibility to actually increase the sensitivity of the measurement. It should be remembered that the ratio of transmissions of the quarter wave plate (r) does not have to be less than one. The other term, adjustment of the quarter wave plate while tracking this adjustment with the polarizer, can yield large changes in sensitivity. Initial experimental verification has been observed on ZT-40M. An increase of about a factor of six was observed when a factor of nine was predicted. Figure 10 shows data taken on two nominally identical discharges, with and without this enhancement. It was very difficult to track the angle of the polarizer with the angle of the quarter wave plate so this discrepancy was not unexpected. This gain in sensitivity may well allow a marginal experiment to become viable. It is not gained at zero cost, however. The magnitude of the signal is decreased almost in direct proportion to the gain in sensitivity. The second cost is loss of self calibration. Before, a fringe shift had a given meaning, in terms of the integral $n_e B \cdot d\ell$. This is no longer the case and a one-time calibration, like rotating a half wave plate a known angle, must be introduced. Because of the signal loss, the utility of this phase enhancement depends on the signal to noise ratio. If the noise is a result of measurement error, due to small signal there is a clear gain. If the phase detectability is limited by noise on the carrier and of the derived quantity, phase shift, scales inversely with available amplitude modulation signal, then there is the customary square root advantage in signal to noise.

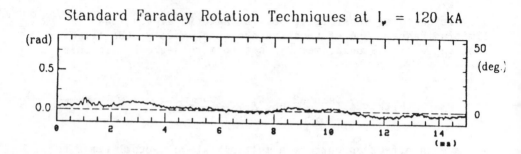

Standard Faraday Rotation Techniques at I_ψ = 120 kA

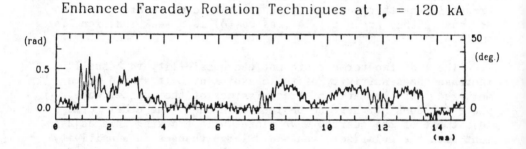

Enhanced Faraday Rotation Techniques at I_ψ = 120 kA

Fig. 10. Faraday rotation data taken on nominally identical discharges with and without enhancement through angular
adjustment of Ψ.

REFERENCES

1. F. C. Jahoda, LA-3963-MS (1968).

2. J. E. Hammel, et al, Nuclear Instruments and Methods, 207,
161 (1983).

3. W. T. Armstrong and P. R. Forman Applied Optics 5, 320
(1979).

4. A. E. Siegman, Optics Communications 31, 257 (1979).

5. L. K. Lam et. al. Optics Letters 6, 475 (1981).

6. Jack Feinberg Optics Letters 7, 486 (1982).

7. G. Dodel and W. Kunz Infrared Physics 18, 773 (1978).

8. M. D. Bausman et. al. Fourth International Conference on Infrared and Millimeter Waves and Their Applications, Miami Beach, Florida (1979).

9. R. M. Erickson, et. al. to be published in IEEE Transactions on Plasma Science, Dec 1984.

TWENTY-PICOSECOND PULSED UV HOLOGRAPHIC INTERFEROMETRY OF LASER-INDUCED PLASMAS

Garlane E. Busch

KMS Fusion
Ann Arbor, MI
USA

1.   INTRODUCTION

Laser fusion research requires the generation and character-ization of high density plasmas ranging from 100-1000 μm in size. Typically, a tiny fueled target is irradiated symmetrically by a multi-terawatt (> 10^{12} W) laser-optical system. The target implodes, compressing and heating the fuel to fusion conditions. This section describes an operational optical diagnostic well suited to characterizing some features of these plasmas. Plasmas of many other types of targets can also be diagnosed by this technique.

2.   UV HOLOGRAPHIC INTERFEROMETRY

This application of holography has many special features. The object is a small plasma, less than one millimeter in cross section, with phase front velocities up to 3 x 10^8 cm/s. The interesting period of the plasma expansion lasts less than one nanosecond. The laser plasma radiates strongly from spectral regions of the near IR to x-rays. Target debris is also present.

A target 100 - 500 μm in diameter is positioned at, or near, the focus of a laser optical system. A high intensity sub-nano-second laser pulse irradiates the target under vacuum. A set of 20 psec, ultraviolet, probing pulses pass transversely through the underdense region of the evolving target plasma. The rela-tive time of probing UV pulses is known to within ± 10 psec and

can be adjusted to arrive at the plasma at any time during its
expansion. A separate optical system relays the UV image of the
plasma to a "camera" outside the vacuum chamber. Figure 1 illustrates the method. A single pulse through the plasma object
would produce a partially opaque shadowgraphic hologram. A
double pulsed set, one before and one during main laser heating
pulse, produces a holographic interferogram. This occurs whenever two holographic images, having phase differences, due to
refraction in the evolving plasma, are spatially superimposed.
When the doubly exposed images are reconstructed they interfere
coherently, producing high contrast fringes. To a first approximation these represent electron isodensity contour lines in the
axial plane of the cylindrically symmetric plasma.

Figure 2 illustrates the experimental layout for one-sided
target shots on disks. Reflective focusing optics, in the laser
heating beam, allow multiwavelength target irradiation. For
spherical fusion type target experiments a novel parabola-
elliposidal system irradiates the Micro-shell targets from two
sides with a high degree of uniformity. The illustrated diagnostic realy optics are compatible with both types of experiments.
The two stage system consists of an f/2 catadioptric system and
an f/4.2 refractive lens. The resolution of this system is 1 µm
at the target. The large collection angle (f/2) relays rays, refracted from the critical density surface in small plasmas.

Two additional features of the system are important: filtration and probe pulse length. The favored filtration method
was, at the time of this presentation, the use of a dichroic
coating on a deep blue, UV transmitting, UG 5 filter glass.
Three-cavity interference filters are not used because of poor
optical quality, low transmittance, and because of low damage
thresholds. More recently a better method has been developed
which shall be published. Secondly, to prevent blurring of the
fringe lines in some experiments, the probing pulse must be about
20 psec or shorter. In order to obtain this pulse length, a regenerative pulse compression amplifier has been developed [1],
with an average temporal compression of four-to-one. The YLF
oscillator in the driver typically produces 100 ps (FWHM) pulses,
one of which seeds the regenerative amplifier cavity. This compressed (25 ps) IR pulse is further shortened in the non-linear
processes of frequency conversions to green and then to UV. The
final UV pulse width is ~ 20 ps.

A multi-frame version has been described [2]. At this
writing, this unit is fully operational and yields four high contrast interferograms, per target shot. Modules have been used
which give 75 ps, 125 ps, and 200 ps frame separations. This

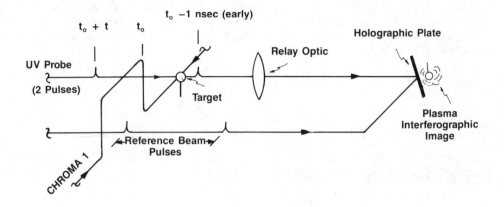

Fig. 1.   Diagram showing sequence of driver and probing pulses to produce holographic interferograms of plasmas.

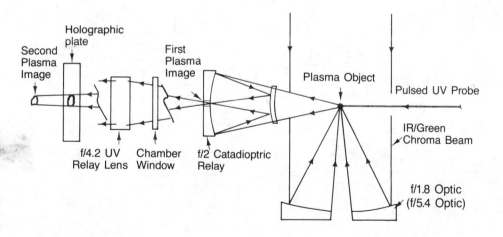

Fig. 2.   Layout of UV probing relay optics with CHROMA 1 plasma experiments.

upgrade permits study of plasma expansion dynamics, while minimizing shot-to-shot data scatter. The method is diagramed in
Fig. 3.

    Finally, Fig. 4a is an example of a single frame holographic
interferogram of a fusion target at 500 ps. Fig. 4b shows the
electron density distribution, along a radius, from an Abel inversion of the photograph [3]. The solid lines in the graph are
computer code simulations of the distributions at various flux
limiting levels.

Frame intervals 75-200 psec

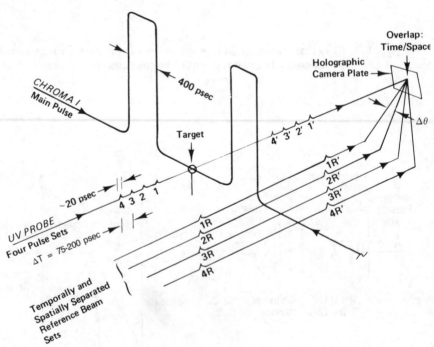

Fig. 3.   Method of time/space multiplexing of four holographic
          frames.

Fig. 4a. Holographic interferogram of spherical target plasma.

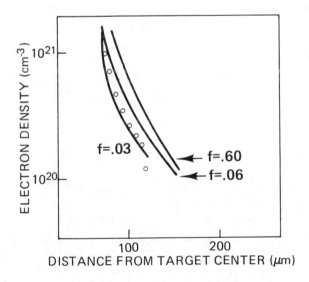

Fig. 4b. Abel inversion provides density profile. Solid lines
          are code simulations.

REFERENCES

1. J. E. Murray, "Temporal Compression of Mode-Locked Pulses for Laser Fusion Diagnostics", IEEE Jour. Quantum Elect., 17, 1713, Sept. 1981.

2. Gar. E. Busch, "Multiframe Holographic Shadowgraphy and Interferometry of Laser Target Plasmas", SPIE vol. 348, High Speed Photography (1982).

3. D. W. Sweeney, J. Opt. Soc. Am. 64, 559 (1974).

X-RAY DIAGNOSTICS

NONDISPERSIVE X-RAY DIAGNOSTICS OF SHORT-LIVED PLASMAS

Robert H. Day

Los Alamos National Laboratory
Physics Department
Los Alamos, NM
USA

1. INTRODUCTION

In this NATO Advanced Study Institute, we have discussed in detail the diagnosis of many pulse power machine properties, including their electrical behavior, grounding and shielding, and related data acquisition techniques. The purpose for many of these machines is to create high-temperature/high-density plasmas and, therefore, the subsequent behavior of these plasmas is of critical concern. The energy density of these plasmas is such that they will naturally radiate in the x-ray regime, and thus the diagnosis of their x-ray emission is a crucial measurement of the entire system performance. In this lecture, I will describe the general techniques used to perform nondispersive x-ray diagnostics of these short-lived plasmas.

First, let's put the issue of x-ray diagnostics into a general framework which will guide our discussions. As in any diagnostic application, there are three basic types of information that can be acquired: 1) the energy distribution of the plasma emission, 2) the temporal history of the x-ray emission, and 3) the spatial distribution of the emission. The precision with which each of these parameters must be measured varies greatly depending upon the application. For example, the plasmas of interest at this Institute have time scales from a hundred picoseconds for laser-produced plasmas, to microseconds for relativistic electron-beam sources. Similarly, the spatial scale can be microns for laser plasmas to centimeters for relativistic electron-beam sources. The requisite energy resolution can also vary from a few parts per thousand to totally energy integrated.

The topic of high-resolution x-ray spectroscopy is an important separate issue and a whole chapter (Chapter 8) is devoted to
its discussion. In particular, the lectures by Hans Griem and
Giovanni Tondello will describe the types of information which
can be deduced from the details of the energy-dependent emission.
Therefore, my comments here will be restricted to broad-band
spectroscopy with or without time and spatial resolution.

The three coordinates of time, energy, and space, which I
have described, are shown schematically as the two, 2x2 matrices
of Figs. 1(a) and 1(b). These matrices represent all classes
of information which can be returned depending upon whether one
integrates over the parameter of interest or resolves information
in that parameter. In the upper left-hand corner of each matrix
element I've tried to indicate some of the major physics issues
which are illuminated by the measurement indicated, while in the
lower right-hand corner of each entry I've tried to indicate some
of the diagnostic techniques which will return this type of information.

As we move from the upper left to the lower right in each
figure, we progress from information integrated over all three
coordinates to information resolved over all three coordinates.
In general, it is not possible nor necessary to return complete
information about all three coordinates; therefore, the majority
of diagnostics will integrate over one or more of them. In this
lecture, we will be primarily concerned with the diagnostics and
physics issues which are illuminated by the spatial and time
coordinates. We will, however, also cover the low-resolution
spectroscopy applications in the lower half of this matrix. In
particular, we will cover time-resolved low-resolution spectro-
copy in some detail and discuss the application of layered synthetic microstructures, LSM's, to x-ray diagnostics measurements.
In section 8, I will also outline the techniques which are available to characterize the instrumentation described below.

2.   BOLOMETRY

Bolometry is the simplest type of x-ray diagnostic and can
only return information about the total amount of x-ray emission
integrated over time, energy, and space. These data yield information on the energy balance and conversion efficiency of the
plasma in turning pulse power into x-rays. Though this is not a
very descriptive indication of detailed plasma performance, it is
frequently one of the most important numbers that can be returned
on the overall performance of a plasma experiment. The two devices which are used to measure this parameter are the flat
energy-response calorimeter and simple bolometers.

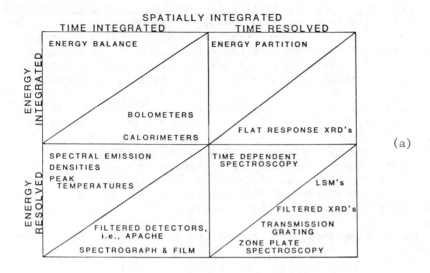

(a)

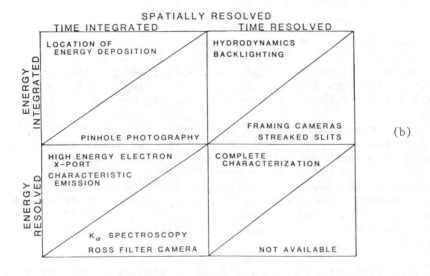

(b)

Fig. 1(a) and 1(b).   One can measure spatially, temporally and/or
                       spectrally resolved information about the
                       x-ray emission from a plasma. The type of
                       physics information that is required will
                       determine what measurement and instrument is
                       needed. The physics information is shown in
                       the upper left corner of each box and the
                       types of instruments to perform these meas-
                       urements are listed in the lower right cor-
                       ner.

Recently, James Degnan [1] has developed a system to perform pulse bolometry to measure the total soft x-ray conversion efficiency of an imploding liner experiment at the SHIVA capacitor bank at the Air Force Weapons Laboratory, AFWL. The advantage of such a system is its simplicity for measuring total radiated power with an integration time of approximately 1 microsecond.

A schematic of this device is shown in Fig. 2(a). The measurement is based upon the principal that as a material is heated, its resistivity change is only dependent upon its temperature. The resistivity change is measured as the change in voltage drop across the foil experienced by a constant current passing through the foil. A magnetic field is frequently applied parallel to the surface of the foil to suppress the emission of photoelectrons created by the incident x-rays.

The voltage drop, ΔV_B, across the foil is given by the expression:

$$\Delta V_B = \frac{\Delta e}{\Delta E} \frac{1}{\rho x^2} \frac{\ell}{w}$$

where $\frac{\Delta e}{\Delta E} = 13.2 \times 10^{-9}$ Ω-cm per Joule-gram. The foil length = ℓ, x = the foil thickness, w = the foil width, and ρ = the foil density. Aluminum foil was chosen for this application because of its availability, well known resistivity, and ease of use.

A cross-sectional view of this device is shown in Fig. 2(b). The foil is mounted on a modified High Voltage Engineering 2-inch diameter vacuum flange with an HN vacuum feed-through connector. The bolometer is suspended between the center conductor and ground of an HN connector. A magnetic field is supplied by external magnets. The device is built as part of a 50-ohm coaxial transmission line, and the coaxial cable connected to the HN connector leads to a current source and oscilloscope recording station.

This device is used as a routine diagnostic to measure total x-ray yield under a wide variety of implosion conditions. The validity of this measurement of total x-ray yield has been tested by comparing the bolometer x-ray yield with the x-ray yield measured by integrating over the spectrum with a filtered detector spectrometer. The filtered detector spectrometer will be described below in Section 4 in greater detail. The results of this comparison are shown in Fig. 3 where the ratio of yields measured by the two techniques is plotted versus the equivalent black-body temperature of the plasma. The temperatures in this plot vary by a factor of five, which means that the x-ray output

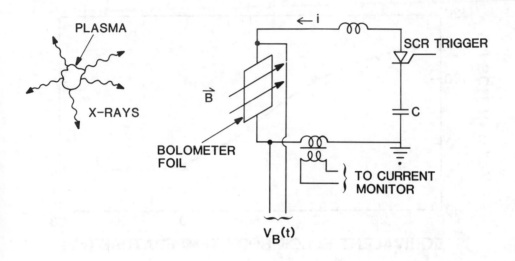

Fig. 2(a).   Schematic of a thin-foil pulse bolometer.

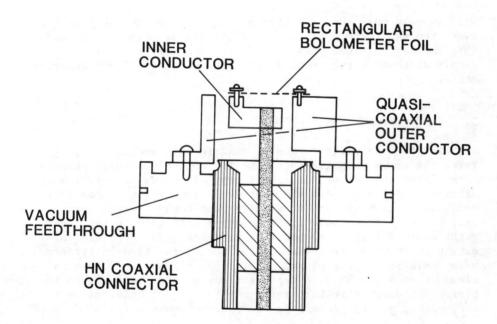

Fig. 2(b).   Cross sectional view of a thin-foil pulse bolometer
             as built by James Degnan at the Air Force Weapons
             Laboratory.

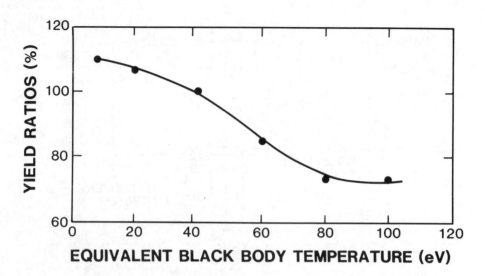

Fig. 3.   A comparison of x-ray yields inferred from a pulse
           bolometer and a filtered detector spectrometer for a
           variety of plasma conditions.

will vary by more than a factor of 600, and yet the agreement between the two techniques is within 15% across this entire range.
This implies that the pulse bolometer is a very reliable and
simple diagnostic for total x-ray conversion efficiency measurements.

3.   FLAT-RESPONSE X-RAY DIODES

       The foil bolometer which is described in Section 2 integ-
rates the x-ray emission over time for most of the high temperature plasma sources of interest today. In order to answer more
detailed questions about the time evolution of the plasma x-ray
emission it is necessary to utilize some sophisticated x-ray
diagnostics. As we move from the upper left cell to the upper
right cell of Figs. 1(a) and 1(b) we are asking to return information on the time evolution of the space- and time-integrated
x-ray emission of the plasma. This type of data answers questions related to the time-dependent energy partition within the
plasma. A diagnostic that returns this type of information for
short-lived pulsed plasmas is the flat-response x-ray diode, XRD.

       We will find x-ray diodes used in a variety of applications,
and it is appropriate to describe this important class of detector before proceeding. In their most basic form, these detectors

are very simple, as shown schematically in Fig. 4. The x-rays
from a plasma source are made to impinge on a metallic photocathode after passing through a transmission filter between the x-ray
source and the detector. An x-ray transparent anode, usually
consisting of a partially transparent metallic mesh, is placed a
short distance from the photocathode so that the photocathode is
biased negative with respect to the anode; photoelectrons are
emission-accelerated away from the cathode. The return current
flowing into the photocathode to replace the electrons emitted
from the surface are then a measure of the incident x-ray flux.
Detailed descriptions of the time and energy response of these
systems are given elsewhere [2,3,4].

    The inherent energy sensitivity of such a system is deter-
mined by the energy dependence of the surface photoemission and
has been shown semiempirically [5] to fall as the energy of the
photon, E, multiplied by the photoelectric absorption coefficient, U(E). This intrinsic response is then modified by the
transmission of the filter placed between the plasma source and
the photocathode, as shown in Fig. 5. It was noted by Harry
Kornblum and William Slivinsky [6] that by appropriately filtering different fractions of a gold photocathode it was possible to
create a response function which was flat to ± 20%, when expressed in units of coulombs per kilovolt, over the energy range
from 100 eV to 1 keV, as shown in Fig. 6. This type of device is

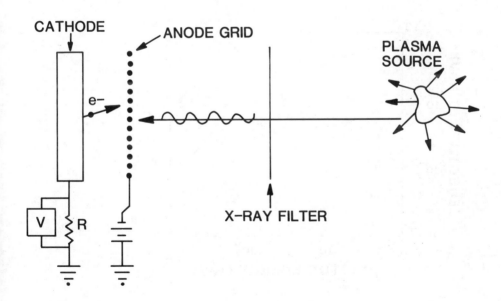

Fig. 4. A schematic of an x-ray diode (XRD) detector.

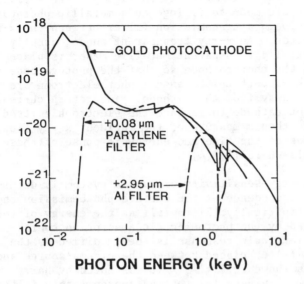

Fig. 5. The intrinsic response of an XRD is modified by the filters placed between the photocathode and the plasma source.

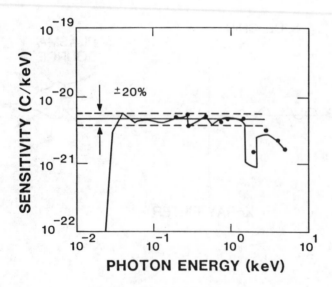

Fig. 6. The response function of a "Flat Response XRD" before being filtered by a critical angle x-ray mirror.

then a flat-response calorimeter, but it is also possible to construct it with a very high time response [4]. Bandwidths have
been achieved with these devices in excess of five gigahertz for
a half-inch diameter diode.

The slow roll off of the high energy response of this system can cause significant error if the spectrum being measured
doesn't also fall fast enough. In several versions built by
Slivinsky [6], this high energy response was further suppressed
by including a critical-angle x-ray mirror between the detector
and the plasma source. For the device shown in Fig. 5 the angle
of incidence and material of the mirror was chosen to give a cut-
off energy of 1.5 keV. This was achieved with a highly polished
beryllium mirror at 1 degree with respect to the incident x-ray
beam. The reflectivity of this mirror is in excess of 80% from
109 eV to 1.5 keV, and then rapidly falls to near zero reflectivity.

The flat-response XRD has been used on a wide variety of
laser plasmas in the Lawrence Livermore National Laboratory,
LLNL, laser fusion program. The inferred flux from this time-
resolved calorimeter has also been compared, like the foil bolometer of Section 2, with the output of a filtered detector spectrometer. As with the folometer, agreement to better than 20%
has been achieved between the two detectors over more than a decade variation in total x-ray output.

4.   FILTERED DETECTOR SPECTROSCOPY

In the last section, I described how it was possible to
compensate for the inherent energy selectivity of the XRD detector by appropriate choice of filters to achieve a constant energy
sensitivity. It is also possible to accentuate this energy
selectivity of the basic XRD detector to build a system which has
low spectral resolution. This type of detector yields low resolution spectral information with high temporal resolution and
begins to return information on the time evolution of the plasma
spectral emission.

Spectral selectivity is achieved by interposing appropriate
filters between the plasma and the photocathode. These filters
are selectively transparent to x-rays in different regions of the
spectrum, and therefore the current measured from that XRD is a
measure of the x-ray fluence in that region of the spectrum. The
various response functions shown in Fig. 7 for an aluminum photocathode filtered by several different materials are typical of
the kind of systems that can be built, and resolving powers of
one to three are achievable by careful selection of filters and
cathodes.

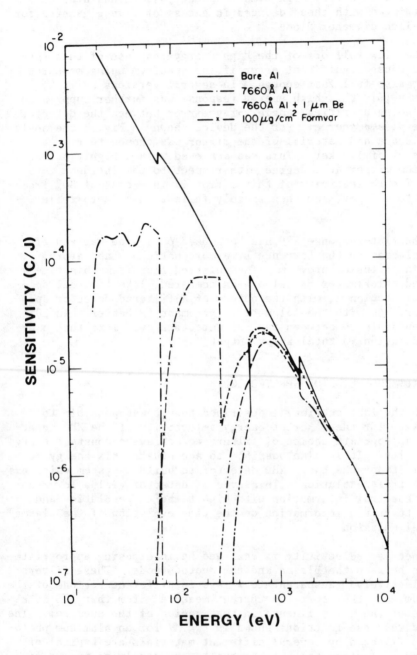

Fig. 7. Spectral selectivity can be enhanced with an XRD by
          appropriate filtering of the plasma emission.

A spectral measurement is made by simultaneously measuring the detector current from a variety of such channels whose response has been optimized in various regions of the spectrum. We have built a seven-channel system of this type at the Los Alamos National Laboratory, LANL, to measure the x-ray emission from laser fusion targets, and similar systems have been used to measure the x-ray emission from a variety of pulsed plasma power machines.

To infer a time-dependent spectrum from this data it is necessary to have a common time reference for the emission from all channels, and then iteratively deconvolve the spectral response of each channel to derive the true spectral nature of the emission. This deconvolution process must be handled with great care and a variety of techniques have been developed to process this type of data [7].

An example of the time evolution of the x-rays emitted from a simple 300-micron diameter gold-coated nickel microballoon illuminated by 2.3 kilojoules of 10.6-micron laser light in a 1-nanosecond pulse is shown in Fig. 8. For the first nanosecond while the laser is on, the temperature of the emission increases quickly to several hundred electron volts. After the energy source is turned off, we see a long decay of the emission from the radiative cooling of the heated target.

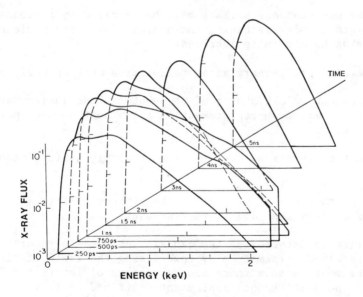

Fig. 8. X-ray emission from a 326-μm diameter Au-coated Ni microballoon illuminated by 2.3 kJ of 10.6-μm laser light in a 1-ns pulse.

The use of XRD's in this type of system is not essential
and similar measurements have been made with PIN solid-state
detectors [8], and photodiode/fluor combinations [9] as the
x-ray-to-current transducers. But XRD's have several advantages
including simplicity, high speed, and moderate to low sensitivity. Their use, however, depends upon measurements of their
absolute efficiency.

The quantum efficiency of an XRD is defined as the number of
electrons emitted from the photocathode for each incident photon.
This parameter has been measured for a variety of materials [4,5]
commonly used in detector systems and typical results are shown
in Fig. 9. The surface of the XRD acts as a solid-state ionization chamber, and primary and secondary electrons, which are
created within an escape depth of the surface of the photocathode, contribute to the signal. This implies that x-ray photoemission is a surface effect and the efficiency of the detector
can be modified by small changes in the surface constituents.

The stability of the response of a typical aluminum XRD
photocathode was investigated in the laser fusion environment.
It was found that by standardizing photocathode preparation and
controlling the handling of the surfaces these systems could be
used reliably. The constraints on using this type of detector as
an absolute instrument are:

a.   The photocathode surface must be prepared by a standard
procedure such as micromachining the photosurface with a clean
tool and using no machining lubricant.

b.   Careful preliminary x-ray calibration must be performed.

c.   Frequent refurbishment of the photocathode surface must
be made. (This may mean replacement on every shot on many pulse
power machines.)

d.   Post calibrations must be made to verify surface stability.

With these precautions it is possible to achieve calibration uncertainties of less than 15% in routine use.

The filtered detector system can also be used to return
spatially resolved information by restricting the field of view
of the instrument to only those source regions of interest. For
example, on the multichannel instrument described earlier a
series of seven lines-of-sight, LOS's, were established by a
series of precisely aligned pinholes to only view a region of
400-micron diameter on the surface of the target. With such a
system it is possible to measure the specific emissivity of the

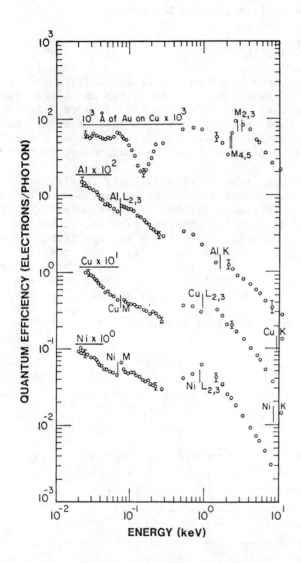

Fig. 9.   The energy-dependent quantum efficiencies of various
surfaces which are useful as XRD photocathodes have been
measured.

source in units of watts per square centimeter and perform detailed comparison of measured x-ray spectra to models of the
emission process.

With a collimated filtered detector instrument, it is possible to return low resolution data on all three parameters of energy, space, and time, yielding the time dependent x-ray spectral emission from a small region on the source.

5.   FILTER/FLUORESCER SPECTROSCOPY

For many applications, such as spectral deconvolution, it is desirable to have simpler response functions and higher spectral resolution than that available from the filtered detector system described in Section 4. For measurements of x-ray emission above 1 KeV, filter/fluorescer spectroscopy is an attractive alternative to significantly increase the measurement resolution.

This type of system is shown schematically in Fig. 10. The x-ray emission from the source is limited by an LOS similar to the type described in Section 4 for the filtered detector system. The detector package on each LOS consists of a filter in the primary x-ray beam to crudely define the x-ray spectrum, and a fluorescer to provide a threshold energy for x-ray detection. The fluorescent photons are then detected through a post filter by a detector which is provided to transduce the x-rays into electrical current. The advantage of this system is improved spectral resolution, while the disadvantage is a reduced sensitivity of at least several orders-of-magnitude.

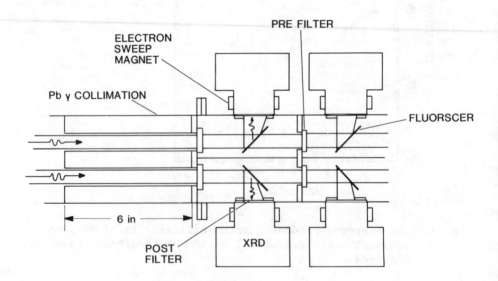

Fig. 10.   Schematic of a generic filtered-fluorescer detector
          system.

The way in which each element shapes the response is shown in Fig. 11. A 1-keV black-body spectrum is shown for reference as the initial x-ray source in the LOS. The effect of the pre-filter is to suppress the spectrum at energies above the absorption edge of the filter. These are usually K-edge filters. The

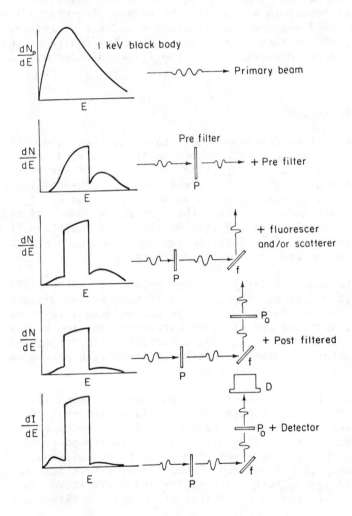

Fig. 11.   This shows the way in which each element of a filter/
          fluorescer channel modifies the incident spectrum.

for the detection of x-rays corresponding to the K-edge of the
fluorescer material. The post filter is used to suppress low-
energy scattered radiation and is usually a K_β filter for the

fluorescent photons. The detector is chosen to optimize the
system response to the fluorescent photons as much as possible.
XRD's, with photocathodes made of elements with Z just below
the fluorescent energy, are a frequent choice because of their
sharply peaked energy response.

Proper optimization of prefilter, fluorescer, postfilter,
and detector can result in a channel with nearly flat energy response between the K-edges of the fluorescer and the prefilter.
The resolving power of this type of system can be as high as ten
and is a significant improvement over the filtered/detector system.

6.   SPATIAL RESOLUTION

All of the systems described above have, at best, poor
spatial resolution, and yet information on the spatial distribution of x-ray emission is an important measurement of energy distribution and evolution within the plasma source. The collimated
LOS geometries can, at best, give an indication of the energy
density at a single location within the source. A number of systems have been devised to return data on the spatial distribution
of x-ray emission and I will describe two such instruments - the
pinhole x-ray camera and the imaging spectrograph.

By far, the simplest x-ray imaging instrument is the pinhole
camera, and this device has been applied in many plasma diagnostics. In its simplest form, the pinhole camera uses x-ray sensitive film for a detector and provides little information on the
time evolution of the plasma, though extremely good spatial resolution, approaching ten microns, can be achieved with precision
pinholes in laser fusion applications.

The sensitivity of a time-integrated pinhole camera is, however, strongly biased to the time of peak plasma temperature.
This results from the fact that for a typical black-body emitter
the intensity of x-ray emission varies as the fourth power of the
plasma temperature and, therefore, the majority of emission will
occur at peak temperature. X-ray pinhole cameras have been used
extensively to diagnose both quantitative and qualitative features of plasmas related to the nature of energy deposition and
transport.

Another class of imaging instrument, the imaging spectrograph, combines good spatial resolution with monochromatic imaging [10]. The system requires that the emission spectrum of the
source have characteristic lines and the source is imaged at the
energy of these lines. A schematic of one such instrument [11]
is shown in Fig. 12.

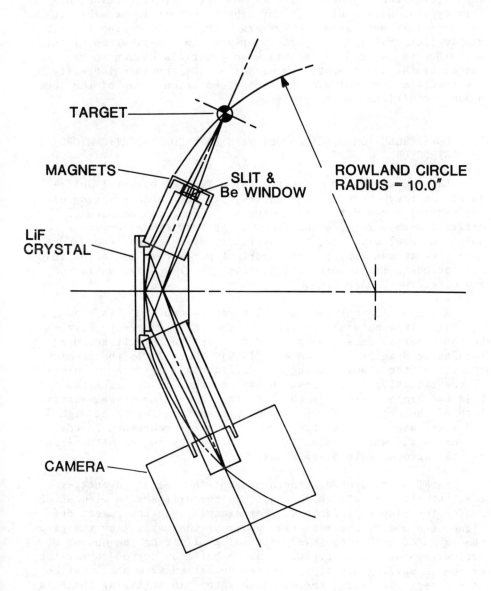

Fig. 12. Schematic of an imaging spectrograph.

For this technique to be successful, the source must produce intense characteristic emission which will diffract from the crystal in the instrument geometry shown. The source is placed on a sphere of radius R, the crystal is bent to a spherical curvature of 2xR, and the surface is ground to the spherical radius R. A real image is then created on the spherical surface beyond the crystal for a source which is dominated by line emission. This type of device has high throughput, but it is possible to build similar monochromatic imagers with lower throughput using simple flat crystals. A particularly useful application of this technique is to seed the plasma with materials chosen to be imaged by the instrument. By measuring the absolute intensity of the resulting spatial distribution of emission a map of the local plasma conditions can be achieved.

7.   LOW-RESOLUTION SPECTROSCOPY WITH LAYERED SYNTHETIC MICRO-
      STRUCTURES

Layered synthetic microstructures, LSM's, or metal multi-layers as they are sometimes called, are the manmade analog of the natural x-ray diffraction crystal. LSM's can be used to diffract x-rays in much the same way as traditional crystals; but unlike natural crystals, the instrument designer has the opportunity to choose many of the important parameters of the diffractor including its atomic constituents, their spatial variation, and effective 2-d spacing.

A comparison between natural crystals and LSM's is shown in Fig. 13. In a natural crystal, the spatially coherent distribution of atoms in a crystal causes x-rays to be diffracted according to Bragg's law: $n\lambda = 2d\text{SIN}(\theta)$. Here 2d is the atomic spacing of the atoms causing the diffraction, λ is the wavelength of the radiation of interest, n is the order of diffraction, and θ is the Bragg angle. In the LSM, the three dimensional structure of the crystal is replaced by alternating layers of high-Z and low-Z atoms. This structure has spatial coherence in one dimension and will diffract x-rays in analogy to the natural crystal according to Bragg's law.

Metal multilayer diffractors have a number of advantages over natural crystals which make them the diffraction element of choice for many applications. For example, the instrument designer can modify the resolving power of the crystal by changing the ratio of high-Z to low-Z material or limiting the number of active layers. These crystals can be built on curved substrates, or the 2d spacing of the system can be varied from one location to another. In short, the major advantage in utilizing LSM's is the ability to engineer the properties of the diffractor to meet

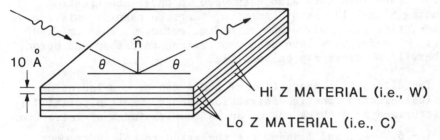

METAL MULTILAYERS
SYNTHETIC CRYSTAL

NATURAL CRYSTAL NaCl

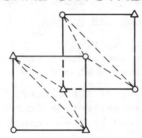

Fig. 13. LSM's are the manmade analogs of the natural diffrac-
          tion crystal. Alternating layers of high- and-low Z
          material take the place of the three-dimensional array
          of atoms in the natural crystal.

the diagnostic requirement rather than to modify the diagnostic
to fit the properties of available crystals.

    The technology for building these systems has been available
for many years [12]; however, it is only recently that thin-film
depositions have achieved the requisite perfection and control
necessary for LSM manufacture. The techniques of sputter deposition and e-beam evaporation have both been used to manufacture
metal multilayers of sufficient perfection to achieve high efficiency x-ray diffraction. Dr. Eberhardt Spiller of the IBM
Watson Research Laboratory [13] uses an e-beam evaporation system
with insitu monitoring of the x-ray diffraction efficiency to
optimize the deposition process. Dr. Troy Barbee of Stanford
University [14] uses the precise control available from sputter
sources to lay down the layers of the LSM.

The present state-of-the-art in LSM construction allows the manufacture of systems with 2d spacing as short as 30 Å. The 2d of the layers has been modified by as much as 10% from the top to the bottom of the stack and a 50% length variation of 2d has been achieved along the multilayer. LSM's have also been deposited on curved x-ray optics with radii of curvature in excess of 1 m. Numerous materials have also been used in their construction including high-Z layers of tungsten, tungsten carbide, silicon carbide, titanium, molybdenum, niobium, paladium, rhenium, and others. Low-Z layers have been constrained to carbon and boron, but beryllium layers may be possible.

The diffraction properties of LSM's can be modeled using classical electromagnetic diffraction theory, where material 1 is characterized by a thickness d_1 and a complex index of refraction $n_1 = 1 - \delta_1 - \beta_1$, and δ and β are the standard unit decrements. Material 2 is characterized by an index of multilayer refraction $n_2 = 1 - \delta_2 - i\beta_2$. Underwood and Barbee [15] have summarized classical diffraction theory to predict a metal multilayer's performance. If x-rays are incident on the material/vacuum interface of the metal multilayer, at a Bragg angle of θ, with s and p polarization components of the electric field of E^S and E^P, respectively, then the x-ray reflection from the boundary between layer ℓ and $\ell + 1$ is predicted by the Fresnel equations:

$$F^S = (g_\ell - g_{\ell+1})/(g_\ell + g_{\ell+1}) \text{ and}$$

$$F^P = (g_\ell/n_1^2 - g_{\ell+1}/n_1^2 + 1)/(g_\ell/n_1^2 + g_{\ell+1}/n_1^2 + 1),$$

where $g_\ell = (n_1^2 - \cos^2\theta)^{\frac{1}{2}}$.

With the Fresnel coefficients we can define a parameter $R_{\ell,\ell+1}$ which describes the reflection, including absorption in the substrate, for an x-ray of wavelength λ:

$$R_{\ell,\ell+1} = a_1^4 \ \frac{R_{\ell+1,\ell+2} + F_{\ell,\ell+1}}{R_{\ell+1,\ell+2} \ F_{\ell,\ell+1} + 1} \ ,$$

where $a_1 = \exp(-i\frac{\pi}{\lambda}g_\ell d_1)$

At each interface within the LSM four waves are present. The wave incident from the surface, the wave scattered towards

the surface, the wave incident from below, and the wave propo-
gating towards the rear surface of the LSM. The boundary conditions on the bottom of the LSM are such that no wave is incident
from below. Therefore, by starting at the lower surface and
solving for the reflectivities recursively, one can solve for the
intensity, $I(\theta)$, of the wave leaving the surface in terms of the
wave incident on the surface, I_o, and thus derive a reflectivity
of the structure:

$$\frac{I(\theta)}{I_o} = (R_{1,2})^2 \; .$$

For systems with uniform structure along the path of the
x-ray, it is possible to sum this recursion relationship explicitly [16] and derive a closed-form solution for the reflectivity
of the LSM.

Use of these theories, however, presupposes accurate knowledge of the optical constants of the materials. A thorough
summary of these parameters has been compiled by Henke [17] in
the energy range from 100 eV to 2 keV; however, there is still
significant uncertainty in these parameters in the regions of
anomolous dispersion near absorption edges. Outside of these
regions, acceptable comparison of theory and experiment is
achieved. For example, Fig. 14 shows a comparison of measurement
and theory for a molybdenum/carbon multilayer at a photon energy
of 1.487 keV. The agreement between theory and measurement is
quite good, with the residual difference probably due to substrate imperfections and surface roughness [14].

These systems are beginning to find widespread application
in the plasma diagnostics community. One of the first systems to
use these devices was a simple multichannel spectrometer built by
Gary Stradling [18] as a front-end dispersion element for an
x-ray sensitive streak camera. This system was a high time-reso-
lution x-ray spectrometer in the LLNL laser fusion program. The
instrument consisted of six discrete channels each with a uniform d-spacing crystal to define the energy range of interest.
The crystals had a resolving power of approximately 40 and the
d-spacings were engineered to cover the energies of several hydrogen and helium-like resonance lines in the plasma spectrum.

At Los Alamos, we have also been developing LSM technology
to build a unique diagnostic instrument which would not be possible with conventional crystal spectrometers. This device is
shown schematically in Fig. 15. An LSM is placed in a beam of
x-rays with a natural divergence of 1 milliradian or less. The
LSM is placed at a 7-degree Bragg angle with respect to the beam
and the d-spacing of the crystal is 50% longer on one end of the

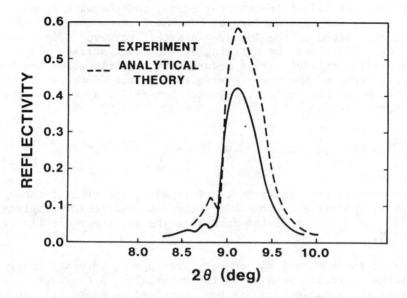

Fig. 14. A comparison of experimental measurement and theoretical reflectivity of a 30-layer pair Mo/C LSM of 110.6-Å
2d spacing. Al K_α radiation was used for the measurement.

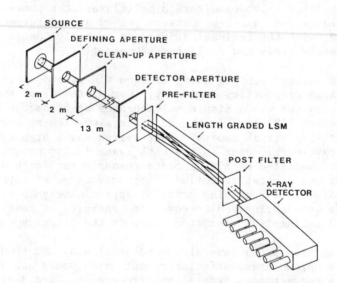

Fig. 15. A unique x-ray spectrometer utilizing LSM diffractors
with 2d spacing which varies from one end to the other.
This instrument covers the energy range from 1.5 to
2.5 keV.

crystal than on the other. The output beam of the spectrometer
is a parallel beam of x-rays with an energy of 1.5 kev on one
side and 2.5 keV on the other.

A detector with eight discrete channels intercepts the output beam and measures the time evolution of the spectrum with a
resolving power of approximately 10. This design is simple and
compact with high throughput and a large signal-to-noise ratio.
This level of simplicity in design would not be possible without
the unique ability to create variable d-spacing crystals provided
by LSM technology.

Obviously, this new class of x-ray diffractors opens up many
new possibilities of instrumentation for plasma diagnostics.
We've only begun to touch on the possibilities of instrument design. Other new instruments will include focusing x-ray optics
with monochromatizing surface coatings, high-collection efficiency optics at high energy, and many others we have just begun
to imagine.

8.   X-RAY CHARACTERIZATION

In this lecture, I have described a variety of diagnostic
instruments to measure the spatial, spectral, and temporal histories of plasma emission. Acquiring absolute measurements of
any of these parameters will usually require that the instruments
have calibrated responses. No single x-ray source is capable of
performing all these calibrations and so a number of systems have
been developed to optimize their properties for these variety of
purposes.

There are five major classes of x-ray generators: the electron-impact source, the ion-impact source, the electrically
driven plasma source, the laser-pulsed plasma x-ray source, and
the electron storage ring synchrotron source. The major parameters of intensity, spectral character, and time structure for
each of these sources is listed in Table 1. In this subsection,
I will outline the major properties and uses of each of these
sources.

8.1   Electron-Impact Generators.

The electron-impact generator is the oldest and simplest
type of x-ray generator and is still in common use today. The
technology of the electron-impact generator is versatile and can
be applied to a variety of applications, including high-intensity
DC sources and short-pulse high-brightness x-ray generators.

Table 1. Typical Source Parameters.

	Electron Impact Generators	Proton Impact Generator	Plasma Sources		Synchrotron
			Electrical	Laser	
Intensity (Photons) sec	~ 10^{12} direct 1" ϕ @ 12" (10^8 fluoresced)	5 x 10^{11} 1" ϕ @ 12"	< 10^9/pulse 1" ϕ 12"	< 10^9/pulse 1" ϕ @ 12"	> 10^{14}
Photon Energy	K L Lines .1 – 100 keV	K L Lines .1 – 100 keV	Lines .1 – 2 keV	.01 – 300 eV	continuously tunable
Purity	> 90%	> 95%	> 90% monochromatized	> 90% monochromatized	> 98%
Divergence	2 π	2 π	4 π	4 π	3 x 10^{-3}
Time Structure	DC, chopped or pulsed to < 50 ps	DC or RF accelerator	~ 10 ns	50 ps – 50 ns	Pulsed < 300 ps – 2 ns

The x-ray generation scheme is very simple and utilizes a source of electrons either from a hot filament or field emission. A potential is established between the cathode and the anode and the electrons are accelerated into the anode. The anode material is chosen to either optimize the continuous bremsstrahlung spectrum or the characteristic line emission. In typical DC applications the anode is water or air cooled to achieve higher power levels. These systems can dissipate tens of kilowatts electrically, resulting in hundreds of watts of x-rays.

The output of such sources can be used directly or further monochromatized. Monochromatization can be enhanced by simple characteristic line filtration or by the use of x-ray monochromators. However, this results in a concommitant loss in intensity.

The parameters of this source are listed in the first column of Table 1; the electron-impact generator is obviously not the system of choice to optimize any one parameter. However, the simplicity of construction and use of this type of source frequently makes it the system of choice for characterization applications.

These systems have been used in many different applications. The quantum efficiency curves shown in Fig. 9, for example, were measured with electrical-impact generators for the data above 100 eV.

8.2 Ion-Impact X-ray Sources.

The ion-impact x-ray generator works in a way analogous to the electron-impact x-ray source when it is used for characteristic line generation. The difference is that the primary vacancy is created by the impact of an ion instead of an electron.

In use the electron source of the electron-impact generator is replaced by a light ion accelerator. The ion energy is typically one to a few MeV per atomic mass unit. Power densities approaching several kilowatts per square centimeter are typically achieved, and characteristic photon energies in the range from 100 eV to several kilovolts are generated.

The advantage of the ion-impact x-ray source is the improved purity which can be achieved in comparison to the electron-impact generator. The knock-on electrons created during ion deceleration will have a maximum energy given by the ion energy, multiplied by the electron mass, divided by the ion mass. Thus, for an MeV ion the maximum electron energy will be below 1 keV. These electrons create less bremsstrahlung background than the typical electron-impact source. The purity of the system is improved from 90% for a typical electron-impact generator to better than 98% for a light ion generator.

The ion accelerators, which are the heart of the ion impact generator, are in use in numerous laboratories for nuclear and atomic physics research. A few dedicated facilities have been commissioned solely for the generation of x-rays. One such facility is the LLNL IONAC source, a picture of which is shown in Fig. 16. It utilizes a beam energy of 300 kV and is used to produce x-rays of energy below 1 keV.

By reviewing the properties of Table 1 it can be seen that when the ion-impact source is optimized for power dissipation the resulting x-ray beams have comparable intensities to traditional electron excited sources. The major advantage of the ion impact source is its improved purity.

8.3 Electrically Driven Plasma Sources.

In order to achieve the highest x-ray intensities, it is necessary to raise a material's temperature until it attains the

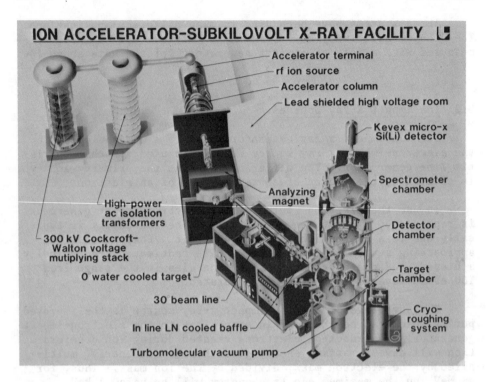

ION ACCELERATOR-SUBKILOVOLT X-RAY FACILITY

Accelerator terminal
rf ion source
Accelerator column
Lead shielded high voltage room
Kevex micro-x Si(Li) detector
Spectrometer chamber
Analyzing magnet
High-power ac isolation transformers
300 kV Cockcroft-Walton voltage mutiplying stack
Detector chamber
O water cooled target
Target chamber
30 beam line
Cryo-roughing system
In line LN cooled baffle
Turbomolecular vacuum pump

Fig. 16. Lawrence Livermore National Laboratory has installed a
          300-keV proton accelerator to excite characteristic
          subkilovolt x-ray lines. This is their "IONAC" low
          energy x-ray facility.

plasma state. Several such plasma x-ray sources are under development [19,20]; however, none have been optimized for x-ray
laboratory characterization purposes.

     These systems usually consist of a Marx generator to create
a high current, and a high voltage source of pulse power which is
then coupled to various plasma loads such as gas Z-pinches or
wire arrays. The characteristic emission of this plasma will
consist of both thermal emission from the plasma as well as characteristic line emission from the optically thin regions of the
plasma.

     We can see by comparing the parameters in Table 1 that this
electrical pulsed plasma source is one of the most intense x-ray
sources and is particularly useful for characterizing systems
with time responses of a few nanoseconds or longer. To date,
this type of machine has been used as a flash radiography source
for backlighting [21] and as a lithography source.

8.4 Laser-Pulsed Plasma X-ray Sources.

High temperature plasmas can also be achieved with short-
pulse lasers. When focused to submillimeter dimensions, lasers
can create plasmas with electron temperatures of hundreds of
volts and 100-ps duration. Such plasmas can generate intense
line radiation useful as a characterization source. On the other
hand, if lower intensity illumination is used it is possible to
create quasicontinuum sources of radiation in the VUV and low-
energy x-ray region using rare earth targets [22].

Again, reviewing the parameters of Table 1 it is seen that
laser plasma x-ray sources can provide the shortest duration
x-ray pulses of 100 ps or less. They also provide the brightest
x-ray sources because of their high energy density and small
size. This type of source is the system of choice where high
bandwidth temporal response is paramount. Such an application
was the bandwidth measurement of the risetime of the high bandwidth x-ray diodes discussed in Sections 3 and 4 [4].

8.5 Synchrotron X-ray Sources.

A new class of x-ray source based upon the synchrotron
emission of electrons in high energy storage rings has become
increasingly important in recent years. This radiation source is
being applied to many of the x-ray characterization needs typically handled by conventional sources. The properties of synchrotron sources are a unique combination of high brightness,
high intensity, continuous energy coverage, tunable polarization,
low divergence, and pulsed time structure.

The intensity of synchrotron radiation is shown in Fig. 17.
The two curves are for the two electron storage rings at the
Brookhaven National Laboratory's National Synchrotron Light
Source, NSLS. One ring has a storage energy of 2.5 GeV and a
maximum current capability of 500 mA. This ring produces the
highest energy photons and is optimized for generating x-rays
between 1 and 25 keV. The second ring has a 1 amp capability and
a storage energy of 750 MeV to optimize it for x-rays of energies
lower than 2 keV. In their range of optimum intensity, each
source is several orders of magnitude brighter than conventional
electron-impact generators.

Synchrotron radiation has been used as an absolutely calibrated x-ray source in only a limited number of experiments.
Some of the first of these included the measurement of surface
quantum efficiencies of x-ray diodes at the Synchrotron Ultraviolet Radiation Facility, SURF II, at the National Bureau of
Standards [23]. With the advent of numerous synchrotron beam

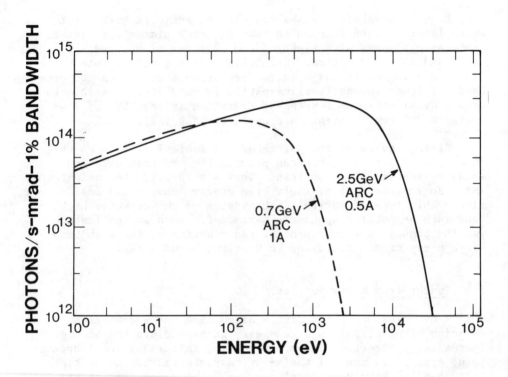

Fig. 17. The source intensity of the two synchrotron storage
rings at the Brookhaven National Laboratory. These
sources are continuous in x-ray emission and several
orders of magnitude more intense than conventional
sources.

lines more are being dedicated to absolute measurements. Facilities being built by Los Alamos at the Brookhaven National Laboratory, and by Livermore at the Stanford Synchrotron Radiation
Laboratory, will also have an absolute radiometric capability to
provide additional facilities for absolute calibration.

REFERENCES

1. J. R. Degnan, "Fast, Large-Signal, Free Standing Foil Bolo-meter for Measuring Ultrasoft X-Ray Burst Fluence," RSI, 50 (10), p. 1223-1226 (1979).

2. R. H. Day, P. Lee, E. B. Saloman, and D. J. Nagel, "Photo-electric Quantum Efficiencies and Filter Window Absorption Coefficients from 20 eV to 10 keV," J. Appl. Phys. 52(11), p. 6965-6973 (1981).

3. G. Beck, "Photodiode and Holder with 60-psec Response Time," RSI, 47, p. 849-853 (1976).

4. R. H. Day, P. Lee, E. B. Saloman, and D. J. Nagel, "X-ray Diodes for Laser Fusion Plasma Diagnostics," LA-7941-MS, (1981), (Los Alamos National Laboratory Report).

5. B. L. Henke, J. A. Smith, and D. T. Attwood, "0.1 - 10 keV X-Ray Induced Electron Emissions from Solids--Models and Secondary Electron Measurements," J. Appl. Phys. 48, p. 8 1952 (1977).

6. H. N. Kornblum and V. M. Slivinsky, "Flat-Response, Sub-Kiloelectron Volt X-Ray Detector with a Sub-Nanosecond Time Response," RSI, 498, p. 1204-1205 (1978).

7. D. G. Shirk and N. M. Hoffmann, "Monte Carlo Error Analy-sis in X-Ray Spectral Deconvolution," to be published in NIM, proceedings of the "5th Topical Conference on High Temperature Plasma Diagnostics," 16-20 September 1984, Granlibakken, Tahoe City, California.

8. D. J. Johnson, "An X-ray Spectral Measurement System for Nanosecond Plasmas," RSI, 45(2), p. 16-19 (1974).

9. P. B. Lyons and D. W. Lier, "Fabrication and X-ray Calibra-tion of Thin Plastic Scintillator Detectors," IEEE Trans. on Nuc. Sci., NS-22, p. 88-92, (1975).

10. D. B. Van Hulsteyn and T. W. Barbee, "Monochromatic X-ray Imaging with a Metal Multilayer Camera," submitted to Appl. Optics.

11. Kenneth B. Mitchell, Los Alamos National Laboratory, private communication.

12. J. Dumond and J. P. Youtz, J. Appl. Phys. 11 p. 357 (1940).

13. E. Spiller, "Evaporated Multilayer Dispersion Elements for Soft X-rays," in Proc. Topical Conference on Low Energy X-ray Diagnostics, Monterey, California (1981), (AIP Conference Proceedings #75) p. 124.

14. T. W. Barbee, "Sputtered Layered Synthetic Microstructure (LSM) Dispersion Elements," in Proc. Topical Conference on Low Energy X-ray Diagnostics, Monterey, California (1981), (AIP Conference Proceedings #75) p. 131.

15. J. H. Underwood and T. W. Barbee, "Synthetic Multilayers as Bragg Diffractors for X-rays and Extreme Ultraviolet: Calculations of Performance," in Proc. Topical Conference on Low Energy X-ray Diagnostics, Monterey, California (1981), (AIP Conference Proceeding #75) p. 170.

16. P. Lee "X-ray Diffraction in Multilayers", Opt. Commun. 37, p. 159-164 (1981).

17. B. L. Henke, P. Lee, T. J. Tanaka, R. L. Shimobukuro, and B. K. Fujikawa, "Low Energy X-ray Interaction Coefficients: Photoabsorption, Scattering, and Reflection. E = 100 - 2000 eV Z = 1-94," Atomic Data and Nuc. Data Tables, 27(1), p. 1-144 (1982).

18. G. L. Stradling, T. W. Barbee, B. L. Henke, E. M. Campbell and W. C. Mead, "Streaked Spectrometry Using Multilayer Interference Mirrors to Investigate Energy Transport in Laser Plasma Applications," in Proc. Topical Conference on Low Energy X-ray Diagnostics, Monterey, California (1981), (AIP Conference Proceedings #75) p. 292-296.

19. G. Dahlbacka, S. M. Matthews, R. Stringfield, I. Roth, R. Cooper, B. Echer and H. M. Sze, "A New Efficient Pulsed Plasma Soft X-ray Source," in Proc. Topical Conference on Low Energy X-ray Diagnostics, Monterey, California (1981), (AIP Conference Proceedings #75) p. 32-34.

20. J. Riordan, J. S. Pearlman, M. Gersten and J. Rauch, "Subkilovolt X-ray Emission from Imploding Wire Plasmas," in Proc. Topical Conference on Low Energy X-ray Diagnostics, Monterey, California (1981), (AIP Conference Proceedings #75) p. 35-43

21. J. Chang, D. L. Fehl, L. Baker, and D. J. Johnson, "Three Frame Flash X-Radiography System for Ion Beam Implosion Studies," in Proc. of 15th International Congress on High Speed Photography and Photonics, San Diego, California, SPIE 348, p. 696-699 (1982).

22. D. J. Nagel, C. M. Brown, M. C. Peckerar, M. L. Genter,
     J. A. Robinson, T. J. McUrath, P. K. Carroll, "Repetitively
     Pulsed-Plasma Soft X-ray Source," Appl. Opt. 23(9), p. 1428-
     1432 (1984).

23. E. B. Saloman and D. L. Ederer, "Absolute Radiometric Cali-
     bration of Detectors between 100-160 Å," Appl, Opt. 14,
     p. 1029 (1975).

HIGH ENERGY X-RAY DIAGNOSTICS OF SHORT-LIVED PLASMAS

Robert H. Day

Los Alamos National Laboratory
Physics Department
Los Alamos, NM
USA

1. INTRODUCTION

High energy x-ray and gamma-ray measurements are a common diagnostic of high-temperature/high-density plasmas. We will define high energy as those energies above 50 kilovolts and discuss the various techniques for returning temporal, spectral, and spatial distributions of this emission.

The study of high-energy x-ray emission from plasmas is important in understanding their energy partition and transport. Terms such as "run-away electrons" and "superthermal electrons" are frequently used to describe the electron component which gives rise to this emission. These terms derive from the fact that this portion of the electron spectrum is not characterized by the same shape or temperature indicative of the bulk of the plasma emission. This component can be generated by run-away electrons as might be observed in a relativistic electron beam experiment, or as an underdense plasma phenomenon in laser-target-interaction experiments. Frequently, these processes are poorly understood and the high-energy x-ray diagnostic is an important aid in interpretating the results of the experiments.

To illustrate the available techniques for measuring emission in this region of the spectrum, I will use examples from the CO_2 laser plasma experiments at the Los Alamos National Laboratory. These systems, however, are typical of the kinds of measurements made at many laboratories of high-energy x-ray emission from short-lived plasmas whether they are from Z-pinches, relativistic electron-beam machines, laser-target interactions,

imploding liner experiments, or the load of some other power
system.

2.   HIGH-ENERGY SPECTRAL RESOLUTION

The high-energy spectral distribution is frequently one of
the most important parameters used in determining the nature and
source of the high-energy electron distribution. One of the
simplest instruments for performing this measurement is the filtered detector system such as that described in Section 4 of my
lecture on "Nondispersive X-Ray Diagnostics of Short-Lived Plasmas."

Such a system was built to diagnose the high-energy spectrum
from the HELIOS laser at Los Alamos [1]. It consisted of ten
filtered detectors as shown in Fig. 1. The spectrum is passed
through a defining aperture and any electron contamination in the
beam is removed by an electron sweeper magnet. The x-rays are
then filtered before being detected by scintillator/photodiodes.
The scintillator type and thickness further modify the response
of the channel.

The scintillators used in this system consist of either
plastic NE111 or CsF. The CsF samples were encapsulated in her-
metically sealed packages to prevent their deterioration. Typical energy-dependent response functions for this system are shown
in Fig. 2. The channels cover the energy range from approximately 10 keV to several megavolts and the system was tuned to
have maximum sensitivity between 50 keV and 500 keV.

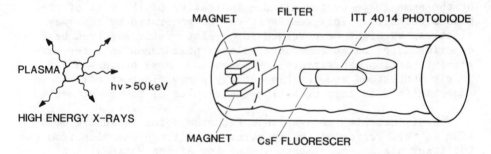

Fig. 1.   Schematic of a filtered detector channel used to measure
          the high-energy x-ray emission from a plasma.

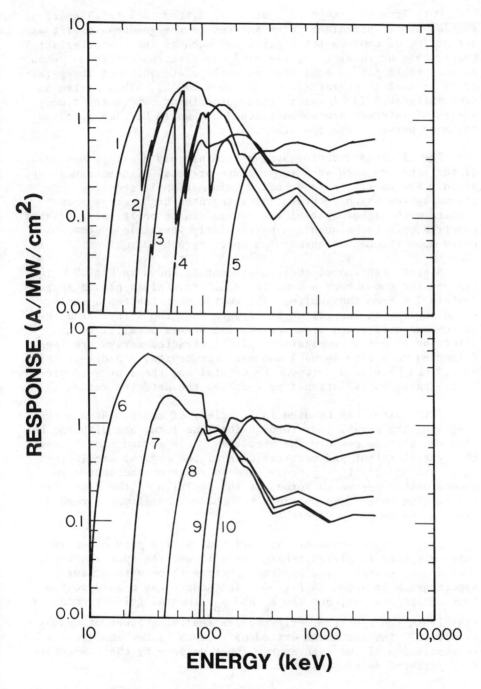

Fig. 2.   Typical response functions of a filtered detector system
for measuring high-energy x-ray spectra.

Data from this type of system are fitted to a trial spectrum such as a bremsstrahlung curve and tested for goodness of fit as a function of some variable parameter such as the characteristic temperature or total x-ray yield of the electron spectrum. When an acceptable fit is achieved the appropriate value of the parameters is used to characterize the spectrum [2]. This system is time integrated for a laser fusion experiment, where the time scales of interest are subnanosecond but would be time resolved for many pulse plasma applications.

The filtered detector system returns only a simple measure of the spectrum such as a temperature or total high-energy x-ray yield. Frequently, more detailed information is required. A second system which returns time-integrated high-energy x-ray spectra with higher spatial resolution can be built using crystal spectroscopic techniques. A particularly versatile system is based upon the Laúe transmission spectrograph [3].

A cross section of this instrument is shown in Fig. 3. In the version shown here a 2-mm thick LiF crystal is placed approximately 1 m from the source. The source must subtend a small solid angle with respect to the crystal and the x-rays from the source are diffracted from the crystal planes parallel to the direction of x-ray propagation. The diffracted x-rays are then deflected to a line focus 1 m behind the crystal. A direct-shine shield can be placed between the crystal and the source to prevent background radiation from reaching the detector region.

This system can be used in a variety of modes. As a spectrograph, the x-rays pass through the line focus and disperse behind the slit to measure the spectrum. As a monochromatic imager the lateral extent and vertical size of the crystal are limited to create an effective pinhole, and an image of the source in monochromatic x-rays is formed at the position of the slit. In the imaging mode it is possible to obtain submillimeter resolution in the geometry described.

For the spectrographic application, a film pack or active imaging system is placed behind the slit and the spectrum is recorded. An example of a typical spectrum taken with a Laúe spectrograph is shown in Fig. 4. The source was a gold-coated laser fusion target, and the K_α and K_β lines of gold are clearly visible in the spectrum verifying its resolving power of approximately 10. The continuum extending to much higher energy is a corroboration of the high energy spectrum seen by the low-resolution filtered detector system.

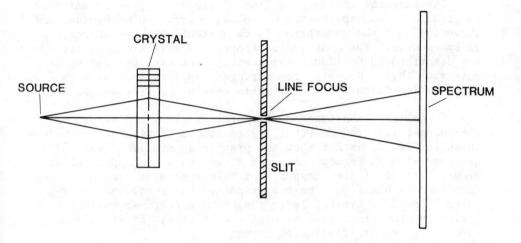

Fig. 3. A cross section of a Laúe-geometry transmission spectrograph.

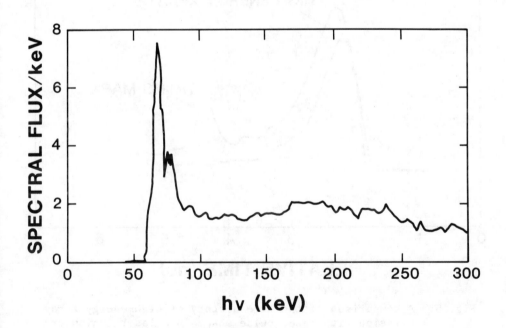

Fig. 4. Example of high-energy x-ray spectra taken with a Laúe spectrometer.

3.   TEMPORAL RESOLUTION

The temporal characterization of high-energy x-ray emission is frequently an important clue to the generation mechanism, and to verify that the generation is coincident with the thermal plasma source. For many applications, the filtered detector system described in Section 2 has adequate temporal resolution to determine this. However, for stringent applications such as laser fusion diagnostics, this time resolution is inadequate.

To achieve subnanosecond temporal resolution a simple single-channel filtered detector system was built similar to the one shown in Fig. 1, except that the plastic scintillator was NE111 quenched with 2% Benzophenone and viewed by a high-speed photo-diode. The final time response of this system, including cable, can exceed 1 GHz [4]. The comparison of the x-ray and laser pulse shape in a typical laser fusion target are shown in Fig. 5. Obviously, the production of high-energy x-rays is coincident with the laser irradiation on target.

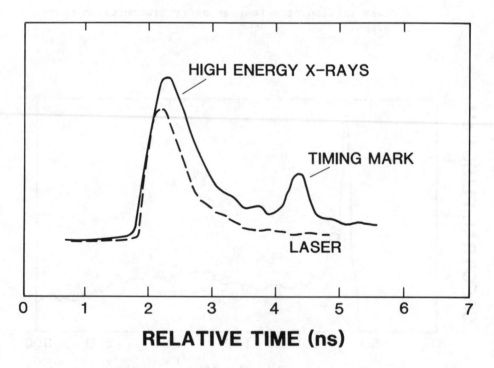

Fig. 5.   A comparison of the time history of high-energy x-ray emission with laser pulse shape in a laser fusion experiment. The high-energy x-ray detection system bandwidth exceeds 1 GHz.

4.   SPATIAL RESOLUTION

The third parameter necessary to characterize high energy x-ray emission from a source is the spatial distribution of the emission.   It is desirable to localize the emission region to verify its coincidence with the thermal source or to localize power flow in some other region.

The technique which is most frequently used is the standard pinhole photograph similar to the type of system used to produce low-energy x-ray images.   The major difference of course is the difficulty of producing small high-contrast pinholes at high energy.   By careful design it is possible to build systems with resolutions of a few tens of microns, though frequently much coarser resolution and larger fields-of-view are desired.   The kinds of advanced x-ray optics systems used in the low-energy x-ray regime are not available and probably will not exist in the foreseeable future.

The detector placed behind the pinhole imager is frequently a simple film pack, but advanced high-energy x-ray-sensitive emulsions using rare-earth intensifier screens have adequate resolution and sensitivity for many applications.   More advanced imagers using electronic intensifiers can also be used.   A system using gated microchannel plates has been built at Sandia National Laboratory and is used routinely on PBFA experiments [5].

The basic instruments necessary to characterize the high energy x-ray emission from plasma experiments are in hand and have been applied.   They provide vital information on the partition and transport of energy in a variety of pulse power applications.   The sophistication of these instruments is not comparable to that of lower energy x-ray devices, but adequate for most present applications.

REFERENCES

1.   W. Priedhorsky and D. Lier, "APACHE: A Low Background Hard X-ray Spectrometer for 10.6-µm Laser Fusion Experiments," RSI 53(8), p. 1189-1197 (1982).

2.   W. Priedhorsky, D. Lier, R. Day and D. Gerke, "Hard X-Ray Measurements of 10.6 µm Laser-Irradiated Targets," Phys. Rev. Lett., 47(23), p. 1661-1664 (1981).

3.   W. C. Priedhorsky, D. W. Lier and R. H. Day, "Laüe Diffraction Hard X-Ray Spectrometer for Laser Fusion Diagnostics," RSI 54(13), p. 1605-1610 (1983).

4. P. B. Lyons, C. R. Hurlbut, and L. P. Hocker, "Subnanosecond
    Plastic Scintillators," NIM 133, p. 175-177 (1976).

5. William Steygar, Sandia National Laboratory, private commun-
    ications.

FIBER OPTICS IN X-RAY DIAGNOSTICS APPLICATIONS

Robert H. Day

Los Alamos National Laboratory
Physics Department
Los Alamos, NM
USA

1. INTRODUCTION

This lecture contains several discussions of the way in which fiber optics have found widespread use in diagnostics apaplications. Several lectures in Chapter 2 outline the way in which fiber optics can be used to measure the current and voltage time histories in this difficult environment. The sections by George Chandler and Jim Chang also discuss fiber optics in plasma diagnostics applications. In this lecture, I will describe how fiber optics can be used as active transmitters of diagnostic data and some ways in which they can be used as passive radiation-to-light converters. I will also discuss the problems of radiation damage and temporal dispersion which are encountered in using fiber optics to transmit diagnostic data and some of the techniques used to mitigate these problems.

In its simplest form, a fiber optic is nothing more than a long thin strand of ultrahigh purity glass. This strand is doped with small concentrations of material to produce a radially dependent variation in the index of refraction of the glass. Light launched nearly along the axis of the fiber will undergo many total internal reflections and be guided along its entire length. These fibers have diameters from a few microns to several hundred microns and can be many kilometers long.

There are two basic techniques for creating this radial variation in the index of refraction. In the first, called step index, a sharp transition in index of refraction is made from an inner core of glass to an outer layer with a lower index of refraction. In the second, called graded index, the index of refraction is made to vary continuously and smoothly from the inner core to the outside of the fiber. This latter type of fiber generally has a smaller core of glass and therefore less time dispersion, and is used almost exclusively for diagnostic data transmission over long distances.

Fiber optics have numerous advantages which make them the transmission medium of choice for many systems. Of special importance is their inherent bandwidth which is many gigahertz-kilometers for a narrow band of visible wavelengths. They are also light weight and small in size which facilitates their installation and routing. Fiber optics are of nonmetallic construction and are almost immune to the traditional sources of electromagnetic interference so familiar in coaxial cable systems. Indeed, an effective technique used to isolate recording instruments from EMI sources is a fiber optic data transmission link. (See the lecture by George Chandler on grounding and shielding.)

2. APPLICATIONS

There are two basic modes - active and passive - in which fiber optics are used in diagnostics data transmission. In the active mode, the fiber is a means by which the information is transmitted from one location to another. The initial signal is usually transduced into an electrical form which is then used to modulate a light source, and this light is then transmitted along the fiber. In the passive mode, the signal is generated directly in the fiber and the fiber then transmits this information to the recording system. Both modes are used in diagnostics systems; however, the passive mode has the advantage of simplicity while the active mode can be coupled to a wider variety of diagnostic instrumentation.

One class of measurements to which the passive fiber system can be applied is the measurement of high intensity gamma-ray sources. An example of such a system is shown schematically in Fig. 1. The gamma ray beam is incident from below and passes through a thin converter plate where a small fraction of the gamma beam is converted into electrons. These electrons are preferentially forward-directed and pass through the fiber. The fiber is placed at the Cerenkov angle for electrons in silica and the resulting signal is transmitted up the fiber as the signal.

Since the intensity of Cerenkov light is proportional to the number of electrons, and the number of electrons is proportional to the gamma-ray intensity, this diagnostic produces a light output which is proportional to the gamma-ray intensity. The dynamic range of this diagnostic is limited by radiation-induced

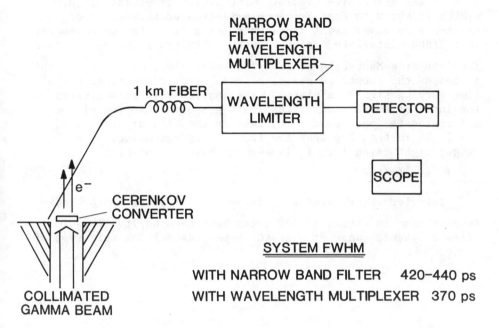

Fig. 1. A schematic of a passive fiber optic system to record
         intense gamma-ray beams.

darkening of the fiber which eventually attenuates the light
created by the electrons. This effect will be discussed in
greater detail below. The dynamic range of this measurement is
usually greater than 40 dB.

Fiber optics can also be used in a variety of active recording systems. The lectures of Chapter 2 by M. Hugenschmidt, G.
Changler, J. Chang and others describe ways in which fiber optics
can be used to record time-dependent voltages and currents.
These systems are indicative of new classes of fiber optics
diagnostics using polarization-preserving fibers and optically
active sections. The optically active sections will be made of a
variety of materials to enhance the sensitivity of the fiber to
either electric or magnetic fields. For example, a fiber made of
Yttrium-doped glass is presently under development for the Los
Alamos National Laboratory, LANL, by Gordon Day of the National
Bureau of Standards in Boulder to have a large Verdet constant to
act as a high sensitivity current sensor in pulse power flow
measurements.

We at LANL are pursuing another type of active system with
MIT Lincoln Laboratories based upon a Mach-Zehnder interferometer

as a voltage-sensitive element. The active component of this
system is shown in Fig. 2. A polarization-preserving fiber
carries a constant light level which is split into two components
in a $LiNbO_3$-integrated optic element. The two light guides form
the legs of a Mach-Zehnder interferometer. One of the two light
guides on the substrate passes between electrodes across which a
time-varying signal is placed. The applied voltage decreases
the index of refraction of the light guide in that leg of the
interferometer and, therefore, retards the time of arrival of
the light in that leg when the two signals recombine. The
output light intensity, I, is seen to have a component

$$I = I_o \sin^2(v(t)/V_\pi).$$

Interferometers with a V_π as small as 5 volts and a breakdown voltage in excess of 100 volts have been built. This implies a dynamic range of over 40 dB per channel for this type of
sensor.

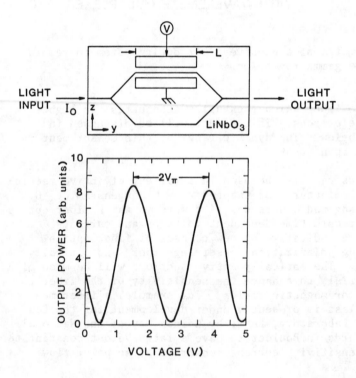

Fig. 2.   A fiber optic voltage sensor built as a Mach-Zehnder
          interferometer.

We have performed detailed tests on the linearity of this instrument by measuring an exponentially rising voltage pulse. The top panel of Fig. 3 shows the voltage-versus-time curve of

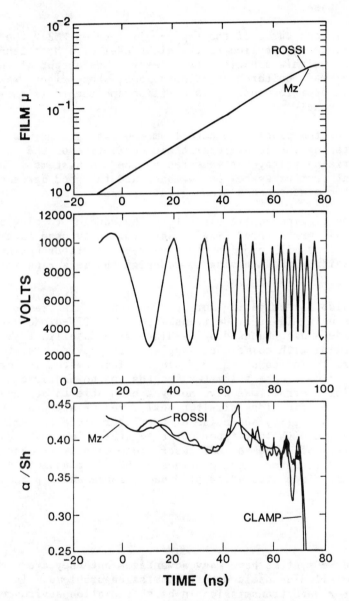

Fig. 3. Comparison of an oscillographic recording with a Mach-Zehnder interferometer measurement of an exponentially rising voltage.

the exponentially rising voltage curve as measured by two techniques which yield essentially identical results. In the curve labeled "Rossi," the voltage was measured by imposing the voltage directly onto an oscilloscope whose horizontal deflection circuit was driven by a high-frequency sine wave. This gives a highly accurate time base.

For the second curve of the top panel, labeled "MZ," the exponential voltage was applied to the electrodes of a Mach-Zehnder interferometer which was excited by a laser. The output of the interferometer was monitored by a photodiode, whose signal was recorded on an oscilloscope. This oscilloscope record is shown in the middle panel.

For the bottom panel, the data from the two systems was analyzed by taking the local logarithmic derivative of the exponentially rising voltage as measured by the two systems. This is a stringent test of system performance and the good agreement between the two measurements shows great promise.

This type of fiber-optic element can be coupled to many of the traditional current-producing sensors routinely used in x-ray diagnostics. This will provide the advantages of fiber optic data transmission without having to redesign the measuring instrument.

We are also investigating more conventional data transmission links using laser-diode transmitters and PIN-diode detectors to provide a high-bandwidth/low-distortion data link. We have been working with commercial suppliers to provide a system with the characteristics shown in Table 1. This system is for transmission of data over kilometer lengths in the underground nuclear testing program. We have not yet been able to achieve all of these specifications simultaneously, and we are presently limited by the stability and noise of the laser-diode transmitters when coupled with long fiber optic lengths. The best system to date consists of a Thomson CSF laser-diode transmitter and a Hewlett Packard receiver which provides a 300-MHz bandwidth over a one-kilometer fiber with 40 dB of dynamic range and 1-percent distortion.

3.   LIMITATIONS

Fiber optic systems have many advantages but they are not the answer to all diagnostic data transmission problems. In particular, for data transmission in harsh radiation environments and over long path lengths some special problems arise which must be considered.

Table 1.   Specifications for Analog Fiber Optic Link.

		1/85 Status
1)	Frequency response - 1.0 kHz to 200 MHz	OK
2)	Gain - 6 dB	OK
3)	Dynamic range - 45 dB	40 dB
4)	Linearity - ± 1% < 100 MHz	@ 1% < 200 MHz
	± 3% < 200 MHz	
5)	Stability - ± .1 dB for 1 hour	XMTR - 1.3X worse
	± .2 dB for 24 hours	OK
6)	Optical cable - 1 km, 600 MHz, 4 dB/km	OK

The problem of radiation-induced darkening of a fiber is shown in Figs. 4(a) and 4(b). Figure 4a is a schematic of a radiation damage experiment where the electron beam from a Febetron 705 strikes a fiber and causes reduced transmission. This reduction in light output is monitored by the extinction of a constant light level passing through the fiber. The resulting absorption in the fiber is plotted in Fig. 4(b).

The fiber transmission is characterized by two attenuation components. The long time-constant component is essentially exponential in time and results from recombination in the fiber. The short time-constant component persists for only a few nanoseconds but creates a higher opacity in the fiber. This effect is not well understood and is under active investigation. The most radiation resistant fiber is a step-index fiber of pure silica which shows onset of radiation-induced opacity of 50% at 10 krads dose. This fiber is acceptable in short lengths but the modal dispersion inherent in long lengths of step-index fiber limits their usefulness for data transmission over long lengths.

Temporal dispersion is a generic problem in all long distance transmission applications. One aspect of this problem is the wavelength-dependent transit time of fibers resulting from the frequency dependence of the index of refraction. This effect is shown in Fig. 5. An impulse of polychromatic light is injected into a kilometer of fiber and the time of arrival of various wavelengths is measured at the other end. By the time the

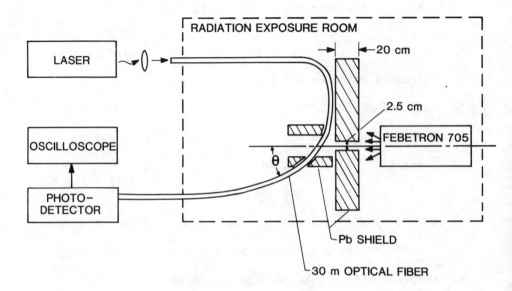

Fig. 4(a).   An intense ionizing radiation field from a Febetron
              705 causes damage in a fiber optic which reduces its
              light transmission properties.

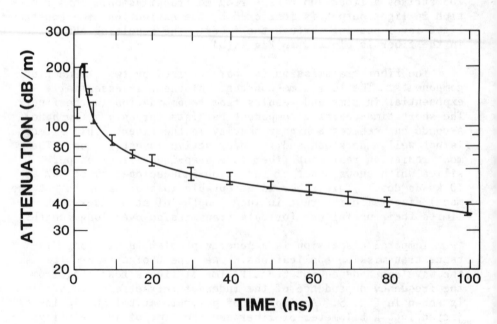

Fig. 4(b).   The light attenuation of the fiber versus time after
              being dosed.

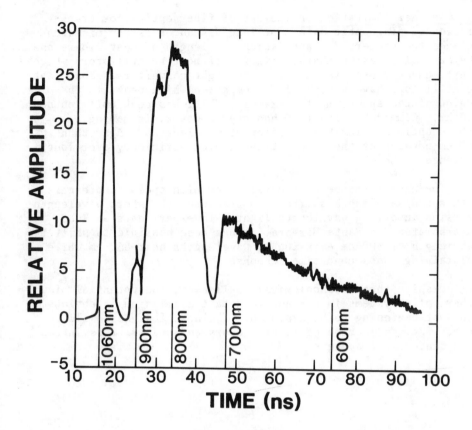

Fig. 5. Long lengths of fiber optic show significant temporal
         dispersion. This figure shows the temporal dispersion
         of a short pulse of polychromatic light in a kilometer
         length of fiber optic.

light traverses a kilometer length, the risetime of the pulse has
been broadened from a nanosecond to greater than 50 nanoseconds
for a 400-nm bandwidth.

     One technique which can be used to restrict the temporal
dispersion in a fiber is to limit the bandwidth of the transmitted light to a narrow wavelength region. A band of a few
nanometers at 600 nanometers will yield a system of gigahertz-
kilometer capability. Unfortunately, this is achieved at the
expense of signal for a polychromatic source. It is possible to
improve this performance by dispersing the light and delaying the
faster wavelengths to compensate for their early arrival. With
this type of correction, it is possible to improve the signal by
up to an order of magnitude.

Another limitation on the use of fiber optics for transmission of diagnostic data over long distances is the mismatch
between the preferred transmission wavelength of most fibers and
the natural emission characteristics of most scintillators which
are frequently used as radiation-to-light converters. Most fast
scintillators have visible light emission which peaks in the blue
region of the spectrum; for example, NE111 has peak emission near
230 nm. Commercial fiber technology, however, is moving to
fibers optimized for transmission at wavelengths longer than
800 nm because of the higher transmission efficiency over long
distances.

The answer has been to develop new high-speed fluorescers
with speed/wavelength shifters to transduce the short wavelength
emission into long wavelength light. In experiments at EG&G,
Incorporated, in Santa Barbara, it has been possible to shift the
wavelength of 300-nm emission to wavelengths near 850 nm while
maintaining nanosecond time response.

Despite these technological challenges, the potential of
fiber optic systems in diagnostic data transmission is obvious
and just beginning to be explored. The next few years will see
increasing application of fiber sensors and transmitters in routine measurement systems.

FLASH RADIOGRAPHY

F. Jamet

Institut Franco-Allemand
de Recherches de Saint-Louis (ISL)
F 68301 Saint-Louis
France

1. INTRODUCTION

Flash radiography is a method for recording radiographic pictures of a non-transparent high-speed phenomenon. Such events occur in a variety of fields: Ballistics, dynamic high pressure physics, detonics, industry, and medical diagnostics.

The phenomena observed by flash-radiography generally move at velocities between 100 m/s and 10000 m/s. The motion blur recorded on the picture is negligible only if the duration τ of the x-ray pulse (or the exposure time) falls in the range 10^{-8} -10^{-7}s. Since the optical density, D, obtained on a film is determined by the product of the x-ray intensity I and τ, the diminution of τ must be compensated by a simultaneous increase of I, which, in turn, is proportional to the term $i \cdot V^2$, where i is the value of the current passing through the tube and V is the voltage applied to the anode. The variations of V cause not only a change in I but also a shift in spectrum. This effect is not always desired. The intensity of the x-rays must therefore be increased mainly by increasing the current i.

With a medical or industrial radiographic device, the correct blackening of a film is obtained with a current of some milliamperes and a mean exposure time of 0.1 s, i.e., $i \cdot \tau \cong 10^3$ A-s. From this it is deduced that the value of the current passing through a flash x-ray tube must be approximately equal to 10^4 A. These values cannot be attained in conventional x-ray tubes with a heated cathode because the number of electrons having a sufficient energy to overcome the potential barrier is not high

enough, and, moreover the current is limited by space charge
(Languir's law).

The production of large currents therefore requires the use
of methods other than thermionic emission. This is mainly
achieved by vacuum discharges and field emission.

2.   PRODUCTION OF X-RAY PULSES

Let us consider two electrodes enclosed in a vessel with gas
pressure below 10^{-6} torr; the mean free path of the electrons is
then in excess of the electrode spacing. When applying a voltage
pulse whose amplitude is higher than the breakdown voltage, we
observe schematically the successive phenomena shown in Fig. 1:

a. Beginning and increasing of a field-emission current,
and emission of x-rays from the anode;

b. Heating of the anode by the impact of the electrons;

c. Development of a vacuum arc which is characterized by
the presence of metallic vapors and other gases emanating from
the evaporation of a thin layer of the anode impacted by electrons.

During the presence of the arc the voltage across the electrodes is low and the x-ray emission is stopped. The design
features of the electrodes currently used in flash x-ray tubes
can easily be derived from the foregoing remarks (Fig. 2). The
cold cathode has a sharp edge or tips to allow a high intensity
field emission current to be built up. The anode generally has a
conical shape in order to keep the apparent x-ray focus as small
as possible. In case of very high voltages the production of
x-rays by the "transmission method" has a better efficiency.
Then the anode is a foil. The metal used for the anode is tungsten or tantalum because they offer both high atomic number
(which is required for high intensity continuous spectra) and
melting temperature.

Generally the tubes are sealed off (Fig. 3), but in some
cases the use of continuously pumped tubes is preferred because
they offer the advantage of allowing an easy change of the electrodes.

The anode of a flash x-ray tube is not cooled because the
local heating due to the very strong electron current densities
causes an immediate vaporization. Consequently, at each pulse
this evaporation of a superficial layer reduces the tube life.
After 100 to 1000 pulses the tube must be changed.

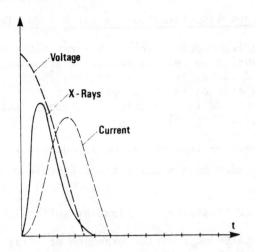

Fig. 1.   Voltage, x-ray intensity, and current vs. time (in
          arbitrary units) in a flash x-ray system (Marx surge
          supply with vacuum discharge tube).

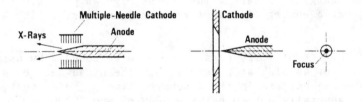

Fig. 2.   Classical electrode shapes used in vacuum discharge or
          field-emission tubes.

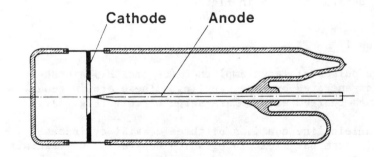

Fig. 3.   Flash x-ray vacuum sealed-off diode tube.

2.1 Marx-surge Generators Associated with a Flash X-ray Tube

The high voltage pulses applied to flash x-ray tubes are in the range of a few kilovolts to several megavolts according to the desired hardness and intensity of x-radiation. Such high values are obtained, in most cases, by a Marx-surge generator. Capacitors, initially charged in parallel, are series-connected with the aid of spark gaps (Fig. 4).

In a first approximation, the theoretical amplitude V_t of the output pulse of the unloaded Marx's multi-stage circuit can be written as follows:

$$V_t = n \cdot V \text{ (n = number of stages, V = loading voltage)}.$$

This theory is incomplete because of the presence of stray capacitances which reduce the value of V_t. The relationship between V_t and $n \cdot V$, taking into account the stray capacitances (C'), shows that the output voltage does not increase further with n greater than 30 to 50, if the ratio C'/C ranges approximately from 10^{-3} to 3.10^{-3} (Fig. 5). These remarks lead to the following conclusions:

a. The charging voltage V must be as high as possible to avoid too great a number of stages (values from 30 kV to 100 kV are nominal);

b. The effect of stray capacitance can be reduced by using a stage capacitor as high as possible, but the disadvantage is the simultaneous increase in the discharge time. The capacity per stage is generally chosen between 2000 pF and 20000 pF.

c. The increase in the number of stages leads to an increase in the inductance of the circuit.

A practical design is shown in Fig. 6.

2.2 Line Supplies

Voltage pulses of large amplitude (several MV) and short duration are generated by line supplies. These are in general Blumlein lines charged with a Marx-surge circuit (Fig. 7).

The Blumlein line consists of three coaxial cylinders (Fig. 7a) or three parallel plates which behave as two transmission lines between which is placed the flash x-ray tube. The study of wave reflections shows the matched Blumlein line (R = $2\sqrt{L/C}$) delivers an output pulse whose amplitude and duration

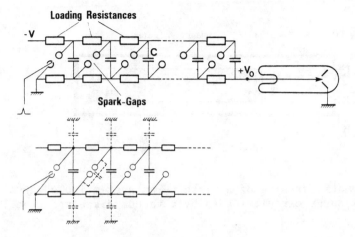

Fig. 4.   a)   Schematic drawing of a Marx-surge pulser supply
          b)   Drawing showing the stray acapcitances

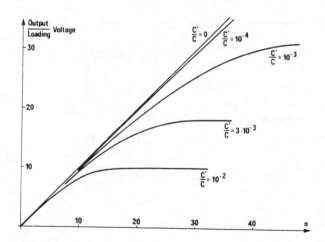

Fig. 5.   Ratio V_t/V of the Marx's output voltage to the charging
          voltage as a function of the number of stages n for dif-
          ferent values of C'/C, where C is the stage capacitance
          and C' the stray capacitance to ground per stage.

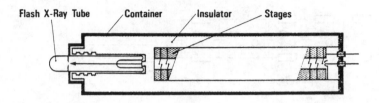

Fig. 6.   Schematic drawing of a flash x-ray system including a
          Marx-surge circuit loaded by a vacuum discharge tube.

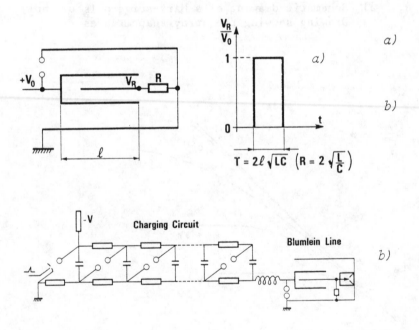

Fig. 7 a) Principle of a Blumlein line consisting of three co-
            axial cylinders.
        b) Charging of a Blumlein by a Marx-surge supply and a
            self-inductance (resonant charging circuit).

equals the charging voltage and $2 \cdot \ell \sqrt{LC}$, respectively. (ℓ is the length of the line, and L and C denote the inductance and the capacitance, respectively, per unit of length.) The loading may be done with either a dc power supply or a Marx-surge generator. In this case it is desirable to introduce a resonant charging circuit allowing for an output voltage which is in excess of the charging voltage (Fig. 7b).

2.3 Other Methods

Among the other methods used to generate x-ray pulses, such as transformer supplies or simple transmission lines connected to a flash x-ray tube, we should like to mention electron accelerators. The electrons reach very high energies (several tens and several hundreds of MeV) and generate, in impacting a tungsten target, very hard x-ray radiation. This method is of particular interest for investigations performed in detonics if large quantities of explosives are involved.

3. MEASUREMENT OF FLASH X-RAY GENERATOR CHARACTERISTICS

3.1 X-Ray Pulse Duration

The measurement of the x-ray intensity vs. time variation is generally carried out as follows: a scintillator transforms the x-radiation into a light pulse detected by means of a high-speed photodiode or a photomultiplier tube. It is necessary that the rise and decay times of the fluorescence of the scintillator used remain below those of the x-ray pulses (plastic scintillators work quite well).

3.2 Dose Per Pulse and Dose Rate

In practice, a very important parameter characterizing a flash x-ray device is the dose per emitted pulse. The dose is generally expressed as "Roentgens" (R)*. The measurement is usually performed with the aid of dosimeters which directly indicate the integrated dose. The dose rate (R/S) is substantially in excess of that delivered by conventional x-ray devices.

---

* The Roentgen is "the quantity of x- or γ-radiation such that the associated corpuscular emission (due to ionization) per 0.001293 g of air produces, in air, ions carrying 1 esu of electricity of either sign". This corresponds to an absorption in air of energy equal to 0.109×10^7 J-cm^3 (i.e., 83 erg/g).

3.3 Quality of the Emitted X-rays

We know that x-ray devices for industrial or medical use are characterized by a network of curves giving the product of the current intensity in the tube by the exposure time (in mA-s) as a function of the thickness of the absorber. These curves allow one to determine the best conditions to attain a specific optical density on a film, the operating voltage thereby being taken into account as a parameter.

Since in flash radiography the x-ray pulse length is fixed for a system, we plot the characteristic curves with variables which differ from those used for conventional devices. The intensity I of the x-radiation impinging on a film placed in contact with the back of an absorber of thickness a, located at a distance r from the intensity source I_o, is given by the following expression

$$I = \frac{I_o}{2} \exp(-\mu a)$$

where the absorption coefficient, μ, is constant if the radiation is monochromatic. It is possible to plot the curves $r = f(a)$ at constant intensity I, i.e., for both a defined voltage and optical density on the film.

The characteristic curves (Fig. 8) drawn in a system of semilogarithmic coordinates are straight lines for absorber thicknesses in excess of a few millimeters. This means that after filtration the spectrum becomes homogeneous and behaves as a monochromatic radiation for which the absorption coefficient can be easily calculated from the slope of the straight line. It is possible to correlate this value of μ with an effective wavelength λ_{eff} which is, for an x-ray device consisting of a Marx-surge generator and a vacuum discharge tube, equal to 50% of the wavelength corresponding to the maximum voltage.

3.4 Source Size

The source size is determined by the well known pinhole camera, including a lead foil with a hole as small as possible placed between source and film. The pinhole camera image of the focus of a vacuum discharge tube shows that the source diameter ranges between 2 and 5 mm.

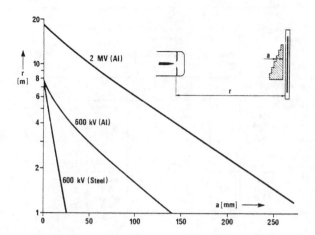

Fig. 8.   Penetration capability of a "Fexitron" flash x-ray sys-
          tem. Kodak Royal Blue film; Dupont industrial screens;
          optical density: 0.7.

4.   CHARACTERISTICS OF SOME FLASH X-RAY SYSTEMS

Table 1 summarizes the characteristics of a few types of
flash x-ray systems.

5.   RECORDING OF THE FRAMES

5.1   Triggering and Synchronization

The following block diagram (Fig. 9) indicates the principle
of the synchronization and the measurement of the time interval
between phenomenon and x-ray pulse.

The signal delivered by the phenomenon under investigation
varies according to the experiment. It can be obtained, for instance, by a short circuit or by a contact between two wires, or
by a light barrier or an x-ray barrier.

5.2   Recording of the Frames with Films

The frames are usually recorded on a radiographic film which
can be placed between two intensifying screens whose function is
to convert the x-radiation either into electrons or into light,
the action of the latter on the emulsion being more efficient.

Table 1.

Flash x-ray system	Electrical characteristics		Characteristics of the x-ray emission			Remarks
			Pulse duration ns	Dose at 1 m from the source R	Source diameter mm	
Marx-surge supplies with vacuum discharge tube	Output voltage : 150 kV-2000 kV Energy : 10 J -1000 J		20-100	0.01-5	2-10	
"SWARF" (UK)	Marx : 24 stages 3 MV 60 kJ	Blumlein line : 60 ns 30-40 kA	60	38	6	Transmission field emission tube
"PHERMEX" (LASL, USA)	Accelerator : Electron beam energy : 27 MeV Target beam current : 30 A		100-200	12-25	3	
"PULSERAD 1480" (PI, USA)	Marx : 80 stages 8.2 MV 500 kJ	Blumlein line : 75 ns 200 kA	150	380	3.5	Transmission field emission tube

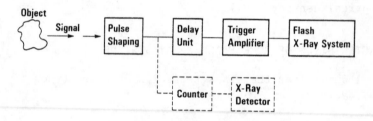

Fig. 9.   Block diagram illustrating the principle of synchronization between phenomenon and x-ray pulse.

In flash radiography fluorescent intensifying screens are generally needed (Fig. 10a).

The image quality is determined by the contrast and the
definition or resolution, the latter depending on the different
blurs, e.g., blur introduced by the film and intensifying
screens, or geometric and motion blur.

5.3 Recording of the Frames with Electro-Optical Systems

In the field of flash radiography, electro-optical systems
are less systematically employed than in industrial or medical
radiography. This is probably due to the destructive effects
of the events under investigation.

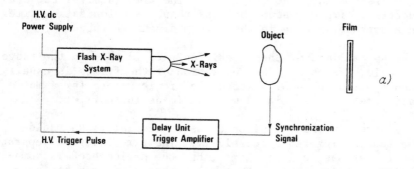

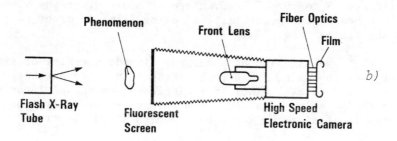

Fig. 10.   Recording of the flash x-ray frames with
           a)   a film
           b)   an electro-optical system

    The principle of image recording by means of such an appa-
ratus is well known (Fig. 10b). The flash x-ray tube emits an
x-ray pulse which, after having passed through the object, is
converted into visible light by means of a fluorescent screen.
The image obtained is then transferred to the photocathode directly if fluorescent layer and photocathode are in contact, or
through a lens having a large aperture. The quality of the image
is not as good as that obtained by direct recording on a film.
But in some cases, especially in flash x-ray diffraction studies,
the use of image intensifiers may be interesting.

6.   FLASH X-RAY CINEMATOGRAPHY

    The information about the evolution of a phenomenon is im-
proved by recording several successive flash x-ray pictures. We
can distinguish between two methods for flash x-ray cinematography.

In the first method, an x-ray pulse train is generated using a single flash x-ray tube (Fig. 11). The time interval between two successive frames cannot be less than the deionization time of the electrode spacing in the vacuum discharge tube.

The upper limit of the repetition rate is then in the range of 10^4 to 10^5 frames/s. The separation of the frames is usually obtained by means of a high speed electronic camera (or a movie camera), sometimes associated with an image intensifier.

The second method consists of using several flash radiographic tubes triggered successively. The time interval separating two frames has no lower limit, but the device used is somewhat bulky. Consequently, the total number of frames has an upper limit of 5 to 10. The advantage of cineradiography with multiple tubes is the possibility of using films in order to record the frames (Fig. 12).

For the investigation of phenomena without a center of symmetry or rotation axis, it is interesting to put the discharge gaps very close to each other in a single vacuum vessel in order to avoid the parallax (Fig. 13). In this case the separation of the frames is obtained by the use of a high speed electronic camera.

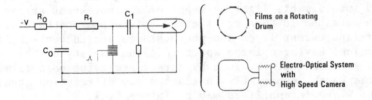

Fig. 11.   Flash x-ray cinematography with a single vacuum discharge tube.

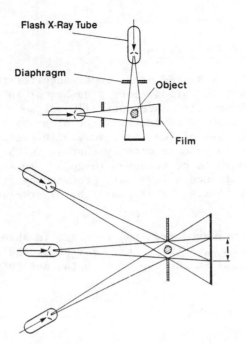

Fig. 12.   Flash x-ray cinematography with multiple tubes.

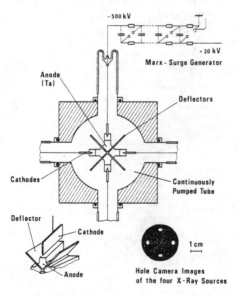

Fig. 13.   Schematic drawing of a flash x-ray cinematographic
           tube: four vacuum discharge gaps are disposed in a
           single vacuum vessel.

7.   APPLICATIONS

7.1   Flash Radiographs of Shock Waves

Flash radiography has achieved great importance for the observation of shock waves in solids since a number of interesting materials - e.g., all metals - are optically opaque.   The compressibility of solids being lower than that of gases (and also slightly lower than that of liquids), very high pressures (p > 10^{10} Pa) are necessary to produce density changes in the shock waves that are great enough to be observed (5-20%).   In most cases dynamic pressures of such a level are produced by direct contact of high explosives with the sample or by hypervelocity impact.

The compression is measured from the increase in absorption of the x-rays in the shock wave (Fig. 14).   For this, the product $\rho \cdot d$ (density and thickness) is introduced into the absorption formula

$$I = I_o \exp (- \mu/\rho \cdot \rho \cdot d).$$

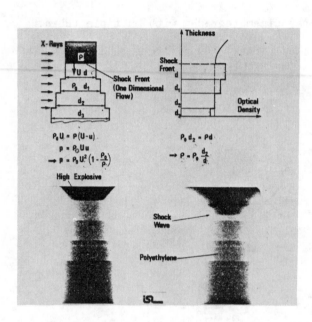

Fig. 14.   Measurement of the density in a shock wave by flash radiography.

A variation of the absorber thickness d by a specific factor affects the absorption in exactly the same way as does a corresponding density variation. So it suffices to record on the same film, together with the shocked sample, calibration steps made of the same material. The photometric analysis - independently of film and development influence - yields the density increase in the shock wave. By measuring the shock wave velocity (two successive flash radiographs are needed), the Hugoniot curve or shock adiabate, which characterizes the behaviour of a material under impulsive loading, is simultaneously determined.

7.2 Flash X-Ray Diffraction

The recording of diffraction patterns by means of generators delivering intense x-ray pulses has attracted increasing interest in the field of physics.

A single x-ray pulse emitted from a flash x-ray system will barely suffice for rough-pattern recording or may even be insufficient, in some cases, for doing so. It should be recalled that the intensity of diffracted rays is only 10^{-2}-10^{-4} as compared to that of the primary beam. When conventional x-ray generators are used, this difference can be compensated by increasing the exposure time. For flash x-ray systems, however, it is necessary to increase the primary intensity and/or the sensitivity of the recording technique.

In the field of solid-state physics in which solids are subjected to high compression generated by shock waves, the recording of flash x-ray diffraction patterns can provide valuable information about changes which occur in crystal structure.

We present, for instance, the detection of distance variation between the (200) lattice planes of a shock-loaded sodium chloride single crystal, the shock velocity thereby being parallel to the (100) direction (Fig. 15). A high intensity flash x-ray source emits the molybdenum K_α characteristic radiation which is reflected from a family of lattice planes. The Bragg angle for these planes is θ. Under the action of shock-wave compression this angle becomes $\theta + d\theta$. and the recorded line is shifted on the film. The measurement of this shift gives the variation $\frac{da}{a}$ between the planes ($\frac{da}{a} = \cot\theta \, d\theta$) (Fig. 16). It is obvious that this method must be used with many precautions. In particular, the depth of penetration of the radiation in the crystal (0.10 mm), as well as the motion of both the free surface and the reflecting lattice planes, must be taken into account.

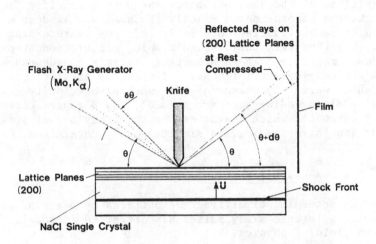

Fig. 15. Measurement of the distance variation between the (200)
          lattice planes of shock-loaded NaCl.

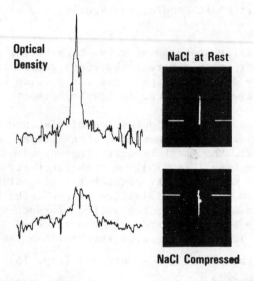

Fig. 16. Shifting of the Bragg line by a sodium chloride single
          crystal submitted to a shock wave.

The measured values of $\frac{da}{a}$ introduced into the relationships given under the assumption of either one-dimensional or isotropic deformation ($\frac{\rho_o}{\rho} = 1 - \frac{da}{a}$, or $\frac{\rho_o}{\rho} = 1 - 3\frac{da}{a}$) allows one to find out the nature of the deformation of the crystal lattice (Fig. 17).

In conclusion we can say that flash x-ray is a useful tool for the study of solids subjected to strong deformations.

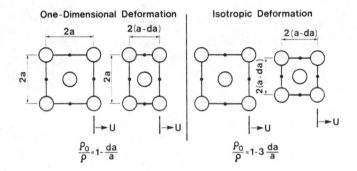

Fig. 17. Deformation models of NaCl by a shock wave.

REFERENCES

1. G. Thomer, Röntgenblitztechnik in Kurzzeitphysik, Springer-Verlag, Wien (1967) pp. 328-366.

2. F. Jamet and G. Thomer, Flash radiography, Elsvier, Amsterdam (1976).

3-NS FLASH X-RADIOGRAPHY

J. Chang

Sandia National Laboratories
Diagnostics Division
Albuquerque, NM
USA

1. INTRODUCTION

In intense particle beam interaction with targets, flash x-radiography (FXR) has become the standard technique used to observe the target dynamic response to particle irradiation. Most of the development was motivated by the particle-beam inertial-confinement fusion program [1,2]. In the inertial confinement approach to controlled thermonuclear fusion, ignition is to be achieved via heating, compression, and confinement of the fuel by an imploding shell. Since the confinement time improves with increasing pusher ρR, target design calculations [1] achieving gains >1 require a $\rho R > 1 g/cm^2$, where ρ and R are the mass density and shell thickness of the pusher, respectively. For such targets the pusher, during its acceleration by the driver and during subsequent deceleration by the compressed fuel, can be unstable, which can lead to reduced compression, fuel heating, and confinement time. An experimental study of this problem is very difficult because of the pusher thickness and requires the FXR technique to observe the hydrodynamic behavior of the critical regions of interest: the interfaces between the target ablator and the pusher, and between the pusher and the fuel.

In making these measurements, the instruments developed must overcome an intense x-ray and EMP background associated with the production of particle beams. In the FXR systems described, microchannel plates (MCP) are used extensively as x-ray converters, gated shutters, and intensifiers to provide signal detection, background discrimination, and signal amplification. The MCP's play an important role in the development of these systems.

In what follows, we will start with a description of the MCP's, and lead into a summary of the FXR development.

2.   THE MICROCHANNEL PLATE

The microchannel plate (MCP) is a 2-D derivative of the earlier channeltron [3], which is a single, continuous dynode formed into a long and narrow hollow tube. When an appropriate bias voltage is applied, secondary electrons generated by intercepted photons or charged particles on the inside surface of the channel tube are multiplied down the tube through additional secondary electron emission. By properly choosing the channeltron material with a secondary coefficient greater than 1, a gain of 10^4 can be achieved in a tube a fraction of a mm long. A close-packed 2-D array of a large number of these channels forms a micro-channel plate.

The initial use of the MCP was for night vision [4]. In this application the MCP is encased in a glass vacuum envelope between a photocathode and a phosphor screen. The high gain and compactness provided by the MCP permits the development of very small and light-weight night vision scopes. It is for this purpose that MCP developed as the mature technology that it is today. Modern MCP's have channel diameters as small as 12 μm, a channel-to-channel spacing of 15 μm, an open area ratio of 50% or better, a gain of 10^4, and can achieve a spatial resolution of up to 40 lp/mm.

The gain in the MCP is achieved through electron multiplication via successive generations of secondary electrons. Secondaries are produced and multiplied as they accelerate down the channel and repeatedly collide with the channel wall (Fig. 1). If the secondary emission coefficient is δ, then after n wall collisions inside the channel the gain achieved is given by $G = \delta^n$. This can be related to physical parameters of the MCP and the bias voltage, V, by the following expression [3]:

$$G = \delta^n = \left(\frac{AV}{2aV_o^{1/2}}\right)^{\frac{4V_o a^2}{V}}$$

A is a material constant defined in the relation

$$\delta = A(V_c)^{1/2} ,$$

with V_c being the average potential gain by the secondaries between collisions; V_o is the average emission potential of the

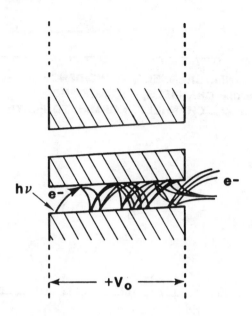

Fig. 1   Schematic of electron amplification in a single channel
of a MCP.

secondaries at right angles to the channel wall; and a is the
length-to-diameter ratio. For a realistic A ~ .2, a bias voltage
of 1kV, a V_o ~ 1 to 2 eV, a length-to-diameter ratio of 42, we
estimate a gain of 500 to 2 x 10^5, which brackets the typically
measured gain of 10^4. The detailed quantitative agreement in
this case is less important than establishing the qualitative
dependence of gain on the length-to-diameter ratio.

Because of the strong gain dependence on a, small variations
in the channel diameter, which is unavoidable due to the MCP production process, will result in large gain variations across the
active area of the MCP. However, it is also due to this strong
dependence that relatively high x-ray image resolutions can be
achieved. This will be made clear presently.

The spectral response of the MCP has been studied and measured from UV to almost 1 MeV [6]. Measurements so far indicate
that the response varies from about 1% at just below 1 MeV to a
peak of 20% at 200 ev [6], and then falls off again to 2% at
1200 Å [7]. The spectral response from near 100 eV to 1 MeV is
shown in Fig. 2. Response for higher energy photons is expected

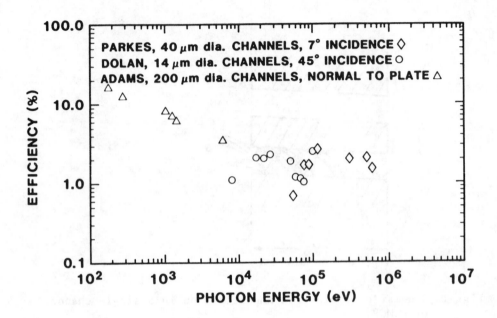

Fig. 2.   Compiled MCP spectral response from three reference
          sources covering the energy range from 100 eV to 1 MeV.

to remain about 1% because the photoelectron production efficiency beyond 1 MeV is relatively constant. Since the MCP thickness is usually less than 1 mm, hard x-rays having energies of a
few tens of keV and higher suffer almost no attenuation going
through the MCP. Therefore, it is equally likely for the incident photons to initiate an event throughout the thickness of the
plate and, furthermore, the photoelectrons produced may have sufficient energy to cross several channels. Therefore, one is lead
to expect poor image resolution due to crosstalk between channels. However, measurements have shown that due to the strong
dependence of gain on the length-to-diameter ratio, events initiated deep inside the channel see almost no gain. In fact, it
is found that nearly all the signal (90%) originates in the first
21% of the 660-µm channel length [6]. This result suggests that
for realistic incidence angles the spatial resolution for hard
x-rays is at least 140 µm, or approximately 4 lp/mm.

     The MCP's sensitivity to x-rays is its single most important
property in the present context and allows the MCP's to function
as a photon converter in addition to an electron intensifier.

3.   FLASH X-RADIOGRAPHY

3.1 System Requirements and Description

Hydrodynamic instabilities have been predicted [8] for particle beam inertial confinement fusion targets during implosion.
Due to the high density target materials used, e.g., gold or
iron, observations of the target implosion dynamics and instabilities are only possible with flash x-radiography (FXR) techniques. The FXR system resolution and penetrating power required
for observing the different instabilities are listed in Table 1.

Table 1.

INSTABILITIES	CHARACTERISTICS LENGTH (cm)	PENETRATING POWER (g/cm^2)
Gross Instability resulting from tar-get and/or irradia-tion assymmetry	.1	<2.5
Ablator-pusher Rayleigh-Taylor	.02	~ .25 - 1.5
Pusher-fuel	.001	~ 1.0 - 4.0

In addition, a unique requirement of the FXR system arises
from the necessity to discriminate against the background bremsstrahlung pulse associated with the particle beam production.
This is a particularly bothersome problem if the target is irradiated by a relativistic electron beam.

In a backlighting arrangement, the required spatial resolution, as dictated by the characteristic lengths in conjunction
with the implosion velocities, specifies the flash x-ray source
duration. Since the implosion velocity may be as high as 20
cm/μs, the corresponding duration of interest for the source is
from 50 ps to 5 ns. Because of the characteristic lengths, the
source spot size also needs to be small compared to the characteristic lengths. This requires the source size to be between
10 μm and 1 mm.

One of the first concerns in developing the desired FXR system is the source spot size. Submillimeter spot size FXR sources
are not available commercially. As a result, we developed a

100-μm spot-size demountable source that is retrofittable to a 3-ns, 600-kV, 10-kA commercially available pulse power unit [9].

There are basically two ways to achieve the system resolution required. One is to place an FXR source of moderate size at a large distance away from the object as shown in Fig. 3. In this arrangement, the limiting geometrical spatial resolution is $\sim S_2 d/S_1$.

Since the image quality is dependent on the ratio of the intensity at the image plane to some noise level (for example, the film-grain noise), the ultimate image resolution is also proportional to I_o/S_1^2, where I_o is the source intensity. Following this line of reasoning, considerable improvement in image quality can be made if d could be reduced and made small. A reduction in d allows a proportional reduction in S_1 without sacrificing resolution. However, the image quality gains by S_1^{-2}. This consideration leads to the second method, e.g., to place a small FXR source close to the object. For a sufficiently small source the distance S_1 can be made less than S_2 and an actual projection-magnified image of the target will result. In our approach the second method is used.

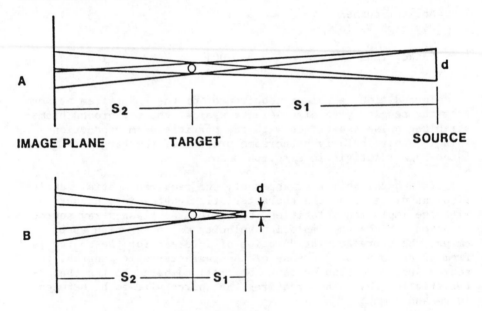

Fig. 3.   Backlighting arrangements for (A) large-spot-size and (B) small-spot-size FXR sources.

3.2   FXR Source

The small spot-size point x-ray source is shown in Fig. 4.
It consists of a small 2-mm diameter aperture cathode centered on
a 50-μm tungsten anode.   This whole assembly is designed to match
mechanically and electrically the output transmission line of a
Febetron 706 unit [9].   During operation a 600-kV, 3-ns pulse
is applied to the diode to generate a peak diode current pluse of
10 kA.   The diode A-K gap is chosen to match the 60-Ω transmission line.

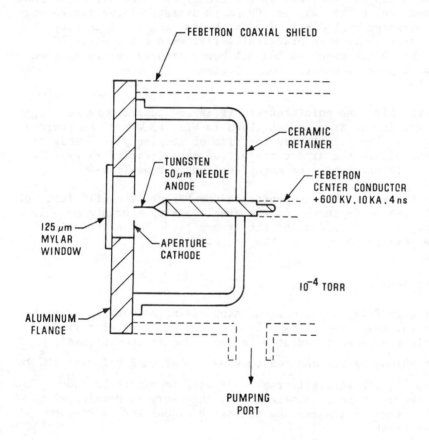

Fig. 4   Schematic of the point flash x-ray source.

The electron trajectories between the cathode and the anode are not determined only by the electric field. There exists a critical current carried by the anode needle beyond which the magnetic field generated around the anode needle is sufficient to turn the electron trajectories such that they miss the anode needle. In fact, the electrons in the presence of the electric and magnetic fields execute (E x B) orbits with a net drift away from both the cathode and anode. The critical current is defined as the current required to cause the gyro-radii of the electrons at peak energy to equal the A-K spacing. For the diode dimension given, $I_c \sim 8$ kA. The virtual cathode formed by the drifting electrons eventually emits toward the axis of the anode needle where the magnetic field is a minimum. Thus, a fine-point focus is achieved.

X-ray pinhole photographs show that only the first 0.5 mm of the needle tip is bombarded by the electrons. The effective spot size achieved in this way is 100 μm in diameter. The source output as measured by TLD's is 7 mR at 66 cm, and the intensity angular distribution is practically flat within a central 30° cone. The output spectrum has not been measured but is assumed to be a tungsten bremsstrahlung spectrum with an end-point energy of 600 keV.

With this fine-point source, it is now possible to backlight the target in the arrangement shown in Fig. B3 where the source is closer to the target than the film or imaging plan. This arrangement allows the direct projection magnification referred to earlier. In practice, the magnification is usually 3-4.

As the x-ray photons penetrate the target, the efficiency of transmission is dependent on the areal density encountered. In this way, a 2-D recording of the transmitted photon intensity gives a detailed mapping of the target.

3.3 REB Applications

Usually film, or film with intensifying rare-earth screens, is used for image recording. However, in the case of relativistic electron-beam irradiated targets, due to the intense bremsstrahlung background, which usually has an R Value of 10^8 to 10^{12} Rsec^{-1}, the straightforward use of film cannot be made. To record the x-ray shadow image it was necessary to develop a special camera to achieve the necessary signal and background discrimination.

This camera is shown in Fig. 5. It consists of two MCP's coupled by a solenoidal magnetic field which serves to maintain the image focus between the MCP's. Photons incident on the first MCP are internally converted to electrons which, after a small amplification, are outputed toward the second MCP. The electrons are accelerated a few hundred volts into an electric-field-free drift region where a longitudinal magnetic field maintains image focus. Initially, we tried a D.C. magnetic field [10]; however, we ultimately found that a pulsed strong magnetic field [11] of 4 kG provided superior focusing between the MCP's. This strong field reduced the electron gyro radii smaller than the resolution achieved by the backlighting arrangement. The second MCP further amplifies the signal and the output electrons from this stage are rendered visible by a phosphor screen. At this point, the image is recorded on polaroid film. The long vacuum cylinder allows the phosphor screen and the polaroid film to be hidden from direct x-ray irradiation as shown in Fig. 6. In addition, in order to keep the scattered x-rays from exciting the MCP's and create a high-noise background, the MCP's are directly gated by pulsing the bias voltages "on" synchronously with the 3-ns FXR source, with both timed to occur after the irradiating beam.

The pulsing scheme is shown in Fig. 7. Here a Krytron-triggered cable pulser is used to generate the needed bias pulses for the MCP's. The pulse lengths used are typically from 10 ns to 40 ns. The exposure or shutter time of the system is controlled obviously not by the MCP gating but by the duration of the FXR source, which is 2.6 ns FWHM as measured by an MCP photodiode with a 200-ps response time. This system has been used successfully to diagnose REB-driven pellet implosions. Fig. 8 shows a 3-mm gold hollow sphere imploded by a relativistic electron beam. The sphere is mounted on a small diameter stalk at the center of the anode. Implosion symmetry and a mass stagnation layer between the target and the anode plane are recorded. In this experiment the symmetry of irradiation is optimized.

An interesting and different implosion configuration is shown in Fig. 9. In this case the pellet is a 3-mm I.D., 200-μm wall hollow gold sphere. Through a novel diode arrangement, the beam is preferentially guided to one pole of the pellet. This asymmetric irradiation pattern caused a jet to form and totally disrupted the implosion process.

To expand the capability of this system, a two-frame system (Fig. 10) was developed [12,13]. In this system, two separate sources were aligned, each with a MCP-gated camera as described above. The two frames could be operated independently and their timing controlled by separate delay and MCP gate pulsers. An innovation in this system is the modification of a straight tube of the camera body to a curved tube, a 76° section of a stainless

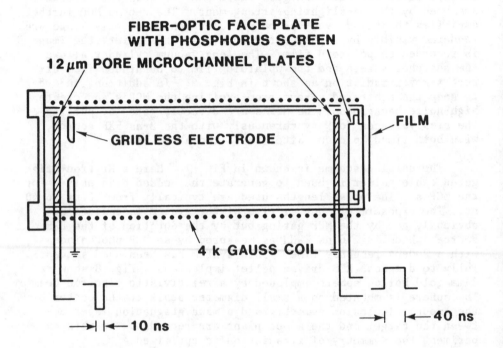

Fig. 5.   Schematic of gated MCP x-ray camera.

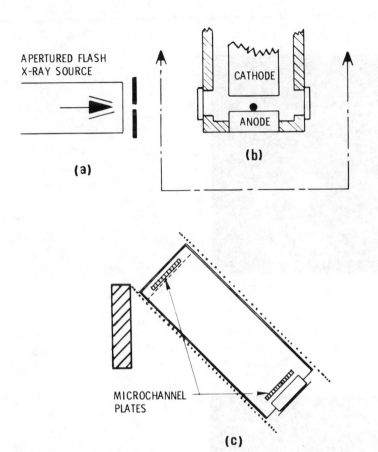

Fig. 6. Experimental backlighting arrangement with an FXR source (a), target mounted in an REG diode (b) and partially shielded MCP x-ray camera (c).

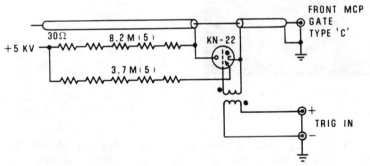

Fig. 7. Circuit diagram of the MCP cable pulser.

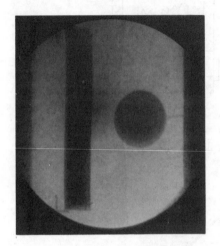

t = 0 ns

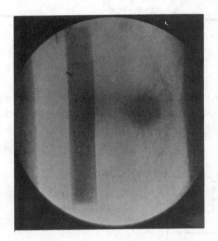

t = 400 ns

Fig. 8.   X-ray shadowgraphs of an anode mounted gold target (3 mm
          I.D., 200 μm wall) before and after beam irradiation.
          The stripe is a gold stepwedge.

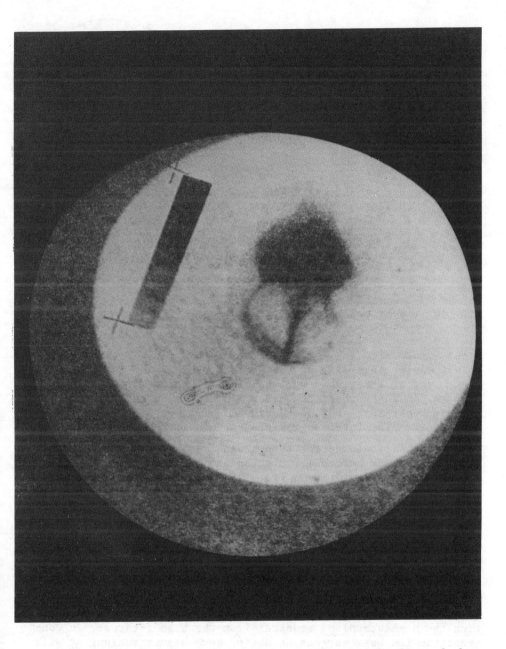

Fig. 9.   An asymmetric implosion caused by polar loading of the target. A stepwedge as well as an instrumental blemish are shown.

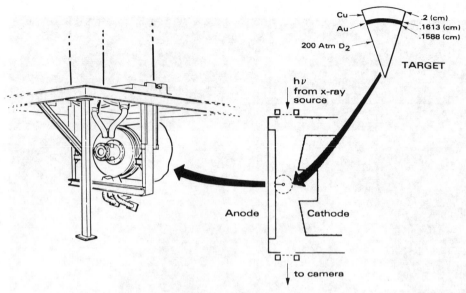

Fig. 10.   Two-frame FXR system.

steel torus (Fig. 11). The purpose of this modification is to
permit easier shielding on the phosphor screen and the film package. Typical x-ray shadowgraphs of a REB target with an ID of
3.18 mm, an inner shell of gold (25 μm thick to enhance radiographic contrast), and an outer shell of copper (390 μm) is shown
in Fig. 12. The interior was pressurized with D_2 gas to 200 atm
to artifically enlarge the turn-around radius. The radiographs
show the targets before irradiation and two times, separated by
110 ns, after irradiation. Fig. 13 shows the imploded and reference target profiles of Fig. 12. The average implosion velocity
measured is $3 \pm 1 \times 10^5$ cm/S.

3.4   Ion Beam Applications

FXR applications for studying intense ion beam interactions
are less difficult. This comes as a result of greatly reduced
x-ray bremsstrahlung background. Whereas in REB target irradiation the bremsstrahlung emanates from the high z-target, in contrast the ion beam targets do not produce bremsstrahlung at all.
However, generally there is a small amount of x-ray bremsstrahlung produced in generating the ion beam, but the location of
x-ray production is removed from the target and not in view of

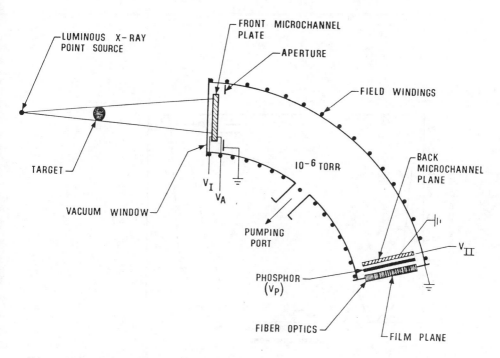

Fig. 11.   Curved tube MCP x-ray camera.

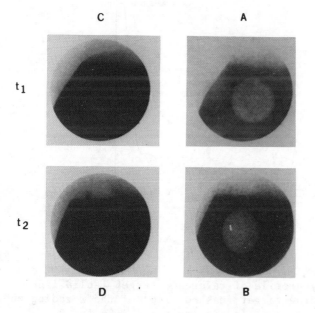

Fig. 12.   Two frames of a single pellet implosion; $t_2 - t_1 = 100$ ns.

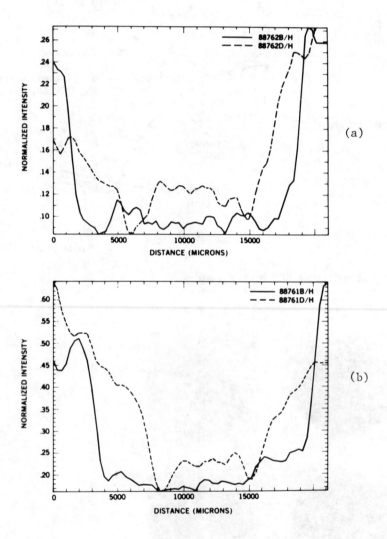

Fig. 13.   Isodensity profile of reference target (solid line)
           and imploding target (dashed line) of the x-radiographs
           shown in Fig. 12.   (a) is for t_1; (b) is for t_2.

the FXR system. For ion beam target applications, the afore-
mentioned FXR system was greatly simplified. The MCP x-ray
camera was not necessary, and only shielded collimating cameras
are needed to shield the film from x-ray background radiation.
This simplification not only improves operational ease but also
enhances image quality.

We have constructed a three-frame FXR system capable of
taking three, 3-ns exposures of an imploding sphere, arbitrarily
separated in time. This system consists of three pulsed FXR
sources and three lead-shielded collimating cameras (Fig. 14)
arranged symmetrically on a biconic surface with a half angle of
12 degrees. Each of the FXR sources has a 600-keV, 10-kA, and
3-ns Febetron 706 pulser that is coupled by an S-shaped, oil-
filled transmission line to a vacuum diode. The x-ray source is
a tungsten anode in this arrangement; it has a spot size of 125
μm, an output spectrum characteristic of a pulsed tungsten bremsstrahlung source, and a dose of 7 mR at 66 cm. In these experiments, the three FXR sources are mounted on a carousel and
can be lifted on and off the top side of the diode. When in position, the sources are 25.4 cm from the target which is located
at the center of the magnetically-insulated radial ion diode.
The cameras are mounted at the bottom side of the diode such that
each camera is aligned with its corresponding FXR source. In
this backlighting arrangement, film, intensified with salt
screens, is located 40.6 cm from the target; a 2.6x magnified
image of the target is produced. The spatial resolution, as
determined by the combination of source spot size and the recording film response, is 125 μm. The system temporal resolution is
determined by the Febetron pulse length which is nominally 3 ns,
but was measured by a fast MCP photodiode to be 2.6 ns. Thus,
implosions with velocities as high as 3-4 cm/μs can be recorded
without motion blur. To protect the film package from off-axis
background radiation, the cameras are shielded with 5 cm of lead
and collimated so that an 8-mm diameter field of view at the target plane is obtained.

Before a shot the system is aligned by taking reference
radiographs which are used later in the data reduction. During
the shot a master trigger pulse fires the Proto-I accelerator
and triggers a series of delayed pulse generators which, in turn,
fire the three Febetrons at preset delay times. The overall uncertainty in setting the exposure times is less than 20 ns, which
is acceptable because the target implosion time in these experiments is approximately 200 ns.

Typical images of ion beam driver implosions of spherical
and cylindrical Au and W targets are shown in Fig. 15 and Fig.
16. The spherical targets were 3 mm in diameter and 25 μm in

3-FRAME FLASH X-RADIOGRAPHY SYSTEM

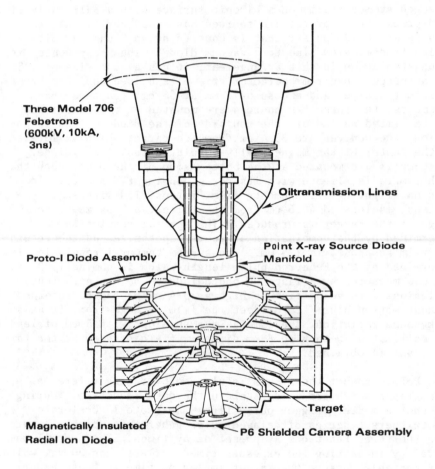

Fig. 14.   Assembly drawing of the 3-frame FXR system.

SHOT 3286 PROTO I
Au SPHERE 25 μm WALL

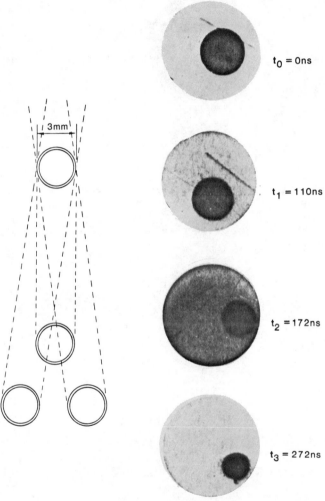

$t_0 = 0ns$

$t_1 = 110ns$

$t_2 = 172ns$

$t_3 = 272ns$

3mm

Fig. 15. X-radiographs of a 3-mm Au sphere with a 25-μm wall taken at three different times during implosion.

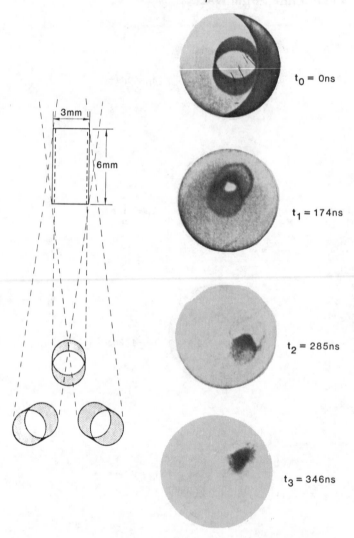

SHOT 3301 PROTO I
W CYLINDER 16.5μm WALL

t_0 = 0ns

t_1 = 174ns

t_2 = 285ns

t_3 = 346ns

Fig. 16.   An imploding cylindrical tungsten target (3 mm I.D.,
16.5-μm wall, and 6-mm high) observed at three different times during implosion.

wall thickness, and the cylindrical targets were 3 mm in diameter, 6 mm in height, and 16.5 μm in wall thickness.

The spherical target was observed to implode with a high degree of symmetry and attained an average velocity of 2×10^5 cm/sec. The cylindrical target, being thinner, imploded faster at 1×10^6 cm/sec.

REFERENCES

1. J. H. Nuckolls, R. O. Bangerter, J. D. Lindl, W. C. Mead, and Y. L. Pan, Proceedings of the European Conference on Laser Interaction with Matter, Oxford, England 1977.

2. J. Chang, M. M. Widner, G. W. Kuswa, G. Yonas, Phys. Rev. Lett. 34, 1266 (1975).

3. Farnsworth, P. T., U. S. Patent issued in 1930 (No. 1969399)

4. Ruggieri, D. J., IEEE Trans. Nucl. Sci. NS-19, No. 3 (June 1972)

5. P. Schagen, Advanced in Imaging Pickup and Display, Vol. 1 (Academic Press, N.Y., 1974) p. 1.

6. K. W. Dolan and J. Chang, X-ray Imaging, Proc. SPIE 106, (1977) p. 178. W. Parker, R. Gott and K. A. Pounds IEEE Trans. Nucl. Sci. NS-17 (1930) p. 360. J. Adams and I. C. P. Millar, Acta Electronica 14 (1971) p. 237.

7. F. Paresce, Appl. Optics 14 (1975) p. 2823.

8. J. R. Freeman, M. J. Clauser, S. L. Thompson, Nuclear Fusion 17, 2, p. 223 (1977).

9. Model 706 Febetron, Hewlett-Packard, McMinnville, Oregon 97128.

10. G. W. Kuswa and J. Change, Adv. X-ray Anal. 18, 107 (1975).

11. J. Chang, L. P. Mix, F. C. Perry, M. M. Widner, and J. W. Poukey, Proceedings of the International Topical Conf. on Electron Beam Research and Technology, Albuquerque, NM, U.S.A., Nov. 3-5, 1975, p. 82.

12. J. Change, D. L. Fehl, M. M. Widner, K. W. Beig, and M. A. Palmer, Appl. Phys. Lett. 39 (3), p. 224, Aug. 1981.

13.   J. Chang and D. L. Fehl, Proceedings the 1st European Conf. on Cineradiography with Photons or Particles, Paris, France, May 18-21, 1981.

14.   J. Chang, D. L. Fehl, L. Baker, and D. J. Johnson, Proceedings of the 15th International Congress on High Speed Photography and Photonics, Aug. 21-17, 1982, San Diego, CA., U.S.A. p. 696.

SPECTROSCOPY

PRINCIPLES OF PLASMA SPECTROSCOPY

H. R. Griem

Laboratory for Plasma and Fusion Energy Studies
University of Maryland
College Park, MD
USA

1. INTRODUCTION

The diagnosis of plasmas using spectroscopic observations has its origins in various older disciplines, including astronomy and discharge physics. As laboratory plasma physics evolved from low-density, low-temperature discharges to higher energy density plasmas, the need for non-interfering diagnostics arose and spectroscopy was applied to determine the physical state and chemical abundance of the plasmas studied.

Such applications of methods originally developed for the quantitative analysis of stellar atmospheres require a quantitative understanding of the interactions between matter in the plasma state and electromagnetic radiation in a large frequency range. Although the principal features of this interaction are well understood in terms of quantum-mechanical perturbation theory, there are nevertheless considerable difficulties in simply transferring experience gained in astronomical research to laboratory plasma physics. These difficulties are connected with the huge factors involved in characteristic densities, and in length and time scales between, e.g., a laser-produced plasma and the atmosphere of a star like the sun. Only if a broader view going beyond diagnostics, e.g., to radiative energy transport is taken, can some similarities be found between line energy transport in stellar interiors and x-ray line spectra from laser compression experiments.

In any event, the range of physical conditions and chemical compositions encountered in spectroscopic plasma diagnostics is large. For fruitful research in this area it is therefore essential to have close interplay between experiments and theory in order to obtain reliable information on the plasma from a reasonably small set of measured intensities. To aid toward this goal, the corresponding theoretical relations are summarized in the following sections.

2. RADIATIVE PROCESSES

There are three distinct ways in which radiation and matter interact so as to change the number of photons present in some beam, namely, spontaneous emission, stimulated emission, and absorption. The corresponding Einstein coefficients, A (for spontaneous emission) and B (for stimulated emission or absorption), for transitions between a given pair of states (of degeneracies or statistical weights g_n and g_m) are related to each other through the celebrated Einstein relations

$$A_{nm} = \frac{\hbar \omega^3}{4\pi^3 c^2} \frac{g_n}{g_m} B_{mn} = \frac{\hbar \omega^3}{4\pi^3 c^2} B_{nm},$$ (1)

which can be derived from the requirement that in case of thermodynamic equilibrium between matter and radiation the spectral intensity $I(\omega)$ should approach the Planck distribution

$$I_p(\omega) = \frac{\hbar \omega^3}{4\pi^3 c^2} \left[\exp\left(\frac{\hbar \omega}{kT}\right) - 1 \right]^{-1} .$$ (2)

The angular frequency obeys Bohr's condition

$$\hbar \omega = E_m - E_n ,$$ (3)

if m and n designate upper and lower states of the transition, while the B-coefficients give the corresponding transition rates, per atom or ion in the appropriate initial state, through

$$dW_{mn} = \frac{1}{4\pi} B_{mn} I_{\Omega}(\omega) \, d\Omega$$ (4a)

$$dW_{nm} = \frac{1}{4\pi} B_{nm} I_{\Omega}(\omega) \, d\Omega$$ (4b)

in terms of the actual intensity which generally depends on the direction Ω (solid angle). Also, if this intensity varies substantially over the actual frequency range covered by the transition (see Sec. 2.2), a corresponding frequency-normalized line shape factor $L(\omega)$ $d\omega$ should be inserted.

2.1 Line Emission

To relate the rate of spontaneous emission to (measurable) intensities, one must first consider the emission coefficient (power per unit volume, solid angle, and angular frequency interval)

$$\varepsilon(\omega) = \frac{\hbar\omega}{4\pi} A_{nm} N_m L(\omega) \ , \tag{5}$$

assuming for the time being that a single line dominates in the frequency range of interest. This formula is self-evident, but relating N_m, the density of atoms or ions in the initial state to the general plasma conditions can be quite difficult (see Sec. 3). Also, the line-shape function appropriate for emission (or absorption) coefficients, $L(\omega)$, is normally not directly reflected by the observed line shape. This complication not only arises because of changes in conditions along the line of sight, but possible also from reabsorption (opacity broadening, see Sec. 2.4).

For diagnostic applications it is usually preferable to use lines for which absorption and induced emission are negligible. In such cases emerging intensities can simply be calculated by integrating along the path taken by the ray traversing the plasma,

$$I(\omega) = \int \varepsilon(\omega)dx + I_i(\omega)$$

$$= \frac{\hbar\omega}{4\pi} A_{nm} \int N_m L(\omega)dx + I_i(\omega) \ , \tag{6}$$

with the incident intensity $I_i(\omega)$ in the proper direction at the far side of the plasma. In dense and pulsed plasmas, observable lines are frequently substantially modified by radiation transport, e.g., in laser produced plasmas [17].

Knowledge of transition probabilities and frequencies (or wavelengths) can not always be taken for granted. Unknown wavelengths must frequently be measured in plasma experiments, with atomic structure theory serving as a guide to level identifications. Measurements of transition probabilities are much more

difficult, and most published results [18] are from atomic structure calculations (for ~5000 lines). Except for atoms or ions
with a very small number of electrons, it is quite likely that
relatively large errors occur, especially for relatively weak
lines.

A natural measure of the strength of a line is the dimen-
sionless oscillator strength or, rather, its product with the
appropriate statistical weight,

$$g_n f_{mn} = g_m f_{nm} = \frac{c}{2r_o \omega^2} g_m A_{nm} \; . \tag{7}$$

Such gf values for some FeX (or Fe^{9+}) and FeXI (or Fe^{10+}) lines
as calculated by Mason (1975) and by Bromage et al. (1977) are
shown in Table 1 to illustrate the sensitivity to approximations
made in typical atomic structure calculations. The remarks in
Table 1 are mostly meant for allowed electric-dipole transitions
involving only single-excited levels. Lines arising from various
so-called forbidden transitions can also be useful for diagnostics, especially for density measurements. Because of their relatively small transition probabilities such lines can only be
seen if their upper levels are substantially over-populated compared to levels giving rise to allowed lines. Such deviations
from essentially statistical populations in turn suggest that
collisional excitation energy transfer is not too important, a
situation consistent with relatively low electron density.

A third general type of line radiation is associated with
doubly-excited states, which involve instead of the usual single
optical electron designated by quantum numbers n and ℓ, two such
electrons $n\ell$, $n'\ell'$, outside of some more or less frozen core of
inner electrons. The additional $n'\ell'$ electron acts mostly as
only a spectator, but of course causes some change in the transitions energies of the $n\ell$ electron, almost always to lower energies. The corresponding lines appear therefore as dielectronic
satellites on the long wavelength side of the parent lines produced by transitions of the $n\ell$ electron in ions without the
additional $n'\ell'$ electron. The ratio of the two ion populations
is mostly a function of electron temperature so that the relative
intensities of satellite and parent lines can be used for temperature determinations.

2.2 Line Profiles

Spectral lines are neither infinitely sharp nor usually
exactly centered at the frequency given by Eq. (3). There are
three general causes of spectral-line broadening as expressed

Table 1.   Calculated gf values for some Iron X and IX lines.

	Mason	Bromage et al.
Fe X $3p^4(^1D)3d\ ^2S_{1/2}$ - $3p^5\ ^2P_{1/2}$	0.34	0.45
$- 3p^5\ ^2P_{3/2}$	2.35	1.51
$3p^4(^3P)3d\ ^2D_{3/2}$ - $3p^5\ ^2P_{3/2}$	0.347	0.115
$- 3p^5\ ^2P_{1/2}$	4.34	3.82
$3p^4(^3P)3d\ ^2P_{1/2}$ - $3p^5\ ^2P_{3/2}$	0.11	0.43
$- 3p^5\ ^2P_{1/2}$	1.76	1.40
Fe XI $3p^3 3d\ ^1F_3$ - $3p^4\ ^1D_2$	7.13	5.64
$^3D_1$ - $3p^4\ ^3P_0$	1.54	1.31
$^1D_2$ - $3p^4\ ^1D_2$	3.03	3.14
$^3P_2$ - $3p^4\ ^3P_1$	0.79	0.60
$3p^3 3d\ ^3S_1$ - $3p^4\ ^3P_2$	1.06	0.16
$^3P_1$	0.3726	0.47
$^3P_0$	0.10224	0.23
$^1D_2$	0.0237	1.13

in terms of the line shape $L(\omega)$ of emission or absorption coefficients. Since at least one of the levels has only a finite lifetime, there is always some broadening according to Heisenberg's uncertainty principle. This natural line broadening gives rise to Lorentzian profiles whose FWHM (full width between half of maximum intensity points) in the ω-scale is given by the sum of the decay rates for upper and lower levels, including absorption and induced emission rates and also any nonradiative rates, like Auger rates in case of inner-shell x-ray transitions. With the exception of the latter transitions, it is normally safe to neglect natural broadening because it typically amounts to only $\sim 10^{-4}$ Å.

The second general line broadening mechanism is also always present. It is due to the Doppler shifts associated with the random velocities of emitting or absorbing atoms or ions. For

nonrelativistic systems only the velocity component along the
line of sight matters, and for thermal plasmas the corresponding
velocity distribution and therefore also the line profiles are
therefore Gaussian. The resulting Doppler width is given by w_D
$\cong v_{th} w/c$, where v_{th} is a characteristic thermal velocity. More
precisely, the broadening due to thermal Doppler shifts is

$$w_D = 2[2kT(\ell n2)/M]^{1/2} w/c \qquad\qquad (8)$$

in terms of the kinetic temperature T and mass M of the emitting
or absorbing species. This FWHM (full width between half of maximum intensity points) width is usually orders of magnitudes
larger than the natural width.

The third general line broadening mechanism is associated
with the perturbations of the emitting or absorbing atoms and
ions caused by other particles in the plasma. If these particles
are charged, they produce electric fields which change with time
more or less radidly but vary in space only slightly over the
region occupied by a given atom or ion in the appropriate states.
Often this spatial variation can be entirely ignored and the
actual interaction between radiating atoms or ions and the rest
of the plasma be replaced by that of a dipole representing the
radiating system and the local electric field. Since all radiators experience different fields and since these are time-dependent, the corresponding Stark effects effectively broaden the
lines, and in the case of the quadratic Stark effect, also shift
them.

Before discussing the various subsidiary approximations made
in Stark broadening calculations, it is worth noting that the
usual monopole-monopole interactions which dominate the scattering between plasma particles do not explicitly cause any line
broadening. The strength of these interactions is the same for
upper and lower levels of the line and the corresponding level
shifts therefore cancel. However, since these Coulomb interactions do cause changes in the relative motions of perturbing
particles and radiators, there is an important influence of these
interactions via the time-dependent electric field. Interactions
with neutral perturbers, on the other hand, tend to be negligible
in plasmas except for very weakly ionized gases [5].

In principle, the response of a given radiator to the various time-sequences of fields must be calculated from time-dependent quantum-mechanical perturbation theory and the corresponding
spectrum then be appropriately averaged. These calculations
would not only be formidable but would also require more information on plasma fields than we now have. In particular, one
would have to carefully join the more or less stochastic particle

produced fields to the wave fields associated with collective
motions of the plasma. Fortunately such great detail is normally
not needed because a multi-time-scale analysis offers itself.

Fields produced by electrons colliding with the radiators
tend to vary so rapidly that corresponding widths and shifts depend only on net changes in the radiator states caused by such
collisions. As in natural broadening, the FWHM width is now
given by the appropriate collision rate from inelastic and (non-
Coulombic) elastic collisions. However, generally there is a
shift which may essentially be viewed as a dynamical generaliza-
tion of the quadratic Stark effect. This collision or impact
theory of Stark widths and shifts has been quiet successful for a
large number of spectral lines the most notable exception being
the far wings of broad lines, especially of hydrogen [6]. Since
line profiles are essentially Fourier transforms of time-dependent wave functions, these far wings correspond to short times
for which details of the collision dynamics are important. Inclusion of these details through relaxation or unified theories
of line broadening indeed improved the agreement with experiment,
without major changes in the structure of the basically
Lorentzian profiles. (Width and shift parameters become functions of the frequency.) In either case, for so-called over-
lapping lines, impact profiles consist of superpositions of
Lorentzians with various widths, and of certain interference
terms. These superpositions are not exactly Lorentzian.

Ion-produced fields vary so slowly that their time-depend-
ence can be neglected in some important cases, e.g., on the
wings of broad hydrogen lines or lines from one-electron ions.
The time interval for which accurate wave functions are needed is
then $\sim\Delta\omega^{-1}$, where $\Delta\omega$ is the angular frequency separation from the
unperturbed line, and this time may indeed be smaller than a
relevant collision time r/v in terms of appropriate radiator-
perturber separation r and relative velocity v. In such cases
the quasistatic or Holtsmark theory can be used, which involves
calculating Stark shifts and intensities for assumed electric
(micro-) fields and then averaging the corresponding Stark patterns. For this weighted average, the probability distribution
of ion micro-fields is required, which for uncorrelated perturbers is given by the Holtsmark distribution. To include such
correlation and shielding effects, Hooper and co-workers have
performed a series of calculations for dense plasmas consisting
of a variety of ions and atoms (see [6] and [8] for references).

In fact, the components of the Stark patterns are also
broadened and shifted by electron collisions. So-called standard
calculations are therefore convolutions of impact-with quasistatic ion-produced profiles. For hydrogen and similar lines these

two contributions tend to be comparable, whereas for other lines electron broadening usually dominates. The need to examine the validity of the quasistatic approximation is therefore greatest for the former type of lines and it was indeed found experimentally that ion-dynamical corrections are important in the line cores of lower series members. Various theoretical correction procedures reproduce these effects [7], [9], [3]. They are also in reasonable agreement with results of the model microfield method [2] in which the actual time-dependent field is replaced by a sequence of constant fields with magnitudes of fields and frequencies of steps chosen to give agreement in the extreme impact and quasistatic limits. This means that the model field has the same autocorrelation and distribution functions as the actual smoothly varying field.

An interesting feature of ion-dynamical effects is their correlation with Doppler broadening. The usual convolutions procedure for combining Stark and Doppler broadening is therefore, strictly speaking, not valid in cases where dynamical corrections to the quasistatic approximation are important. Finally, there are situations in which higher-than-dipole multipole interactions are noticeable. Quadrupole interactions caused by spatial inhomogeneities in the field acting on hydrogen atoms are responsible for some asymmetry in the otherwise almost symmetrical profiles of lines subject to linear Stark effect. The corresponding shifts, however, tend to be smaller shifts caused by electron collisions due to interactions involving states of different principal quantum numbers.

2.3 Continuum Radiation

Dense plasmas produce spectra which over substantial frequency ranges are continuous, i.e., have no pronounced maxima or minima, i.e., no strong emission or absorption lines. However, unless the spectrum observed is the result of a number of emission and absorption processes in an optically thick plasma, it is not simply described by the blackbody, or, Planck formula, Eq. (2). For a quantitative description of optically thin continua one must extend the relation for line emission coefficients, Eq. (5), from bound-bound to free-bound and free-free transitions. A very physical way of doing this is through the use of a line shape function given by the inverse of the absolute value of the frequency separation between adjacent lines in a series, and of transition probabilities or f-numbers which are asymptotically valid for large quantum numbers [5] [14]. Thirdly, the density N_m of the upper level of the line is expressed in terms of the electron density and the density in the next higher ionization stage using the appropriate Saha equation (see Sec. 3.2 for the corresponding ionization equilibrium).

The result of these replacements is an approximate expression for the contribution of, e.g., substantially broadened and therefore merged high series members of the Balmer series to the emission coefficient at frequencies below the Balmer edge, down to frequencies corresponding to the Inglis-Teller limit [10]. By now replacing the energy of the electron in the upper level, $-E_m/n^2$, by the Bohr relation, Eq. (3), this relation can be extrapolated beyond the usual series limit to calculate free-bound transitions into a given lower level. Summing over these levels gives the total free-bound continuum, and extension of the sum over lower levels through an integral over positive energy states also yields the free-free, or bremsstrahlung continuum.

Except for the Gaunt factors $g(\omega)$, which are often close to 1 ([5], [8]), the above procedure gives a continuum emission coefficient of

$$
\varepsilon(\omega) = \frac{32(\alpha\, a_o)^3 z E_H}{3^{3/2} \pi^{1/2}} \left(\frac{z^2 E_H}{kT}\right)^{3/2} [\sum \frac{g_n(\omega)}{n^3} \exp(\frac{z^2 E_H}{n^2 kT})
$$

$$
+ \frac{kT}{2z^2 E_H} g(\omega) \exp(\frac{z^2 E_H}{n_\ell^2 kT})] \, N_e N_z \, \exp(-\frac{\hbar\omega}{kT}) \tag{9}
$$

for recombination into and bremsstrahlung on fully stripped ions of density N_z and nuclear charge z. The temperature in this relation is that of the electrons whose density is N_e, and E_H is the ionization energy of hydrogen (13.6 eV). Also, α and a are the fine-structure constant (1/137) and the Bohr radius (0.529Å), respectively. The sum over bound-free or recombination continua involves lower state principal quantum numbers such that

$$
\hbar\omega \gtrsim z^2 E_H (\frac{1}{n^2} - \frac{1}{n_\ell^2}) , \tag{10}
$$

and n_ℓ must be chosen according to the merging of upper levels.

For continua arising from electron interactions with incompletely stripped ions, Eq. (9) must be modified more or less drastically, especially for recombination into the ground state [11]. There is not only a change in the degeneracy of the final

state or, in the extreme of closed shells, even a strict rule
(Pauli principle) against such transitions, but also the possibility of a stronger dependence on frequency, e.g., the Cooper
minimum in some recombination or photoionization cross sections.

We should also not overlook the fact that in partially
ionized plasmas electron-atom interactions can cause continuum
emission as well, both by recombination, i.e., formation of negative ions, and by bremsstrahlung. The most famous example is the
role of H^- (negative hydrogen) in the solar atmosphere.

2.4 Radiative Transport

Only if absorption is negligible can emission coefficients
be directly converted into observable intensities according to
Eq. (6). Usually the underlying assumption of negligible optical
depth first fails in the center of resonance lines. We must
therefore consider the magnitude of the line absorption cross
section,

$$\sigma_{mn} = 2\pi^2 r_0 c f_{mn} \, L(\omega), \tag{11}$$

where f_{mn} is the absorption oscillator strength first introduced
in Eq. (7) and r_0 the classical electron radius. For strong
lines, $f \cong 1$, and $L(\omega)$ may be as large as the inverse of the
Doppler width in Eq. (8). A numerical estimate for the maximum
cross section is

$$\sigma(\omega_0) \cong 3 \times 10^{-17} \left(\frac{AE_H}{kT}\right)^{1/2} \lambda f_{mn} \, [cm^{-2}]. \tag{12}$$

Here the wavelength λ is an $\overset{o}{A}$ units and A is the atomic weight of
the absorbing atom or ion. Depending on the densities N_n in the
lower state, the corresponding mean free paths $(N_n\sigma)^{-1}$ can be
quite short. If Stark broadening is important as well, cross
sections are smaller according to the ratio of actual and Doppler
line shape functions. Absorption into the continuum, if at all
important, would be mostly controlled by the photo-ionization
cross section, which for hydrogen and one-electron ions of nuclear charge z in state n obeys

$$\sigma_n = \frac{64\alpha}{3^{3/2}} \pi a_0^2 \left(\frac{E_H}{\hbar\omega}\right)^3 \frac{z^4}{n^5} g_n(\omega) \, , \tag{13}$$

where g_n is again the Gaunt factor ($g \cong 1$, $64\alpha\pi a_o^2/3^{3/2} \cong 0.88 \times 10\text{-}16 \text{ cm}^2$). This process is, of course, the inverse of recombination, whereas there is no special term for absorption by inverse bremsstrahlung which is important in laser-matter interactions. The corresponding cross section can be obtained from the free-free emission coefficient using Kirchhoff's law.

In addition to absorption and spontaneous emission, it is necessary to consider induced emission. This is often done via the effective absorption coefficient, namely, in the case of a single line

$$k' = 2\pi^2 r_o c f_{mn} (N_n - \frac{g_n}{g_m} N_m) L(\omega) \ , \tag{14}$$

where Eq. (7) was used to relate emission and absorption oscillator strength. Also, we assumed that the same line shape could be used for both processes which is usually but not always true. The change of intensity due to the three processes considered, i.e., neglecting scattering, along the line of sight is in terms of ε and k' given by

$$\frac{dI}{dx} = \varepsilon - k' I \tag{15a}$$

or, introducing the optical depth τ through

$$d\tau = k' dx \tag{16}$$

by

$$\frac{dI}{d\tau} = \frac{\varepsilon}{k'} - I \equiv S - I, \tag{15b}$$

where S is called the source function.

In general, the source function, i.e., the ratio of Eqs. (5) and (14), requires a self-consistent solution of the integral of the radiative transfer Eq. (15) and the set of rate equations (see Sec. 3) which control the populations in the upper and lower levels of the line. Such calculations are very demanding; they both require a large atomic data set and efficient numerical methods, in particular if they are coupled with a plasma code which predicts input parameters like electron densities and temperatures. Calculations of this kind are very important for the understanding of the radiation dynamics of non-LTE plasmas, i.e., plasmas for which local and instantaneous thermodynamic equilibrium cannot be assumed in calculations of level populations and charge state distributions. On the other hand, if densities are sufficiently high and spatial and time variations reasonably

weak, then we can use Boltzmann factors to relate level populations to each other,

$$\frac{N_m}{N_n} = \frac{g_m}{g_n} \exp\left(-\frac{E_m - E_n}{kT}\right) .$$ (17)

Criteria for the validity of this relation can be obtained using
approximations for the various rate processes [5], and are often
reasonably met for resonance lines from dense plasmas. In such
cases, Eqs. (5), (7), and (14) reveal that the source function S
reduces to the Planck function I_p in Eq. (2), as to be expected
from Kirchhoff's law.

Returning to the radiative transfer equation, we note that
its formal solution is

$$I(\tau) = \int_o^\tau S(\tau) \exp(-\tau)d\tau + I(0) \exp(-\tau) ,$$ (18)

if the optical depth is calculated from Eqs. (14) and (16) along
the ray entering the plasma with intensity $I(0)$ and leaving with
intensity $I(\tau)$. Only if the traversed optical depth is large and
if S is constant and equal to I_p does the plasma radiate like a
blackbody. In the other extreme, i.e., $\tau \ll 1$, Eq. (18) is easily seen to reduce to Eq. (6), the optically thin solution.

3.   COLLISIONAL PROCESSES

Although transitions in excitation and ionization states are
usually not directly observable, they nevertheless play a crucial
role in the formation of the spectrum emitted by a plasma because
they tend to control population of radiating states. Except for
photo-excitation, with a cross section corresponding to Eq. (14)
without the density factor, radiation-induced transitions are
usually less frequent than transitions caused by electron-ion
collisions which may or may not involve photon emission as in
radiative or dielectronic recombination.

Knowledge of collisional rates is required at two levels of
accuracy. At the lower level of accuracy it is sufficient to
find whether or not collisional rates for some process are significantly larger than radiative rates. For Maxwellian distributions of the colliding electrons the steady-state solution for
the system of coupled rate equations for level populations in a
homogeneous plasma then gives thermodynamic equilibrium populations. Such low accuracy estimates of collisional rates can also

be used to estimate permissible rates of plasma parameter variations in time and space to which the atomic states could respond
without substantial deviations from steady state populations.

The purpose of the following sections is to describe the
physics of the collision processes with enough realism to allow
us to make estimates of the various rates sufficient for the
above purposes. For more demanding applications, e.g., for use
in plasma modelling codes, one would have to use quantum mechanical scattering theory.

3.1 Excitation and De-excitation

To calculate electron-atom (or ion) cross sections it is
necessary to consider the full bound electron-free electron
Coulomb interaction,

$$H = \frac{e^2}{|\underset{\sim}{r} - \underset{\sim}{x}|} ,$$ (19)

where $\underset{\sim}{r}$ and $\underset{\sim}{x}$ represent the coordinates of the free and the active bound electron relative to the nucleus, respectively. (This
is in contrast to atom or ion interactions with the radiation
field, for which the dipole approximation is usually an extremely
accurate description.) Fermi's golden rule gives

$$Q_{fi} = \frac{2\pi}{\hbar} \left| <f_t|H|i_t> \right|^2 dn_f/dE$$ (20)

for the transition rate between degenerate states i_t and f_t of
the total system consisting of a free electron and the atom or
ion to be excited or de-excited. (Since the free electron gains
or loses the atomic transition energy, the states of the total
system involved here are indeed degenerate.) Moverover, there is
actually a continuum of final states whose number per energy interval is dn_f/dE. The states of the total system are assumed to
be well approximated by products of bound state wave functions
|i>, |f>, etc., and free electron wave functions, $|k_i>$, $|k_f>$,
etc. In other words, the interaction is assumed to be weak on
the average.

If we normalize the free electron wave functions in a volume
V, the matrix elements between plane wave states are

$$\langle k_f | H | k_i \rangle = V^{-1} \int \exp[i(\underset{\sim}{k_i} - \underset{\sim}{k_f}) \cdot \underset{\sim}{r}] \frac{e^2}{|\underset{\sim}{r} - \underset{\sim}{x}|} d\underset{\sim}{r}$$

$$= \frac{4\pi e^2}{Vq^2} \exp(i\underset{\sim}{q} \cdot \underset{\sim}{x}) \qquad (21)$$

in terms of the momentum transfer

$$\underset{\sim}{q} = \underset{\sim}{k_i} - \underset{\sim}{k_f} . \qquad (22)$$

Equation (21) suggests that small q values are most effective in causing transitions and we therefore expand the exponential. The first term contributing to the transition matrix element is the linear (dipole) term,

$$\langle f | \langle k_f | H | k_i \rangle | i \rangle \cong \frac{4\pi e^2}{V^2 q^2} i\underset{\sim}{q} \cdot \langle f | \underset{\sim}{x} | i \rangle \qquad (23)$$

The absolute value squared of this matrix element, averaged over relative orientations of $\underset{\sim}{q}$ and $\underset{\sim}{x}$ is

$$\langle f_t \, H \, i_t \rangle^2 = \frac{16\pi^2 e^4}{3V^2 q^2} | \langle f | \underset{\sim}{x} | i \rangle |^2 = (\frac{4\pi e^2 a_0}{Vq})^2 \frac{E_H}{\Delta E} f_{fi}, \qquad (24)$$

where we have used the definition of the oscillator strength to obtain the second version. (The factor $g_i \hat{=} g_n$ is omitted because we require an average over initial states, while $g_f \hat{=} g_m$ appears on both sides of the defining equation.) Also, ΔE is the energy transfer or transition energy, corresponding to $\hbar\omega = E_m - E_n$ in the radiative case.

The density of final free electron states is as usual, with $E = \hbar^2 k^2 / 2m$,

$$\frac{d^2 n}{dE d\Omega} = V \frac{k_f^2 dk_f}{(2\pi)^3 dE} = V \frac{m k_f d\Omega}{(2\pi)^3 \hbar^2} , \qquad (25)$$

where Ω is the solid angle associated with the directions of $\underset{\sim}{k_f}$. Substitution of Eqs. (24) and (25) into Eq. (20) gives

$$Q_{fi} \cong \frac{4e^4 a_0^2 mk_f}{h^3 V} \frac{E_H}{\Delta E} f_{fi} \int \frac{d\Omega}{q^2} . \tag{26}$$

To relate solid angle and momentum transfer, one considers

$$q^2 = (\underset{\sim}{k_i} - \underset{\sim}{k_f})^2 = k_i^2 + k_f^2 - 2k_i k_f \cos\theta \tag{27}$$

which, with $d\Omega = - 2\pi(d\cos\theta)$, leads to

$$d\Omega = \frac{2\pi}{k_i k_f} q dq. \tag{28}$$

The integral in Eq. (26) therefore becomes $\ln(q_{max}/q_{min})$ where q_{min} corresponds to $|k_i - k_f|$. However, use of $k_i + k_f$ for q_{max} would often be inconsistent with the replacement of $\exp(\underset{\sim}{iq} \cdot \underset{\sim}{x})$ by $1 + \underset{\sim}{iq} \cdot \underset{\sim}{x}$, and Bethe therefore suggested $q_{max} \cong r_{if}^{-1}$, r_{if} being representative of the range of the bound state wave functions. Using this estimate for the integral and replacing k_i by $mv/\hbar$ and V^{-1} by N_e (remember that there was exactly one electron in the normalization volume) we obtain for the collisional excitation or de-excitation rate

$$Q_{fi} \cong \frac{8\pi}{v} \left(\frac{\hbar}{m}\right)^2 N_e \frac{E_H}{\Delta E} f_{fi} \ln |(k_i - k_f) r_{if}|^{-1} \tag{29}$$

and for the corresponding cross section

$$\sigma_{fi} \cong 8\pi \left(\frac{\hbar}{mv}\right)^2 \frac{E_H}{\Delta E} f_{fi} \ln |(k_i - k_f) r_{if}|^{-1} . \tag{30}$$

The Maxwell-averaged excitation rate coefficient is

$$X_{fi} \cong \frac{16\pi}{\alpha c} \left(\frac{\hbar}{m}\right)^2 \left(\frac{E_H}{kT}\right)^{1/2} \exp\left(-\frac{\Delta E}{kT}\right) \frac{E_H}{\Delta E} f_i \ln | \ldots |^{-1} ,$$

if we neglect the velocity dependence of the logarithm. (In the case of de-excitation, there is no exponential factor.)

To derive a similar formula for electron collisions with ions, the plane wave states used in Eq. (21) must be replaced by

free Coulomb states. This results in a replacement of the logarithmic factor by Gaunt factors $g(k_f, k_i)$ multiplied by $\pi/3^{1/2}$,
in Eq. (31).

$$X_{fi} \cong \frac{16\pi}{\alpha\ c} \left(\frac{\hbar}{m}\right)^2 \left(\frac{\pi E_H}{3kT}\right)^{1/2} \exp\left(-\frac{\Delta E}{kT}\right) \frac{E_H}{\Delta E} f_{fi} \bar{g} \ . \tag{32}$$

Often the averaged Gaunt factors are close to unity, and Eq. (32)
then tends to give a reasonably accurate estimate for dipole
transition rates. However, it would be wrong to conclude that
other transitions are negligible. Higher order terms in the ex-
pansion of $\exp(i\underset{\sim}{q}\cdot\underset{\sim}{x})$ are generally not small enough for this
conclusion to be valid. Moreover, close coupling effects that
come in if the interaction is not actually weak as assumed here
may suppress the dipole transition rate.

3.2   Ionization and Three-body Recombination

Estimates of the collisional ionization rates can be made in
the same way, but it is more convenient to extrapolate our result
for the excitation rate. The first step is to consider excitation into a group of states in some interval E, E + dE and to
replace the oscillator strength by (df/dE)dE. The derivative
is, for hydrogenic spectra, of order E_H^{-1}, and assuming df/dE =
$16\ E_{\infty n}^2 /\ 3^{3/2}\pi n E^3$ (Kramers approximation, [14]) for atoms or ions
in the n-th level of ionization energy $E_{\infty n}$, this estimate for the
ionization rate coefficient is

$$S^z \cong \frac{50v}{\alpha cn} \left(\frac{\hbar}{m}\right)^2 \left(\frac{E_H}{kT}\right)^{1/2} E_{\infty n}^2 E_H \int_{E_{\infty n}}^{\infty} g\ \exp\left(-\frac{E}{kT}\right)\frac{dE}{E^4}$$

$$\cong \frac{50\ v}{\alpha\ cn} \left(\frac{\hbar}{m}\right)^2 \frac{n}{z^2} \left(\frac{E_H}{kT}\right)^{1/2} \frac{\exp(-\ E_{\infty n}/kT)}{3 + (E_{\infty n}/kT)}\ g_{eff} \ . \tag{33}$$

The second version is obtained by assuming the Gaunt factor to be
essentially constant and by using an interpolation formula between the low and high temperature limits of the remaining inte-
gral. With $g_{eff} = (\sqrt{3}/\pi)[1 + \ell n(1 + kT/E_{\infty n})]$ this estimate is
then seen to be consistent (for n = 1) with the classical Thomson
cross section if $kT \lesssim E_{\infty n}$ [16]. It also comes close to the semi-
empirical relation of Lotz [12], although again only for n = 1.
(Thomson [16] and Lotz [12] have $S^z \sim n^2$ instead of $S^z \cong n$.)

Inner shell ionization is normally small, but can be estimated
similarly using appropriate ionization energies, etc.

To estimate the rate coefficient for the inverse process, we
invoke the principle of detailed balance for the process in
question, namely,

$$e + e + I_z \rightleftarrows e + I_{z-1},$$ (34)

where I_z and I_{z-1} stand for the z and (z-1) charged ion of the
element in question. In steady state we have

$$C\, N_z\, N_e^2 = S\, N_{z-1}\, N_e$$ (35)

in terms of the various densities, and the ionization rate coefficient S, and writing $C\, N_e^2$ for the three-body recombination
rate. In thermodynamic equilibrium, the densities are related by
the Saha equation, i.e., by the extension of Boltzmann factors
Eqn. (17) to continuum states,

$$\frac{N_z N_e}{N_{z-1}} = \frac{g_z}{4a_o^3 g_n}\left(\frac{kT}{\pi E_H}\right)^{3/2} \exp\left(-\frac{E_{\infty n}}{kT}\right) .$$ Saha Eqn (36)

Here a_o is the Bohr radius, and g_z and $g_n < 2n^2$ are statistical
weights of the initial and final ions. From Eqs. (33), (35), and
(36) follows

$$C_n \cong \frac{740 v n^3 a_o^3}{\alpha c g_z z^2}\left(\frac{\hbar}{m}\right)^2\left(\frac{E_H}{kT}\right)^2 \frac{g_{eff}}{1 + (E_{\infty n}/3kT)}$$ (37)

for recombination into the (unoccupied) states of principal quantum number n. To obtain an effective rate coefficient for recombination leading to lower states of the recombined ion, the C_n
should be multiplied by a suitable branching ratio and then be
summed over n. Since for $kT \gg E_{\infty n}$, further collisional excitation is certainly dominant over collisional de-excitation, we
will simply assume that for $E_{\infty n} < kT/10$ there is no effective
recombination, while recombination into lower levels contributes
fully. With $\Sigma n^3 \cong \frac{1}{4} n_{max}^4$ and $n_{max}^2 = z^2 E_H/E_{\infty n} \cong 10\, z^2 E_H/kT$ this
estimate for the effective three-body recombination coefficient
is then

$$C_{eff} \cong \frac{3.5 \times 10^4 vz^2}{\alpha c g_z} \left(\frac{\hbar}{m}\right)^2 a_o^3 \left(\frac{E_H}{kT}\right)^4 , \tag{38}$$

which is typically in order of magnitude agreement with more detailed calculations or measurements [4]. To apply our estimate
to recombination into atoms or ions, g_z must be interpreted as a
sum over all low-lying states which have significant populations.

3.3 Radiative and Dielectronic Recombination

Because in many plasmas the collisional ionization rate is
not balanced by its inverse in the sense of the principle of
detailed balancing (three-body recombination), but rather by
radiative and dielectronic recombination (corona ionization equilibrium), it is important to calculate the corresponding rate
coefficients as well. As one might expect, this causes no particular difficulty, the radiative recombination process being an
extension of line emission. In the same way as we can obtain the
number of atoms or ions undergoing a particular radiative decay,
per unit time and volume, from the corresponding line emission
coefficient by integrating it over frequency and solid angle and
dividing by the photon energy, we can also find the volume rate
of recombinations. Basically, this leads to Eqn. (9) for the
recombination continuum, divided by the photon energy and integrated over the angular frequency. Taking all these steps for
the partial recombination rate into hydrogen or hydrogenic level
n we obtain

$$\alpha_n N_z N_e \cong \frac{2^7 \pi^{1/2} (\alpha a_0)^3 z^4}{3^{3/2} n^3} \omega_L \left(\frac{E_H}{kT}\right)^{3/2} \exp(z^2 E_H / n^2 kT)$$

$$\cdot N_z N_e \int_{z^2 \omega_L / n^2}^{\infty} \exp(-\hbar\omega/kT) d\omega/\omega. \tag{39}$$

Here $\omega_L = E_H/\hbar = 2.1 \times 10^{16} \sec^{-1}$ is the angular frequency corresponding to the limit of the Lyman series of hydrogen. The corresponding radiative recombination rate coefficient is therefore

$$\alpha_n \cong \frac{\alpha}{6c} \left(\frac{\hbar}{m}\right)^2 z (z^2 E_H/n^2 kT)^{3/2}$$

$$\cdot \exp(z^2 E_H/n^2 kT) \int_{z^2 \omega_L/n^2}^{\infty} \exp(-\hbar\omega/kT) d\omega/\omega , \qquad (40)$$

written in a form suggestive of the z dependence remaining after the required approximate scaling of the temperature with z^2 is accounted for, and combining the various constants as was done for the excitation and ionization rate coefficients in Eqs. (32) and (33).

As in case of recombination continua, Eq. (40) should be multiplied, under the integral, with appropriate Gaunt factors g $\cong$ 1 to obtain exact results for recombination with bare ions. In the case of other ions, further corrections are required. The most important of these are factors <1 that are of order $v_n/2n^2$, where v_n is the number of electrons already present in a given principal quantum number shell. These factors take care of the Pauli principle, and other corrections are required to allow for deviations of dipole matrix elements, etc., from hydrogenic behavior.

We finally observe that the remaining z scaling of collisional excitation and ionization rates, after also allowing for the approximate temperature scaling, is $\sim z^{-3}$. In other words, the ratio of ionization and radiative recombination rate coefficients goes approximately as z^{-4} which shows that the temperature for maximum abundance of a given ion in corona equilibrium actually increases faster than z^2. These effects are even stronger than is implied here because of the elementary process corresponding to dielectronic recombination which should perhaps more properly be called radiationless capture. The name dielectronic, on the other hand, indicates that the transition requires two electrons in an essential way: while the originally free electron is captured, a bound electron is promoted to a higher excited state and takes up the energy set free in the capture process. Normally the captured electron is in a highly excited state of the resulting ion, while the originally bound electron is more likely than not promoted to an excited state corresponding to the upper level or resonance lines of the recombining ion. These doubly excited states can, of course, auto-ionize, i.e., undergo the microscopic inverse of the capture process. In that case, there is no effective recombination. However, it is also

possible that the originally bound electron returns to its ground
state through a radiative (stabilizing) transition. The corresponding photon corresponds in frequency almost to the resonance
lines in the spectrum of the recombining ions, and the slightly
displaced and often unresolved lines are called dielectronic
satellites. The other electron, at sufficiently low densities,
eventually cascades down to complete the process.

The usual procedure for calculations of effective dielectronic recombination rates begins with an estimate of the population in the doubly excited states in terms of the density of free
electrons and the ground state population of the recombining ion
through the appropriate Saha equation. This estimate is corrected by multiplication with the ratio of auto-ionization to
total decay rate, including radiative decay. Such a factor allows for the fact that radiative processes are not in balance and
is sometimes called the Saha decrement. The auto-ionization
rate, on the other hand, can be expressed in terms of the elementary capture rate through the principle of detailed balancing.
To finally obtain the recombination rate, the doubly excited
state populations are multiplied with the spontaneous radiative
decay rates, and these products are summed over all doubly excited states.

Because of the large number of states involved and because
of the uncertainties in cross sections, etc., it is difficult to
estimate the accuracy of such calculations. Moreover, especially
at laboratory plasma densities, a number of additional processes
affecting the doubly excited state densities may have to be included. For example, although the primary capture process will
usually only result in small orbital angular momenta ℓ for the
captured electron, both electron and ion collisions may result in
a redistribution over ℓ-values and therefore a net reduction of
the auto-ionization rate and increase of the effective recombination rate. On the other hand, electron collisions may also cause
further excitation and even ionization of the $n\ell$-levels of the
captured electron and thereby decrease the effective dielectronic
recombination rate.

Rather than attempting a quantitative description of the
current theoretical research on these complex processes, we shall
now return to the description of the capture process as below-
threshold excitation of the recombining ion. We estimate the
corresponding rate coefficient by taking the difference of the
excitation rate coefficients according to Eq. (32) for a fictitious reduced excitation energy and for the actual excitation
energy for the "promoted" electron,

$$d \cong \frac{16\pi}{\alpha \, c} \, (\frac{\hbar}{m})^2 (\frac{E_H}{kT})^{3/2} \, \exp(- \frac{\Delta E}{kT}) \, \frac{E_n}{\Delta E} \, \bar{fg} \; . \tag{41}$$

Here $E_n \cong z^2 E_H/n^2$ is an estimate for the extra energy available for excitation because of the binding energy of the captured electron. (We assumed $E_n \lesssim kT$ and expanded the exponential accordingly.)

Since auto-ionization rates decrease as $1/n^3$ while radiative stabilization rates are nearly independent of n, we impose a lower bound on n corresponding to near quality of these competing processes and obtain as an order of magnitude estimate for the effective dielectronic recombination coefficient

$$d \cong \frac{10^3 v}{2\alpha \, c} \, (\frac{\hbar}{m})^2 \, (\frac{E_H}{kT})^{3/2} [\alpha(z + 1)/n_v]^{8/3}$$

$$x \; (z^2 E_H/\bar{E}) \, \exp(- \bar{E}/kT) \; , \tag{42}$$

where n_v is the principal quantum number of the ground state (with v equivalent electrons) of the recombining ion of charge z. The relative values of radiative and auto-ionization rates were chosen as $\sim 20[\alpha(z+1)/n_v]^4$ and $\sim (20 \, n^3)^{-1}$, and $\bar{fg}$ replaced by 0.2 v to obtain this formula, in which $\bar{E}$ is an average excitation energy for the resonance lines of the recombining ion.

Comparison with Eq. (40) suggests that dielectronic recombination is much more probable than radiative recombination, especially at relatively high temperatures, i.e., kT values close to $\bar{E}$. (Note, however, that there is obviously no dielectronic recombination on fully stripped ions.)

4.   DIAGNOSTIC METHODS

Although spectroscopic methods can be used to determine a number of macroscopic and microscopic plasma parameters from non-interfering measurements, there is the difficulty of obtaining local parameters from measurements which are integrated along the line of sight. In principle, this difficulty can be overcome, especially when the symmetry of the plasma is known and simple, by taking data through various lines of sight and inverting these data (Abel transform, tomography). Even for optically thin plasma, data must be of high quality for such inversions to be practical, and for pulsed plasmas this approach is therefore rarely used. A more fruitful approach often is to assume or calculate

plasma densities and temperatures as functions of time and space
and then to calculate synthetic spectra. By trial and error,
iteration of these calculations will usually lead to reasonable
agreement with observed spectra so that the corresponding spatial
variation of plasma parameters used in the calculation can be
assumed realistic. In the following, we take it for granted that
such an analysis of spatial variations was either performed or
considered not necessary.

4.1 Density Measurements

From absolute line and continuum measurements one can determine densities of atoms or ions in the appropriate initial state,
or in the case of continuum emission the products of these densities with the electron density and a function of temperature,
see Eq. (9). This dependence must be kept small by measuring at
long wavelengths, i.e., at $h\omega < kT$. There is an analogous expression for line emission, which may be obtained from Eq. (5)
and the Saha Equation (36) and can also be used to determine the
product of electron and ion densities, as long as the conditions
for LTE of the initial state of the line with the respect to
higher states and the continuum are valid [5]. In the case of
single ion species plasmas, the neutrality relation helps reduce
errors in the electron density because the two intensity relations are then quadratic in N_e.

In the above measurement of line intensities, integration
over the profile was assumed ($\int L(\omega)d\omega = 1$), i.e., total line
intensities were used. In dense plasmas, a more direct way to
determine N_e is via line profiles of Stark broadened lines. Some
lines, especially hydrogen and lines from one-electron ions, tend
to show significantly more than Doppler broadening, and their
widths are very closely proportional to $N_e^{2/3}$. Their Stark profiles are quite characterisitc so that profile fitting is preferable to measurements of halfwidths. Besides possible corrections
for other line broadening mechanisms, one must also make sure
that the measured intensity distribution $I(\omega)$ is not broadened or
otherwise distorted from the line shape $L(\omega)$ by radiation transport effect in optically thick plasmas (see Sec. 2.4). Otherwise, one sometimes speaks of opacity broadening.

4.2 Temperature Measurements

For lines and conditions such that Doppler broadening is the
dominating line broadening process, the kinetic temperature of
the emitting species can be determined from the measured Gaussian

profile or its width by inverting Eq. (8). High experimental
accuracy is very desirable here because the temperature is quadratic in the width and because considerable accuracy may be lost
by corrections (deconvolutions) for other than Doppler broadening, e.g., instrumental broadening.

Other spectroscopic methods for temperature measurements
basically give electron temperatures. In plasmas consisting
mostly of fully stripped ions and approximately for other plasmas
also, the continuum emission for $\hbar\omega > kT$ is dominated by the
exponential factor so that a measurement of the slope of the continuum in this region can be used to infer the electron temperature. Similarly, relative line intensities can be used, provided
the densities in the initial states can be related to each other
by a Boltzmann factor, Eq. (17). To make this a valid temperature measurement, it must be ascertained that electron-ion collisions control the populations of the two initial levels. Also,
their energies should fulfill $(E_m - E_m')/kT > 1$ for the intensity
ratio to be sensitive to the temperature. The latter requirement
is rarely met, and it may be advantageous to chose lines from
adjacent ionization stages. However, one then has to use Saha's
Equation (36) as well, which is often not valid in high temperature plasmas and must be replaced by a more detailed rate equation approach.

A special case of lines from adjacent ionization stages is
the ratio of dielectronic satellite to parent line ratio. Under
some conditions, this ratio only depends on electron temperature.
Experimentally the proximity in wavelengths facilitates relative
intensity measurements greatly.

4.3 Fluctuations

Density and electric field fluctuations are important microscopic plasma properties that can be studied spectroscopically.
To measure density fluctuations, one may observe a small volume
from different directions and determine correlations between the
two signals. Depending on the frequency of these fluctuations
and the characteristic relaxation times, the underlying fluctuations in initial state fluctuations may be interpreted also in
terms of electron density or temperature fluctuations.

Measurements of electric field fluctuations are usually
based on the high-frequency Stark effect [1]. This effect produces to lowest order in the field-atom (or ion) interactions in
addition to the usual forbidden Stark component (corresponding to
$\Delta\ell = 0$ or ± 2 transitions) two plasma satellites symmetrically

located with respect to this forbidden component at separations near ω_p, the dominant plasma frequency. The relative (integrated) intensities of these satellites, relative to the intensity of the allowed line, are

$$S_\pm \cong \frac{3}{8} \frac{\ell_>}{2\ell_i + 1} (\frac{a_o e}{\hbar z})^2 \frac{n^2 - \ell_>^2}{(\omega_{ii}' \pm \omega_p)^2} <F^2>$$ (43)

according to second-order perturbation theory [6]. In this relation, $\ell_>$ is the larger of ℓ_i and ℓ_i', the orbital quantum numbers of initial levels of the allowed line and forbidden component, respectively, and n is their common principal quantum number. The two levels are assumed to have an intrinsic splitting of ω_{ii}', and $<F^2>$ is the average of the square of the wave electric field. To estimate the dipole matrix elements, hydrogenic wavelengths for an effective nuclear field of charge z were assumed, e.g., z = 1 for HeI.

Fields have to be quite large (~10 kV/cm) for plasma satellites to be observable in emission. Such fields typically correspond to wave energy densities of two or more orders of magnitude above thermal levels.

4.4 Magnetic Fields

Since Zeeman splittings, which are of the order of the electron cyclotron frequency ω_{ce}, tend to be smaller than Stark effects, spectroscopic measurements of magnetic field fluctuations have so far not been possible. Zeeman splittings in strong and reasonably homogeneous fields have, however, been observed in a plasma focus experiment [15] and in laser-produced plasmas [13]. Since broadening due to thermal Doppler effects and mass motion and also from opacity are competing with Zeeman effects, it is necessary to perform line profile and structure measurements with polarizing optics and to look for small differences in the corresponding spectra. For the high fields encountered, the theoretical analysis must allow for deviations from the low field (anomalous) Zeeman effect.

5. SUMMARY

Spectroscopic methods for the diagnostics of plasmas rely on quantum-mechanical results for atomic and ionic structure, for radiative processes, and for collisional interactions. Through

intensive laboratory and theoretical research the general framework originally transferred from astronomical research has been
used to acquire a considerable body of required atomic data and
analytic methods. Present research is aimed towards the extension of the range in physical conditions and at improved accuracy.

REFERENCES

1. Baranger, M., and B. Mozer, 1961, Phys. Rev. 123, 25.

2. Brissaud, A. and U. Frisch, 1971, J. Quant. Spectrosc.
    Radiat. Transfer 11, 1767.

3. Cauble, R. and H. R. Griem, 1983, Phys. Rev. A 27, 3187.

4. Eletsky, A. V. and B. M. Smirnov, 1983, Ch. 1.2 in Handbook
    of Plasma Physics, Eds. M. N. Rosenbluth and R. Z. Sagdeev,
    Volume 1: Basic Plasma Physics I, edited by A. A. Galeev
    and R. N. Sudan (North-Holland Publishing Company).

5. Griem, H. R., 1964, Plasma Spectroscopy (McGraw-Hill, New
    York; Univ. Microfilms International 212 00000 7559, Ann
    Arbor).

6. Griem, H. R., 1974, Spectral Line Broadening by Plasmas
    (Academic Press, New York).

7. Griem, H. R., 1979, Phys. Rev. A 20, 606.

8. Griem, H. R., 1983, Ch. 1.3 in Handbook of Plasma Physics,
    Eds. M. N. Rosenbluth and R. Z. Sagdeev, Volume 1: Basic
    Plasma Physics I, edited by A. A. Galeev and R. N. Sudan
    (North-Holland Publishing Company).

9. Griem, H. R. and G. D. Tsakiris, 1982, Phys. Rev. A 25,
    1199.

10. Inglis, D. R. and E. Teller, 1939, Astrophys. J. 90, 439.

11. Kim, Y. S. and R. H. Pratt, 1983, Phys. Rev. A 27, 2913.

12. Lotz, W., 1968, Z. Physik 216, 241.

13. McLean, E. A., J. A. Stamper, C. K. Manka, H. R. Griem,
    D. W. Droemer, and B. H. Ripin, Phys. Fluids (in press).

14. Omidvar, K., 1982, Phys. Rev. A 26, 3053.

15. Peacock, N. J. and B. A. Norton, 1975, Phys. Rev. A 11,
     2142.

16. Thomson, J. J., Phil. Mag. 23, 419.

17. Tondello, G., E. Jannitti, and A. M. Malvezzi, 1977, Phys.
     Rev. A 16, 1705.

18. Wiese, W. L. and G. A. Martin, 1980, Wavelengths and Transi-
     tion Probabilities for Atoms and Atomic Ions, NSRDS-NBS (US
     Govt. Printing Office, Washington, D.C.).

SPECTROSCOPY OF LASER-PRODUCED PLASMAS

G. Tondello

Instituto di Elettrotecnica e di Elettronica
University of Padova
Italy

1. INTRODUCTION

The study of the irradiation of solid targets by laser light is a very interesting subject for the following investigations: a) the absorption of laser light by matter and the physical processes leading to plasma heating; b) the extension of laboratory plasma parameters to very high densities ($n_e \geq 10^{24} cm^{-3}$) and pressures ($n_e kT_e \cong 10^8$ atmospheres); c) the generation of highly stripped ions; d) controlled fusion research. Spectroscopic analysis play an important role in all of these fields of study. [1]

Modern laser installations can easily produce pulses of energy ranging from 10J to 3KJ with pulse duration from 100 psec to 10 nsec. When focused on solid targets, power density in the range $10^{12}-10^{18}$ W cm^{-2} can be obtained on irradiated areas whose typical dimensions are 0.1-1mm. The electric field associated with the electromagnetic wave is related to the power density by:

$$E \cong 20 \; \phi^{\frac{1}{2}} \tag{1}$$

with E in $V \cdot cm^{-1}$ and ϕ in $W \cdot cm^{-2}$;E ranges from 2×10^7 to 2×10^{10} $V \cdot cm^{-1}$. If one remembers that the electric field on the electron in the hydrogen atom at the Bohr radius is 5×10^9 $V \cdot cm^{-1}$ one can easily understand the great effects that laser irradiation can produce on matter in general and in the formation of energetic plasmas in particular.

   This lecture is concerned with the main aspects of spectroscopy of laser produced plasmas (l.p.p.) with some more emphasis
on the optical field ($\lambda \geq 20\overset{\circ}{A}$) since X-ray diagnostics have already been treated [2].

   Also, time-resolved spectroscopic techniques will be mentioned only briefly since they have been covered separately [3].

2.   PRODUCTION OF PLASMA BY LASER LIGHT

   We now consider, in a very simple way, the formation and
characteristics of plasmas created by laser irradiation [4]. A
thin sheet of plasma close to the target surface is quickly
formed (often within few optical cycles of the laser field) by
the leading edge of the laser pulse. In case of conducting target materials this is caused by interaction of the radiating
field with the conduction electrons in a very shallow region on
the surface and in the case of insulators, electrons are produced
possibly through such mechanisms as multiphoton ionization, surface effects or presence of impurities and non linear effects
(harmonic production). Once the priming plasma is produced, the
incoming laser radiation is absorbed mainly by inverse bremsstrahlung. The absorption coefficient is proportional to

$\dfrac{Z n_e^2}{T_e^{3/2}}$ where Z is the charge of the ions, n_e is the election

density and T_e the electron temperature. At the beginning, the
frequency of the laser light ω_L is $>> \overset{.}{\omega}_p$ the plasma frequency

$\omega_p = \dfrac{(n_e e^2)^{\frac{1}{2}}}{m_e \varepsilon_o}$, but quickly, as a result of energy absorption,

there is an increase of the electron temperature and consequently
of the ionization and of the electron density. The process esca-
lates rapidly until the so called critical density n_c is estab-

lished across a surface some distance into the plasma. The
critical density is such that $\omega_p = \omega_L$, i.e., the plasma frequency
equals the laser field frequency. For 1 μm laser light (Nd^{3+})
$n_c = 10^{21}$ cm^{-3}; for 10 μm laser light (CO_2) $n_c \cong 10^{19}$ cm^{-3}. For 0.3 μm
(third harmonic of Nd^{3+} or typical excimer laser) $n_c \cong 10^{22}$ cm^{-3}.

   At the surface, the plasma becomes opaque to the incoming
radiation which is then partially reflected outwards. It can be
easily seen that the absorption coefficient of the plasma becomes

very large in the layers immediately in front of the critical
density surface. However, because of the heating which follows
the absorption of energy, the plasma is driven rapidly away from
the target surface (expansion) with a corresponding decrease of
the electron density: the laser radiation can again reach the
target. The various processes quickly adjust themselves to an
equilibrium situation in which generation, heating and expansion
of plasmas take place throughout the length of the laser pulse.

Often and particularly so for high irradiation intensities,
the situation is considerably more complicated with other phenomena such as resonant abosrption, ponderomotive forces due to the
light pressure and self-generated magnetic fields playing an
important role [1].

The velocity distribution of the electrons can in such cases
depart from the Maxwellian and develop a high energy component:
the so called hot electrons (with $T_e > 10 KeV$).

3.   HYDRODYNAMIC DESCRIPTION OF THE PLASMA

An idealized picture of the physical states of a laser
produced plasma on plane targets at a given time t is shown in
Fig. 1 assuming a semi-infinite block irradiated by a beam of
constant power density ϕ coming from the right [4] Fig. 1 shows
the initial position of the target; the interaction region F
which travels into the target with velocity v_F and the shock wave

S that moves ahead into the umperturbed solid with velocity v_S.

To the right of the front F there is the expanding plasma. In

Fig. 1a the electron density starts from n_c and the narrow region

(between n_c and say $0.9 . n_c$) is where most of the energy is de-

posited. The variation of intensity of the incoming and the out-
going radiation after reflection at the critical layer are shown
in Fig. 1b. Fig. 1c shows the temperature profile at the time t.
It rises sharply at the ablation front and stays almost constant
(in this approximation) in the streaming plasma. The density
illustrated in Fig. 1d shows a strong increase in the region between the shock front S and the interaction front F. It is precisely this effect that is used in spherical targets to induce
implosion and to increase the overall density several times over
solid state values. The density of the expanding plasma is much
less and it falls off with increasing distance from the target.
Finally in Fig. 1e, the velocity profile of the system is shown.
The shock front travels with a constant velocity into the target
while the plasma expands with increasing velocity into the vacuum. The scales on the figures are arbitrary and depend on the

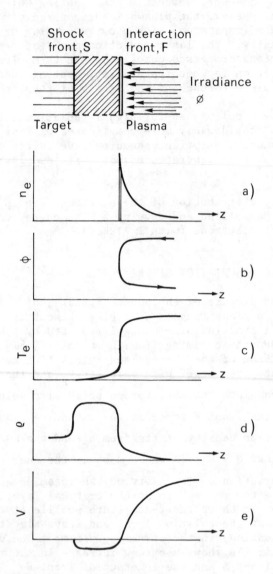

Fig. 1.   Schematic of the conditions in a laser produced plasma at
           a given time t:
           a) electron density
           b) intensity of incoming and reflected radiation
           c) electron temperature
           d) density
           e) velocity.

pulse used for creating the plasma (energy and time duration being the most important parameters). Typical scale lengths are, for pulses of a few nsec duration, 1 mm for the expanding plasma and 0.1 mm or less for the shock front penetration. The temperature peak is of the order of 100-1000 eV. Typical values for the expansion velocities are a few times 10^7 cm sec^{-1}. Several modifications to the idealized picture have to be taken into account.

    a. In reality the irradiation is limited laterally and two-dimensional effects are very pronounced. The plasma expands as a plume as is shown in Fig. 2. In general it is reasonable to assume rotational symmetry around the laser axis. Also the ablation front into the solid is curved. (This has been proved by some beautiful experiments analyzing the properties of the back-scattered light [5]).

    b. For relatively high irradiation intensities strong preheating of the target is produced by several mechanisms like fast electron (created by several non-linear processes) preheating, by X-ray radiation transport, etc.

    c. In plane targets for low Z materials most of the absorbed energy goes into kinetic energy of the expanding plasma with only a few percent converted into radiation emitted by the hot plasma. The situation can be different for high Z targets where copious soft X-ray production has been observed (up to ≅30% of the observed energy).

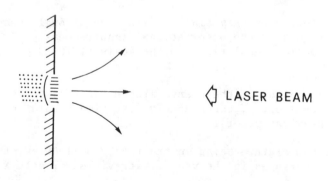

Fig. 2. Scheme of a laser produced plasma with a finite focal spot.

For spherical targets irradiated uniformly (with several beams) the situation is somewhat different although in this case also a one-dimensional treatment is valid.

Two regimes can be distinguished [1]: the exploding pusher implosion and the ablatively driven implosion. Exploding pusher configurations have been applied to implosions of thin-walled microballoons ($r \cong 50$ μm, $\Delta r \cong 1$ μm) with hot electrons in the plasma whose range of energy deposition is $\lambda_H > \Delta r$. The laser pulse (rise time and duration) is short (<100psec) to give a high value of hot electron preheat in the shell before it relaxes by explosive expansion. The explosion of the walls drives approximately half the mass inward and half outwards. The inward part compresses the gas that has however been preheated by hot electrons and shock waves. Electron temperature at maximum compression is of the order of 0.1-1 keV.

Ablatively driven implosion applies to spheres with higher Δr in order to stop significant preheating of the core and achieve significantly higher compression ratios. The laser pulse irradiates the pellet during the whole implosion process (1 nsec or longer). A fuller discussion of this important, for fusion, mode of implosion is given by Key and Hutcheon [1]. For imploding spherical targets, due to the high density, spectroscopic observations are mainly concentrated in the X-ray regime either in emission or in absorption with the help of an auxiliary laser produced plasma acting as a background source [6].

Spectroscopic observations and diagnostics of laser produced plasmas are strongly dependent on the physical nature of l.p.p. as sketched before:

a. The source is very small in size and strongly non-homogeneous; consequently the observational requirements put a strong emphasis on spatial resolution (on the scale of 10-30 μm or even smaller).

b. The duration of the events are in the time scale of few nanosec or even shorter; consequently fast time resolution is required (see previous paper).

c. The temperature being in the range 100-1000 eV, the bulk of the emission ranges in the vacuum ultraviolet or soft X-ray regions.

d. The density range being very high, with density approaching or exceeding solid state density, there are severe problems in the trapping of the radiation. The spectrum emitted is considerably more complicated than in the "thin" conditions and for the extreme of high density only x-rays could ultimately escape from the plasma.

4.   SPECTRA EMITTED BY LASER PRODUCED PLASMA

The emission by laser-produced plasmas consists, as in general that from any plasma, of continuum and discrete spectral components [7]. We review them here looking mainly to l.p.p applications.

4.1 Continuum-Emission

Continuum emission can arise from two different effects: Recombination radiation or free-bound radiation corresponding to the process of recombination of an ion of charge Z with a free electron to give an ion of charge Z-1 with the emission of a photon:

$$A^{z+} + e \rightarrow A^{(z-1)+} + h\nu \tag{2}$$

The calculation of the emitted recombination spectrum is relatively straightforward [8] in the case of a fully stripped ion recombining to give a hydrogenic ion and can be given by:

$$\frac{dE_{fb}}{d\lambda} = \frac{\kappa}{4\pi} \, n_e n^{z+} \, \frac{C}{\lambda^2} \left(\frac{\chi_H}{T_e}\right)^{2/3} \left(\frac{\chi_{in}}{\chi_H}\right)^2 g_{fb} \frac{\int_n}{n} \exp\left\{\frac{\chi_{in} - h\nu}{T_e}\right\} \tag{3}$$

where E_{fb} is the emitted intensity, κ a constant $(= 1.7 \times 10^{-55}$ joule $cm^4 \, °^{-1})$, n_e and n^{z+} the densities of electrons and ions of charge Z, $\chi_H(eV)$ is the ionization potential of hydrogen, $T_e(eV)$ is the electron temperature of the plasma, $\lambda(cm)$ and $h\nu(eV)$ are the wavelength and the energy of the emitted photons. χ_{in} is the ionization potential of the ion into which there is recombination counted from the state n of multiplicity $\int_n$. Expression (1) can be written also as:

$$\frac{dE_{fb}}{d\lambda} = n_e n^{z+} \, F_{fb}(\lambda, T_e) \tag{4}$$

where the dependence on the wavelength and temperature is contained in the function F_{fb}. Recombination radiation has a step-like structure in wavelength with discontinuities corresponding

to the recombination of the electron on the various shells of the
ion. The largest one corresponds to the recombination to the
ground state and is at the shortest wavelength. Often is the
only discontinuity clearly recognizable in the emitted spectrum
of a plasma.

Bremsstrahlung or free-free radiation arises because of the
deceleration of an electron in the field of an ion and can similarly be calculated. The spectral intensity is given by:

$$\frac{dE_{ff}}{d\lambda} = \frac{\kappa}{4\pi} n_e n^{z+} \frac{C}{\lambda^2} Z^2 \frac{\chi_H}{T_e} y2 \ g_{ff} \ \exp \frac{-h_\nu}{kT_e}$$

$$= n_e n^{Z+} F_{ff} \ (\lambda_1 T_e) \tag{5}$$

Free-free radiation has, apart from the discontinuities, a similar dependence on wavelength than recombination. It reaches a
maximum at $\lambda_m(\overset{o}{A}) \cong 6200/T_e(eV)$, decaying at long wavelengths as
λ^{-2} and at very short wavelengths as: $\exp\{-h\nu/kT_e\}$ [9]. Free-
free and free-bound continua both depend on the product $n_e \cdot n^{z+}$,
i.e. on n_e^2: in laser produced plasmas due to the very high density they appear quite strongly. Both continua depend also on
Z: the free-free on Z^2 and the free-bound on Z^4. Consequently
the emitted energy increases rapidly with the atomic number.

Free-free and free-bound continua always exist together;
at relatively high temperature, long wavelengths and high Z the
free-free radiation is dominant. Fig. 3 shows the calculated
continuum spectrum emitted by a plasma of fully ionized beryllium
(Z=4) [10] for two temperatures: T_e =50 and 100 eV. The func-
tional dependence of the free-bound continuum at short wave-lengths
$\frac{1}{\lambda^2} \exp \left(\frac{\chi_{in} \ -h_\nu}{T_e}\right)$ lengths is the basis for a useful method of meas-

uring the temperature. If one measures the slope of the intensity (i.e., only a relative measure) of the continuum as it
falls off at wavelenghts shorter than the Lyman edge, one can
derive the electron temperature. The measurement can be done
with spectrographs or more simply if there are no disturbing
lines present, with absorbing foils. In this very simple method
one is relaying on the preferential absorption by metals in thin
foils of wavelengths shorter than the K(or L) edge of the atoms
[11]. In this way a quite coarse (but highly efficient in terms
of photon detection) spectral resolution can be achieved.

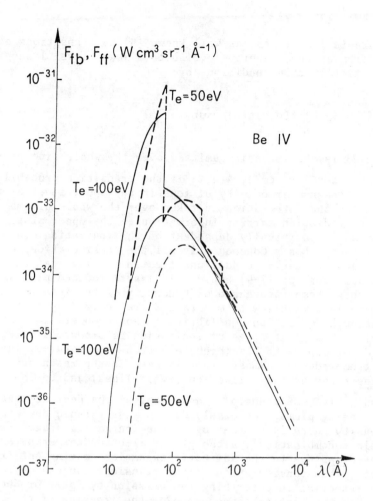

Fig. 3. Calculated spectrum of free-free (thin lines) and free-
bound (thick lines) emission for a fully ionized beryllium plasma for two temperatures.

On the other hand, a measure of the absolute intensity of
the continuum, particularly at long wavelengths where there is
no dependence on T_e, can be used to derive the quantity $\int n_e^2 dl$,
i.e., the integral of n_e^2 along the plasma depth if one assumes
that the plasma is optically thin in the continuum.

4.2 Line Emission

Line radiation arises, as well known [8] by transitions of the atoms from an upper energy level u to a lower level 1 and is given in optically thin conditions by

$$I(u,\ell) = \frac{1}{4\pi} \int_L A(u,\ell)n(u)h\nu \, dL \tag{6}$$

where I (u,ℓ) is the intensity emitted over the whole line profile (photons $sec^{-1} sr^{-1} cm^{-2}$), $A(u,\ell)$ is the transition probability, $n(u)$ is the number density of ions in the upper level of the transition and the integration is taken over the whole plasma depth. Line radiation carries information on the upper state population and is critically dependent on the ionization equilibrium of the plasma. Consequently all the considerations on ionization and excitation conditions treated in classical texts of plasma spectroscopy [7-9] apply. In laser produced plasmas, due to the high temperature and high density of the interacting region, materials with low to moderate Z are ionized up to the K shell, although ionization equilibrium is not complete in most cases i.e., the highest stage of ionization is produced is less than would correspond to the (high) electron temperature. The line spectrum reduces (almost) in this case to the resonance 1s-np transitions of the H-like ion (Lyman lines) and to the corresponding $1s^2 \, ^1S- 1snp \, 1p^0$ transitions of the He-like stage. In the expanding plume, conversely due to a very rapid decrease of the density there is a freezing of the ionization stages (with very little recombination), although the electron temperature in most cases drops due to the adiabatic expansion, moving out from the target. Thus consideration of the transient aspects of ionization is essential in describing the emission by laser produced plasmas. One interesting consequence of the freezing of the population in rapid plasma expansion is in the possibility of producing population inversion between levels with (generally) moderate to high values of the principal quantum number n. This effect is quite important in the search to obtain soft x-ray laser action. Fig. 4 shows population inversion obtained in the rapid expansion of an aluminium plasma [12]. The line originating from n=4 and decaying to the ground state in He-like Al XII is clearly stronger than the line coming from n=3 of the same ion. The corresponding possible laser transition 4^3F-3^3D is at 129 Å. Another case extensively studied is the population inversion on the Balmer α line of CVI at $\lambda=182$ Å found in the rapid expansion of a plasma produced irradiating a 5 μm diameter carbon fibre with a 150 psec, 2J laser pulse [13].

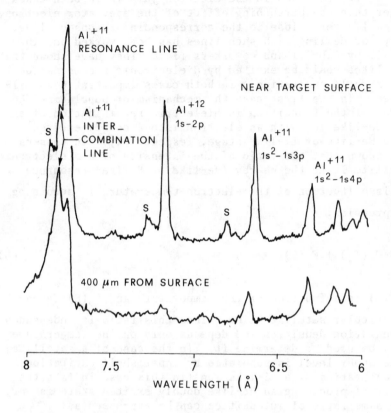

Fig. 4. Aluminium spectrum recorded with spatial resolution
         showing inversion of the (1s3p,1s4p) pair.

In addition to resonance lines, corresponding to single
electron transitions (in a normal atomic scheme), the so-called
satellite lines are often very prominent in l.p.p.. They correspond to transitions of the type 1sn1-2pn1 or 1s^{2}n1-1s2pn1. They
appear as a group at the long wavelength side of the first resonance line of, respectively, the H-like or the He-like ion.
These lines correspond to transitions between doubly excited
states of the He-like and Li-like stages of ionization, but can
in fact be thought as screened transitions, i.e., normal one-
electron transitions in the presence of a "spectator" electron:

$$2p(n\ell) \rightarrow 1s(n\ell)$$

$$n \geq 2$$

$$1s2p(n\ell) \rightarrow 1s^2(n\ell) \qquad (7)$$

Since the energy involved in the electron jump is in both cases
much larger than the perturbing effect of the spectator electron,
these lines lie very close to the corresponding resonance line.
The theory for dealing with such lines has been worked out in
great detail by Gabriel and coworkers [14]. They have shown that
such satellites could be excited by dielectronic recombination
or by inner-shell excitation. In both cases important possibilities arise. In the first case the mechanism of populating the
upper level of the transition is dielectric recombination of an
H-like or He-like ion with an electron into a doubly excited
state of a He-like or Li-like stage, respectively. In this case
it can be shown that the ratio of the intensity of a dielectronic
type satellite I_s to the nearby (He-like or H-like) resonance

line, I_R is a function of the electron temperature T_e according
to the expression:

$$\frac{I_S}{I_R} = F_1(T_e) + F_2(S) \tag{8}$$

where $F_1(T_e)$ is a function of the temperature and $F_2(S)$ depends
on the particular satellite line. Thus this ratio is independent
of the population densities and depends only on the temperature,
and so may be used to determine it. In the case of a satellite
line whose upper level is populated by inner shell excitation,
i.e., by excitation of a Li-like ion (in this case in fact the
probability of producing an He-like doubly excited state is very
low), the same ratio of intensities can be expressed as:

$$\frac{I_2'}{I_R} = \frac{N^{Li}}{N^{He}} \beta F_2'(S) \tag{9}$$

where β and $F_2'(S)$ are functions of the satellite line and N^{Li},
N^{He} are the population densities of the Li-like and He-like ions,
respectively. Thus the ratio I_s'/I_R gives the ionization temperature T_z through the ratio N^{Li}/N^{He}. If the plasma is not in a
steady state $T_z \neq T_e$ and one therefore gets an indication of the
departure of the plasma from the state of ionization equilibrium.
It is to be noted, however, that with respect to the theory of
Gabriel, developed originally for low density plasmas (tokamaks
and solar corona), several high density effects have to be included in l.p.p. like the opacity effect on the resonance line,
and the collisional mixing of the closely spaced upper levels of

the satellites. However, the method is quite useful as a diagnostic tool, taking into account that satellites and resonance
line lie close together which simplifies calibration problems.

Fig. 5 shows a simplified term scheme of the carbon ions
summarizing the various transitions of interest in l.p.p.: Lyman
lines, He-like resonance lines, and satellites.

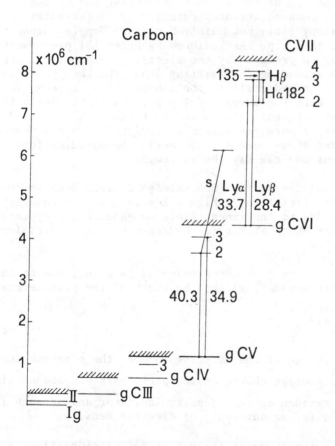

Fig. 5. Term diagram of the ionization stages of carbon with the
main lines observed in laser produced plasmas.

For high Z materials ionization is far from complete and the emitted line spectrum could be highly complicated. In fact the field of line identification using l.p.p. as a light source (with considerable advantages over other competing sources for spectral purity etc.) has grown considerably. Very high stages of ioniza-tion have been reached (up to W^{45+}) and important contributions to the knowledge of atomic structure in general have been made.

5.   LINE BROADENING IN L.P.P.

A quite general feature of line radiation in l.p.p. is the remarkable broadening of the lines. Broadening can be due to various physical reasons, treated extensively in textbooks on plasma spectroscopy [16], but mainly Stark and Doppler broadening are of interest in l.p.p. As reviewed by Griem [8] previously, Stark broadening is produced by the electric fields of the ions and electrons surrounding the emitting ion. The theory of Stark broadening has been and still is the subject of an intense re-search effort. With the advent of l.p.p., and particularly for imploding pellets, research has moved to treat highly charged ions for diagnostic purposes (whereas previously, mostly neutral or singly ionized atoms were of interest). Summarizing for l.p.p. conditions one can say the following:

a.   Stark broadening for (moderately) charged ions is re-markable only for (relatively) high n levels and particularly so for H-like ions due to the linear field dependence (as opposed to quadratic) of the Stark shifts for hydrogenic (degenerate) lev-els.

b.   In the latter case the broadening is mainly due to the ions (quasistatic broadening) and the width of the profile scales according to:

$$\Delta\lambda \propto Z^{-5} \, n_e^{2/3} \, Z_p^{1/3} \tag{10}$$

where Z is the charge of the emitting ion, n_e the electron den-sity and Z_p the average charge of the perturbers. Consequently, as known, line broadening is a density indicator and as such is extensively used for measurements of electron density.

c.   Several improvements to this simple considerations are to be mentioned: correlation between perturber-emitter pairs are increasingly important as the density goes up and reliable full profile calculations are very complex, particularly for non-hy-drogen like ions. Fig. 6 shows an example of a profile calcu-lated by Lee [17] of the $1s^2$-1s4p transition in BeIII. The profile has been calculated for an electron density of 5×10^{20} cm^{-3} and T$_e$=90eV). As evident, due to the unshifted central component, the profile departs remarkably from a simple Lorentzian shape which, however, applies very well to the transition 7→6 at 3433 Å of CVI as shown in Fig. 7 [18]. The latter is an extreme example of Stark broadening (often seen in l.p.p.) due to the hydrogenic nature of both levels involved in the transition (a full width of 35Å for an electron density of 10^{19}cm^3). Indeed only instruments of moderate to low resolition are needed to perform measurements on this line.

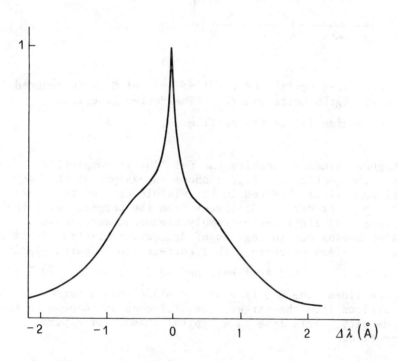

Fig. 6.   The line is 1s^2-1s4p of Be III calculated for laser produced plasma conditions of n$_e$ = $5 \cdot 10^{20}$ cm^{-3} and T$_e$=90 eV.

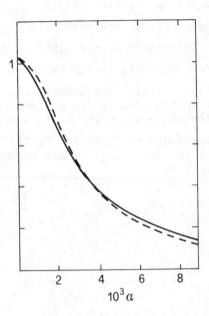

Fig. 7.   Calculated profile of the line 3434 of C VI in reduced
wavelengths units $\alpha = \Delta\lambda / F_o$.   The dotted line is a
Lorentzian fit to the profile.

At higher values of density the CVI line is completely
merged into the continuum. Fig. 8 shows an example of the behaviour of some lines observed in the visible part of the spectrum in a l.p.p. at various distances from the target. Near the
surface almost all lines are extremely broadened and almost
merged; when moving out to regions of decreasing electron density
the widths the lines decrease. This correlation is particularly
evident on the lines 4673 Å of BeII and 4d-5f of BeIII. In Fig.
8 one of the lines, the BeI line at λ = 4572Å, has a narrow width
at all positions from the target, i.e., it does not depend on the
electron density. This line is a typical example of a pure
Doppler profile.

Doppler broadening is very classical in nature and in l.p.p.
can be divided between thermal Doppler due, as known, to the Maxwellian distribution of the velocities of the ions, giving a

width $\Delta\lambda \propto \lambda \sqrt{\dfrac{T_i}{M}}$ where M is the mass of the radiating ion and T_i

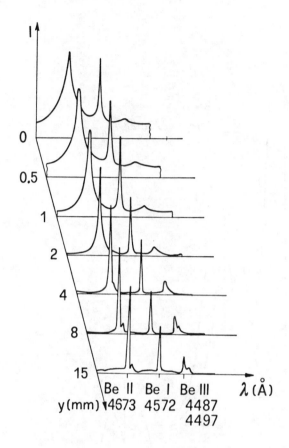

Fig. 8. Spectrum of a beryllium laser produced plasma in the region 4400-4700 Å recorded at various distances from the target. The peak intensity is normalized at the same value in order to show the changes of the profiles.

the ion temperature, and motional Doppler width (or shift) due to the relative motions of the single plasma elements as they expand or implode in the l.p.p. The latter effect can produce shifts or splittings of the lines (depending on the distribution of velocity) as shown in Fig. 9 where a profile of the resonance line of BeIV at $\lambda = 75$ Å is recorded at (relatively) large distances from a plane target where the optical depth is negligible [19]. Indeed a noticeable splitting is present which becomes increasingly clearer as one moves far from the target. When the optically

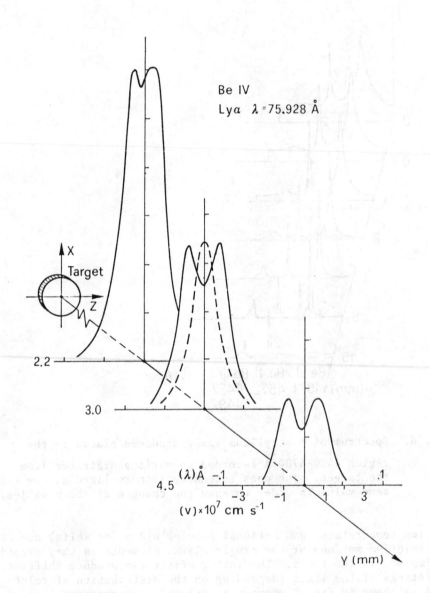

Fig. 9.   Profiles of Be IV Ly α λ = 75.93 Å for three distances
          y from the target (full lines). Corresponding values
          for expansion velocities by assuming that the lines are
          Doppler dominated are also shown.

thin condition is not fullfilled, motional Doppler shifts can
lead to asymmetries of the lines.

6.   EXAMPLES OF OBSERVATIONS FROM L.P.P.

In addition to the examples already shown it is interesting
to examine some more cases of observations of l.p.p. Fig. 10
shows a spectrum of a carbon l.p.p. recorded in two regions: the
expanding plume and the interaction region [20].   It is possible
to see the various components of the spectrum treated up to now:

a.   the recombination continuum at and beyond the limit of
CVI and CV resonance series whose slope is a function of the
electron temperature and whose intensity is proportional to n_e^2;

b.   the resonance lines whose broadening - see inset where
a comparison with a model spectrum is made - is a function of the
electron density;

c.   the satellite lines to both the CVI and CV first resonance lines; in particular the latter are more prominent in the
interaction region where the contribution due to the inner shell
excitation is very important for these lines. These Li-like
satellites are nearly absent in the expanding plasma.

Observations with finer spatial and spectral resolution of
the expanding plasma is reported for a beryllium l.p.p. in
Fig. 11. Here the Lyman series of BeIV and the recombination
continuum are shown for distances from the target from 0 to
1.4 mm. The various spectra refer as before to conditions of
decreasing density and temperature exemplified by the above
mentioned phenomena of the intensity of continua and widths of
the lines and by the slope of the continuum. In addition another
effect of broadening is evident: the merging of the high series
members at progressively higher density: the so-called Inglish-
Teller effect. In fact [9] the highest resolvable quantum number
n_s for ions of charge Z is given by

$$7.5 \log n_s = 23.26 - \log n_e + 4.5 \log z \qquad (11)$$

Obviously the Inglish-Teller effect can be used as a density
indicator. For considerably greater irradiation intensities and
higher densities we refer to the example shown in Fig. 12 [21].
Here the spectrum recorded in the soft-X-ray region of an imploded pusher-type microballon is shown. The glass microballon
is filled with two pressures of neon: 2 and 8.6 atmospheres.
Several interesting features are readily apparent. The neon
lines (H-like and He-like) are considerably more broadened than

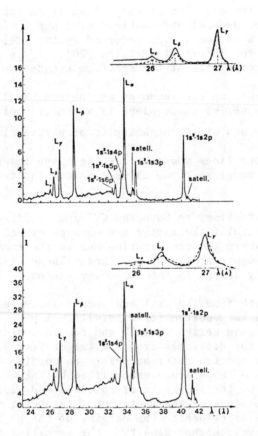

Fig. 10.   Spectrum of a carbon laser produced plasma in the re-
            gion 24-42 Å. Upper: recorded observing the expanding
            plasma side-on at 0.16 mm distance from the target.
            Lower: observing end-on inside the crater. In each of
            the inserts an enlargement of the spectrum near the
            series limit of C VI is shown. The dotted lines are a
            best fit theoretical spectrum corresponding to values
            of $n_e = 2.2 \cdot 10^{20}$ cm^{-3} (upper) and $1.2 \cdot 10^{21}$ cm^{-3}
            (lower).

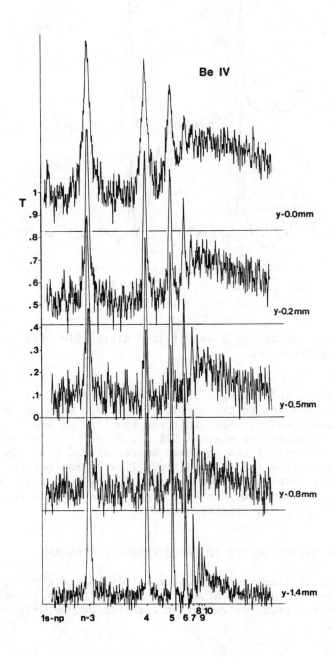

Fig. 11.   Spectrum of Be IV near the series limit recorded side-
on for various distances from the target.

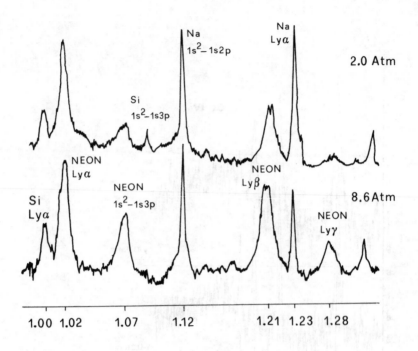

Fig. 12. Spectrum emitted by a neon filled microballon for two
          filling pressures.

the Na and Si lines, indicating a higher density of the compressed core with respect to the shell. Comparison between the
2 and 8.6 atmosphere cases, shows that the Lyα line of NeX is
almost constant in intensity whereas Lyβ increases by a factor of
two and Lyγ and the NeIX lines increase by factors of 3-4. This
is due to the saturation (at the blackbody limit) from optical
opacity that first sets in for Lyα and less for Lyβ and Lyγ. Indeed the influence of optical depth in l.p.p. is essential in in-
terpreting practically all spectra recorded.

7. THE ROLE OF OPTICAL DEPTH IN LASER PRODUCED PLASMAS

    Reabsorption of line radiation is a fairly common event in
l.p.p. due to the high density of the plasma itself. As is well
known [8,9], the equation for radiative transfer can be written
as:

$$\frac{dI(\nu,x)}{dx} = -\chi(\nu,x)I(\nu,x) + J(\nu,x) \tag{12}$$

where I (W cm^{-2} sr^{-1} Å^{-1}) is the intensity at frequency v and at the position x in the plasma, χ(cm^{-1}) and J (W cm^{-3} sr^{-1} Å^{-1}) are respectively the absorption and emission coefficients.

If one considers several simplifying assumptions like the fact that the populations of the states are not affected by the reabsorption process (as for example in LTE), and that the ratio $S(x) = I(v,x)/\chi(v,x)$, known as the source function S, is not a function of the frequency, the equation for a homogeneous plasma of depth D can be solved as:

$$I(v) = S(v)\left[1 - e^{-\tau(v)}\right] \tag{13}$$

where $\tau(v) = \int_D \chi(v,x)dx$ is the optical depth of the plasma at frequency v. It can easily be demonstrated that the highest value of optical depth appears on discrete transitions. If one considers an isolated line in a plasma:

$$\tau(v) = \int_o^D N_1(x)\ \phi(v,x)\ \frac{hv}{4}\ B_{\ell n}\ dx \tag{14}$$

where $N_\ell(x)$ is the number density of the atoms in the lower level ℓ of the transition, B is the Einstein coefficient of absorption and $\phi(v,x)$ is the normalized local profile of the line; $\phi(v,x)$ is a function of the intrinsic plasma broadening causes: it can be a convolution of several mechanisms, the most common being Stark and Doppler broadening (thermal and mass motion).

It is instructive to consider the variation of a spectral line shape as it emerges from a plasma for various values of the optical depth. Referring again to a very simple example of a homogeneous plasma whose optical depth increases (for instance by increasing the physical length D of the layer), the line appears as in Fig. 13a). For small values of τ as τ_1 the line is optically thin - given by Eq. (6) and its profile is given by the function $\phi(v)$. Increasing the optical depth, the line broadens progressively until it saturates at $I(v) = S(v)$, i.e. at the value of the source function. This can also be expressed in terms of the escape probability [22] of the photons emerging from the plasma with a mean free path given by $\ell = \chi(v)^{-1}$. The escape probability approaches unity in the far wing of a saturated line but is very small at the center.

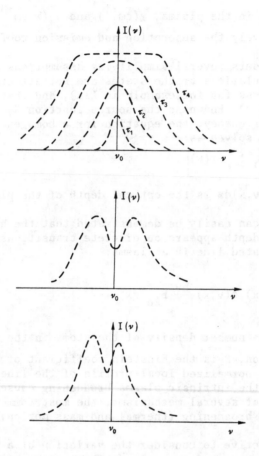

Fig. 13. a) Behavior of the profile of an isolated line for
various values of the optical depth; b) the profile for
a source function with a maximum near the center of the
plasma; c) the profile for a streaming plasma and a
source function as in b).

In most laser produced plasmas there are strong local spatial variations of the physical parameters and the profile of a
line can depart markedly from the one shown in Fig. 13a, depending on the integrated effect of the non-stationary population of
the upper and lower levels throughout the plasma. A quite common
situation is that the source function $S(v,x)$ is not any more
uniform in the plasma, but presents a maximum near the center.
This can happen if for example the temperature (and sometimes
also the density) is higher in the center of the plasma with
respect to the outside. (A very similar effect is the one shown

in a common spectral lamp). In this case the profile for large
values of the optical depth is shown in Fig. 13b, showing a self-
reversal of the line with a minimum of intensity at the central
frequency.

This is the consequence of the predominance of the reabsorption over emission mechanisms in the outer cooler layers of the
plasma.

If the plasma is also moving like in an expanding plasma
out of a plane target or a compressing spherical pellet, the profile of a thick line takes the shape shown in Fig. 13c (positive
v to the right for an expanding and to the left for a contracting
plasma).

Suppose we have a situation like the one shown in Fig. 14
for an expanding plasma whose various elements are streaming out
from a plane target in a cone-like fashion. For an observer
looking side-on the various elements along a line of sight have
velocities whose components (along the line of sight) are zero
for the central element and reach a maximum of $\pm v_x$ at either

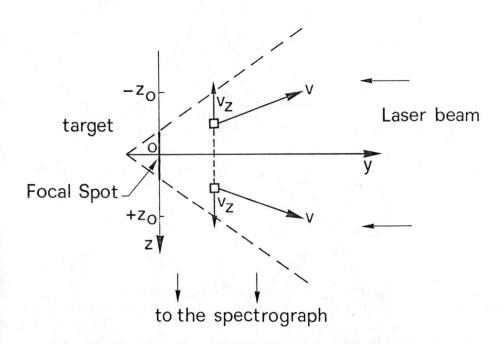

Fig. 14.   Schematic of an expanding laser produced plasma on a
          plane target.

boundary. If one assumes also that the source function has a
maximum at the center (i.e. for x=0) as if the plasma is, e.g.,
hotter in the center than at the boundary, then in terms of emission and absorption coefficients we have a situation like the one
shown in Fig. 15 [23]. It is clear how the red photons, i.e.,
the ones emitted by the plasma layer moving away from the observer, have a higher escape probability than the blue photons.
The opposite is true for a collapsing plasma. This situation
presents obvious similarities with the well known problem in
astrophysics of expanding or contracting stellar atmospheres.

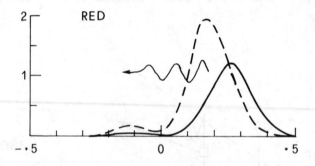

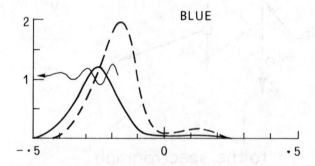

Fig. 15. The origin of the asymmetry: with dotted lines are
          shown the distribution of the emissivity coefficients;
          with solid lines the absorption coefficients inside the
          streaming plasma.

   An example of observations dealing with such profiles is
shown in Fig. 16 [24]. Here the Lyα line of BeIV is observed at
various distances from the target with space resolution. One can
clearly see the asymmetric self-reversed profile near the target
evolving to a single (almost thin type) profile far from the target. A model calculation taking into account various physical
parameters of the plasma and atomic data for the actual line,
predicts very well the profile of the line. Another example of
such asymmetry is the Lyα line of NeX observed in an imploding
plasma as shown in Fig. 17 [21]. Here the blue and red peak intensities are reversed. It happens that the asymmetry in the
profile is very sensitive to the value of the velocity of the
plasma elements and therefore constitutes a useful measure of
this quantity [24]. Note finally that since the optical depth
depends, through χ, on the number density of the lower level of
the line (i.e., for a resonance line on the ground state density
n_g) the fitting of a thick line profile can provide a direct

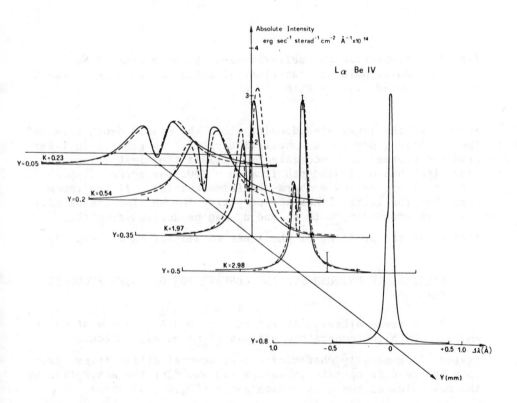

Fig. 16.   Profiles for the Be IV Lyα line at various distances y
           (in mm) from the target. Dashed curves: experimental
           profiles.

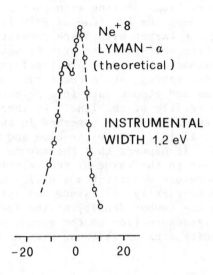

Fig. 17.   Asymmetric and self-reversed Lyman- α line of Ne^{9+}.
          Solid points: experimental data; dotted line: com-
          puted profile with τ = 86.

measure of the integrated density along the plasma depth (i.e. of
the quantity ∫ ρdr for a spherical target).   In general in laser
produced plasmas with moderate Z materials the first resonance
line (Lyα) has optical depth from 10 → 100, the second (Lyβ) ≅ 1,
and the higher series members τ<1 (almost thin).   If one remembers that the latter lines are sensitive to the electron density
via Stark broadening both n_g and n_e can be derived from the
fitting of the profiles of the lines to all series members [25].

8.   SPECTRAL TECHNIQUES FOR THE OBSERVATION OF LASER PRODUCED
     PLASMAS

     The extreme ultraviolet region λ < 1000 Å, where most of the
emission by high temperature plasmas (T_e > 30 eV) is concentrated, is generally characterized by several difficulties, among
them the need to operate in vacuum for avoiding the absorption by
the molecules of the most common gases (CO_2, N_2, O_2 etc.).

This spectral region can be naturally divided into three parts: [26]

a. The normal incidence region: $1000 > \lambda > 300$ Å

b. The grazing incidence region: $300 > \lambda > 10$ Å

c. The X-ray region: $30 > \lambda > 1$ Å

In region a) the reflectivity at near normal incidence of the usual optical components is good to low (80%) at 1000 Å to 10% at 300 Å) but still sufficient, provided the number of re-flecting surfaces in the whole optical system is kept small (2 or 1) and provided suitable reflecting layers (mainly of high Z materials) are used. A typical spectrograph in this region uses a concave grating employed in the Rowland configuration, i.e. with entrance slit, grating and detector lying on a circle of diameter equal to the radius of curvature of the grating like the scheme of Fig. 18. Due to the relatively low amount of aber-rations of a spherical surface used at small angles of incidence ($\alpha \lesssim 10°$), the aperture of such a spectrograph can be comparable to visible or near u.v. type instruments (f number = 10 or 20). However, spherical surfaces, when used far from normal incidence, suffer from astigmatism, i.e. the image of a point source like one on the entrance slit is not a point but consists of two images in form of two narrow lines perpendicular to each other, located according to the equation:

$$\frac{1}{p} + \frac{1}{p'} = \frac{2}{R\cos\alpha}$$

$$\frac{1}{p} + \frac{1}{p'} = \frac{2\cos\alpha}{R} \tag{15}$$

where p is the distance source-grating, p' and $\bar{p}'$ are the dis-tances from the grating of radius R to the first and second astigmatic images, respectively. In fact the Rowland circle is the locus of the first astigmatic image which is perpendicular to the plane containing the Rowland circle itself, for point sources on the same circle. It is then natural that for spectroscopic purposes the Rowland circle mounting is well suited, pro-viding sharp spectral lines even for long slits. Resolutions $\lambda/\Delta\lambda$ at 500 Å of 5000-10000 have been achieved (and even more for larger installations).

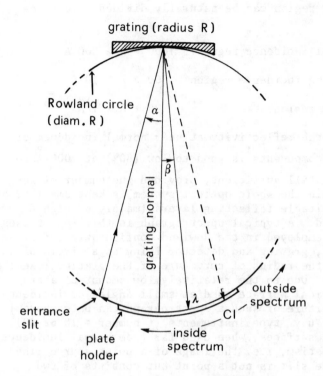

Fig. 18.   Optical layout of a basic normal incidence spectrograph.

However, such a system cannot provide space resolved images
of a source located - or whose image is projected - on the entrance slit.

A careful analysis shows [27] that stigmatic conditions can
be realized - with a toroidal grating, i.e. a grating ruled - or
recorded - on a toroidal surface having two radii of curvature,
one in the plane of the Rowland circle R_h and one smaller R_v in a
plane perpendicular to it.   Provided the relation

$$\frac{R_v}{R_h} = \cos\alpha \cdot \cos\beta \tag{17}$$

is fulfilled, the spectrograph will be stigmatic for point
sources on the entrance slit seen by the grating at an angle of
incidence α for wavelength λ corresponding to a diffraction angle
β.

Another approach is sometimes more convenient, for l.p.p.,
because it includes the necessary focussing of the source on the
entrance slit.   It involves the combined use of a toroidal mirror
and a conventional spectrograph with a spherical grating.

In this approach as shown in Fig. 19, the toroidal mirror
compensates the astigmatism of the spherical grating by producing
of a point source two astigmatic images in "complementary" positions with respect to the spherical grating, i.e. one at the
entrance slit and one some way ahead of it (Sirks position [26]).
In this way stigmatic images with resolution of the order of $\cong$ 20

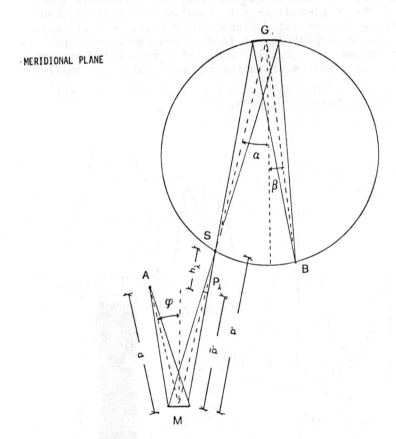

Fig. 19.   Schematic diagram of the use of a toroidal mirror M
           to correct for the astigmatism of a spherical grating
           G in the normal incidence mounting.

μm can be obtained as shown in Fig. 20. The figure shows the CIV
resonance lines 2s-2p recorded in an expanding laser produced
plasma looking side-on (as shown in Fig. 14). The structure of
the emitting plasma is clearly seen to vary from the center
through the edges, and the self-absorption on the lines is also
evident.

　　　At wavelengths shorter than 300 Å the reflectivity of all
materials at near normal incidence falls below any useful limit.
Recourse is here made to the reflecting properties of surfaces of
conductors when used at extreme angles of incidence. As known
from the electronic theory of metals [26], the refractive index
(real part of the complex dielectric constant) is smaller than
one (by a very small amount). Consequently total reflection can
be obtained in the propagation of an electromagnetic wave passing
from air (or vacuum) to a metal. Such total reflection happens
only for very small angles of incidence, especially for smaller
wavelengths (for λ = 10 Å and for a platinum layer the limit
angle for total reflection is 86°).

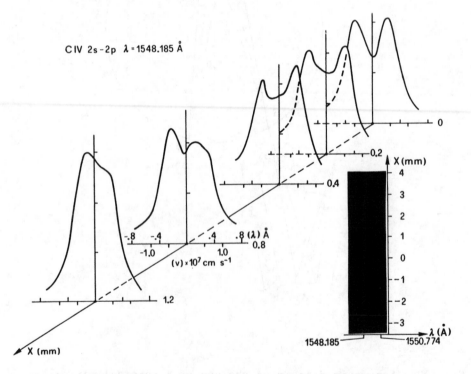

Fig. 20. The line 2s-2p of C IV in a laser produced plasma re-
           corded side-on in stigmatic conditions in the expanding
           plasma.

Spectrographs for the range $\lambda < 300$ Å use again as disperser the reflecting concave grating with all its focusing properties but with grazing angles of incidence ($\alpha > 80°$) in order to reach an acceptable degree of efficiency. A scheme of such spectrograph is shown in Fig. 21. Unfortunately there are important restricting consequences to this mounting. As apparent, due to the grazing geometry, all the positional tolerances of the various optical components must be very tight. In fact, a positional tolerance in normal direction of say 10 μm leads in a grazing incidence (88°) setting to a defocusing of $\cong 300$ μm which is quite noticeable. Another unpleasant consequence of working at grazing incidence is the fact that, as the aberrations of spherical reflecting surfaces increase dramatically when working very far from orthogonality, the aperture of the spectrograph must be kept very small: values of numerical aperture of 100 are very common in grazing incidence applications. This offsets the increase in efficiency due to the total reflection. Typical spectral resolutions obtained at $\cong 80$ Å are of the order of 3000 or smaller.

The third consequence is the extreme (total) astigmatism of the grating used in grazing incidence. The effect is qualitatively the same as explained for the normal incidence configuration of the spherical grating, but for angles $\alpha > 80°$ reaches extreme values. The image of a point source on the entrance slit of a grazing incidence spectrograph is practically a line, i.e., is brought by the grating to a focus only along the Rowland circle and is unfocused in a perpendicular direction. The grating in this case resembles a cylindrical lens with no focusing power in a perpendicular direction to the plane of the Rowland circle.

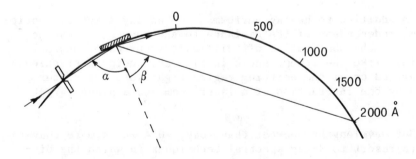

Fig. 21. A grazing incidence spectrograph.

The consequence is that no stigmatic images can be obtained
in grazing incidence and this fact cannot be corrected by the use
of a torodial grating because for the latter, due to the extremely small value of its minor radius R_v needed to correct the
astigmatism (e.g. R_h = 2000 mm, R_v = 20 mm) the aberrations are
very severe and degrade the spectral resolution as well.

For obtaining spatial resolution one can use a pinhole technique, i.e. by using an additional slit perpendicular to the main
one placed half way between the grating and the plasma. In this
way as shown in Fig. 22 a spatial resolution can be obtained in
one dimension but at the expense of reducing the aperture (and
the flux) in one direction to very small values. However, this
is a quite common technique in the grazing incidence region, and
most of the spectra presented before have been obtained in this
way. Another disadvantage of the method is the fact that the
plasma must be located very near the entrance slit for inter-
cepting it with the angular aperture of the field of the spectrograph and yet preserving a good spatial resolution.

A different approach for obtaining spatial resolution is
based on the same technique as the use of a toroidal mirror mentioned before but working at grazing incidence for compensating
the astigmatism of the grating. The scheme of such method is
shown in Fig. 23. The toroidal mirror produces of a point like
source as a l.p.p. two astigmatic images (like two narrow segments), one at the entrance slit of the spectrograph and another
at the position of the diffracted wavelength λ where stigmatic
conditions are required. A complete optical analysis of the system [28] shows that it is considerably more advantageous over the
pinhole technique in term of the related quantities: spatial
resolution - luminosity.

In addition to having performances, as any stigmatic optical
system, independent of the distance to the source, it can be
applied to the observation of "difficult" sources or unusual
positions like the scheme shown in Fig. 24, where observations
are carried on of the emitting region right into the crater
formed by the interaction of a laser beam and a plane target
[20].

For wavelengths shorter than, say, 20 $\overset{\text{o}}{\text{A}}$ it is more convenient to resort to X-ray spectral techniques in which the dis-
persing element is a crystal. Since this topic has already been
covered, only very few remarks are made here in connection with
observations of laser produced plasmas. Focusing or non-focusing
spectrographs have been employed, spatial resolution being obtained most often with a slit. Fig. 25 shows an example of

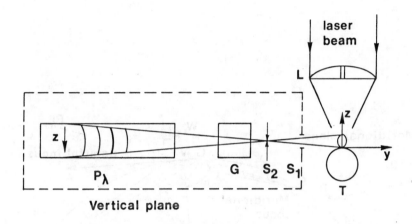

Vertical plane

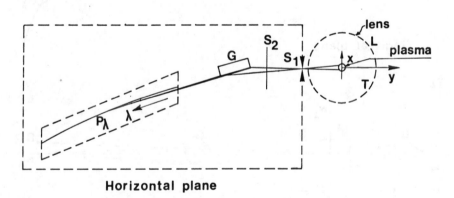

Horizontal plane

Fig. 22. Method to obtain spatial resolution. The plasma pro-
         duced on target T is observed along the z axis by a
         grazing incidence spectrograph. S_1 is the entrance
         slit of the spectrograph and the slit S_2 perpendicular
         to S_1 provides spatial resolution on the plate. G is
         the concave grating.

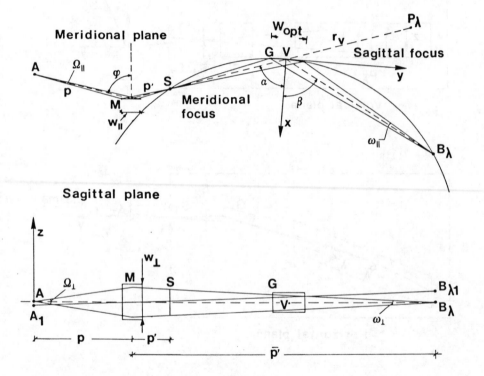

Fig. 23.   Schematic diagram of the use of a toroidal mirror M to
           correct for the astigmatism of a grating G. Upper
           part: view in the plane of dispersion of the grating-
           the meridional plane; lower part: view on the sagittal
           plane.

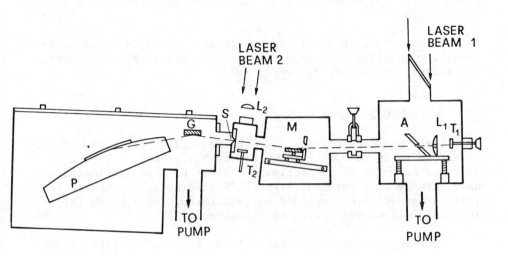

Fig. 24.   Scheme of an experiment for observing inside the interaction region in a laser produced plasma. The XUV
light is collected by the toroidal mirror M.

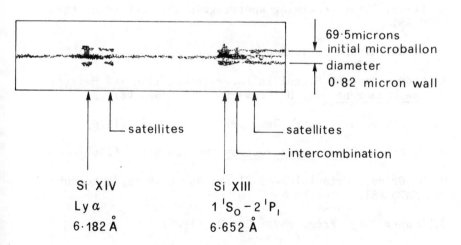

Fig. 25.   Space resolved X-ray spectrum from a laser irradiated
glass microballoon filled with neon.

observations of an imploding sphere taken with a crystal spectro-graph [29]. Values of spectral resolution $\lambda/\Delta\lambda$ are of the order of 500-1000.

Use of crystals as dispering elements can also be advanta-geous for coupling to streak cameras in order to obtain spectral resolutions as already described [3].

9. CONCLUSIONS

Laser produced plasmas, due to their high brightness, can be considered also as powerful sources of XUV radiation either in discrete emission or in continuum radiation. Few applications have already been pursued, like EXAFS studies of shocked matter [30], diagnostics of imploded targets through absorption [31], and ionic plasma absorption spectroscopy [32].

Some applications are under study, such as XUV lithography.

REFERENCES

1. M. H. Key and R. J. Hutcheon, Adv. Atom. Molec. Phys. 16, 201 (1980).

2. N. J. Peacock, "X-ray Spectral Dispersion Techniques", Lec-ture at the NATO ASI.

3. R. Price, "Time resolving spectrographs", Lecture at the NATO ASI.

4. P. K. Carrol and E. T. Kennedy, Cont. Phys. 22, 61 (1981).

5. K. Eidmann and R. Siegel in Laser Interaction and Related Plasma Phenomena eds. H. Schwarz and H. Hora Vol. 3.

6. M. H. Key et al., Phys. Rev. Letters 41, 1467 (1978).

7. H. R. Griem, Plasma Spectroscopy, McGraw-Hill, (1964).

8. H. R. Griem, "Principles of Plasma Spectroscopy", Lecture the NATO ASI.

9. J. Cooper, Rep. Prog. Phys. 29, 35 (1966).

10. P. Nicolosi, E. Jannitti and G. Tondello, Appl. Phys. B26, 117 (1981).

11. F. T. Stratton in Plasma Diagnostic Techniques, eds.
    R. H. Huddlestone and S. L. Leonard, Academic Press (1965).

12. V. A. Bhagavatula and B. Yaakobi, Opt. Commun. 24, 331
    (1978).

13. D. Jacoby, G. J. Pert, L. D. Shorrock and G. J. Tallents,
    J. Phys. B15, 3557 (1982).

14. A. H. Gabriel, Mon.Not. R. Astr. Soc. 160, 99 (1972).

15. J. Reader and G. Luther, Phys. Rev. Letters 45, 609 (1980).

16. H. R. Griem, Spectral Line Broadening by Plasmas, Academic
    Press (1974).

17. R. W. Lee, Unpublished Results (1981).

18. P. C. Kepple and H. R. Griem, Phys. Rev. A26, 484 (1982).

19. D. Santi, E. Jannitti, P. Nicolosi and G. Tondello, Nuovo
    Cimento B65, 198 (1981).

20. G. Tondello, E. Jannitti, P. Nicolosi and D. Santi, Opt.
    Commun. 32, 281 (1980).

21. B. Yaakobi et al, Phys. Rev. A19, 1247 (1979).

22. R. W. P. McWhirter in, Plasma Diagnostic Techniques eds.
    R. H. Huddlestone and S. L. Leonard, Academic Press (1965).

23. F. E. Irons, J. Phys. B8, 3044 (1975).

24. G. Tondello, E. Jannitti and A. M. Malvezzi, Phys. Rev.
    A16, 1705 (1977).

25. J. D. Kilkenny, R. W. Lee, M. H. Key and J. G. Lunney, Phys.
    Rev. A22, 2746 (1980).

26. J. A. R. Sampson, Techniques of Vacuum Ultraviolet Spec-
    troscopy, Wiley (1967).

27. M. C. E. Huber and G. Tondello, Appl. Opt. 18, 3948 (1979).

28. G. Tondello, Opt. Acta 26, 357 (1979).

29. R. G. Evans et al., Proc. 7th Int. Conf. on Plasma Physics
    and Controlled Fusion Research, Vol. III, p. 87 (1979).

30. P. J. Mallozzi, R. E. Schwerzel, H. M. Epstein and
     B. E. Campbell, Science 206, 353 (1979).

31. M. H. Key et al., J. Phys. B12, L 213 (1979).

32. E. Jannitti, P. Nicolosi and G. Tondello, Physica C, to be
     published (1983).

ACTIVE OPTICAL TECHNIQUES

LASER INDUCED FLUORESCENCE TECHNIQUES

S. J. Davis

Air Force Weapons Laboratory
Kirtland AFB, NM
USA

1. INTRODUCTION

Of all the applications that have arisen from the invention of the laser, laser induced fluorescence (LIF) has been one of the most fruitful. A relatively recent technique, LIF has proven itself to be an extremely valuable research tool. Many areas of atomic and molecular physics have utilized various forms of LIF spectroscopy. Radiative and collisional phenomena for a multitude of ions, atoms, and molecules have been successfully investigated [1]. Energy distributions and product species resulting from chemical reactions have been studied. It has found wide use in combustion studies. Medical and biological researchers are also utilizing LIF. LIF is exceedingly sensitive for species detection (fluorescence from a single ion has been seen.) Experiments previously thought to be unrealistically difficult are now routinely performed. Using LIF, extremely high spectral, temporal, and spatial resolution can be obtained. These properties are, of course, attributable to the laser itself and LIF takes full advantage of them. Most LIF techniques are non-intrusive and allow nearly ideal experiments to be performed. For example, an atom or molecule can be prepared in a selected pure quantum state with essentially a delta function pulse of radiation.

In the following we will cover some of the important features of several LIF techniques. The approach is tutorial and basic principles are emphasized over specific details. It is hoped that the reader will gain some appreciation of various LIF techniques and applications so that he or she may utilize them as diagnostics (or invent some new ones!). Of course, not all areas and applications can be covered in this brief presentation.

Nevertheless, most LIF experiments will have features in common
with those presented here.

2.   THEORETICAL CONSIDERATIONS

The theory behind LIF is relatively straightforward and many
reviews of the subject have appeared in the literature. In what
we develop below the mathematics are minimized in order to empha-
size the salient features of LIF as a diagnostic.

Laser induced fluorescence is closely related to the perhaps
more familiar technique of absorption spectroscopy. In absorption spectroscopy a beam of light of frequency ν is passed
through a cell containing N absorbers/cc as shown in Fig. 1.

For simplicity we assume that the absorbers in the cell have
only two energy levels E_1 and E_2. If the frequency of the radiation matches the difference between two energy levels of the particular species in the cell, i.e., $h\nu = E_2 - E_1$ then absorption
will occur and the transmitted intensity $I_\nu(T)$ is described by
Beer's law:

$$I_\nu(T) = I_\nu(0)e^{-N\sigma_\nu L} \tag{1}$$

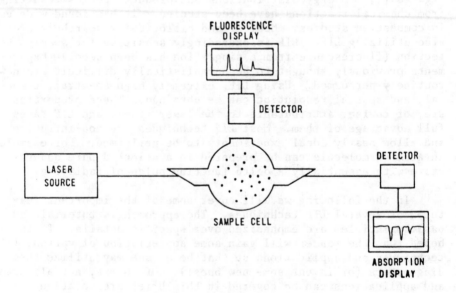

Fig. 1.   The typical laser absorption/fluorescence experiment.

where I_v = power/(cm^2 - frequency interval), σ_v is the optical absorption cross section and L is the cell length. A common convention is to define an absorption coefficient $k_v = N \sigma_v$. By measuring the absorption, the number density of the absorbers can be determined from equation (1). (Note that N is actually the number density of the absorbers in level 1. For our simplistic case we assume that negligible population resides in level 2.) When the absorption takes place the atoms are excited to level 2 from which they can emit spontaneously. This emission is the source of LIF.

A rate equation analysis of absorption and fluorescence is useful for gaining insight into the process. For simplicity in our analysis we assume a fictitious two level system as indicated in Fig. 2.

The rate equation for N_2 is given by Eq. (2).

$$\frac{dN_2}{dt} = (I/c)B_{12}N_1 - (I/c)B_{21}N_2 - N_2A_{21} - N_2Q \qquad (2)$$

where A_{21} = radiative decay rate, B_{12} and B_{21} are the Einstein "B" coefficients for absorption and emission, and Q is the non-radiative collisional quenching rate. The quenching can be caused by a bath gas, M, or by the absorbing species itself.

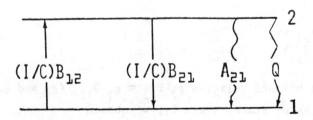

Fig. 2.   Processes of importance in laser induced fluorescence.

In steady state

$$N_2 = \frac{(I/c)B_{12}N_1}{(I/c)B_{21} + A_{21} + Q} \tag{3}$$

or, since $N_T = N_1 + N_2$ we can write Eq. (3) in terms of N_T

$$N_2 = \frac{(I/c) B_{12}N_T}{(I/c)(B_{21} + B_{12}) + A_{21} + Q} \tag{4}$$

It is instructive to look at two limiting cases of the laser intensity I. In the low power limit, i.e., $(I/c)(B_{21} + B_{12})$ $\ll A_{21} + Q$, Eq. (4) reduces to

$$N_2 = \frac{(I/c)B_{12}N_T}{A_{21} + Q} \tag{5}$$

The fluorescence power density, P_F, is given by:

$$P_F = N_2 A_{21} h\nu = \frac{(I/c)B_{12} N_T A_{21} \; h\nu}{A_{21} + Q} \tag{6}$$

Although in principle a determination of P_F can be used to find the number density N_T, knowledge of the quenching rate is required, we see that P_F depends upon the quenching term Q. As described later the quenching coefficient can be measured but its presence in Eq. (6) can be troublesome when trying to obtain absolute number densities of the absorber.

This difficulty can be overcome by increasing the laser power to the limit $(1/c)(B_{12} + B_{21}) \gg A_{21} + Q$. Eq. (4) then reduces to

$$P_F = \frac{B_{12} N_T A_{21} \; h\nu}{(B_{21} + B_{12})} \tag{7}$$

Using the detailed balance relation $g_2 B_{21} = g_1 B_{12}$, (g_1 and g_2 are the degeneracies of states 1 and 2), we obtain

$$P_F = \frac{N_T A_{21} \; h\nu}{(1 + g_1/g_2)} \tag{8}$$

We note that P_F is now independent of both quenching and laser power. This is because, at high intensity, absorption and stimulated emission dominate the other kinetic processes. The laser holds N_2/N_1 at a constant ratio, e.g., if $g_1 = g_2$ the steady state population of N_2 is equal to $N_T/2$ or N_1. This phenomenon has led to a very active area of research called laser saturation spectroscopy.

Since this Institute is concerned with short time diagnostics we will now examine the time dependence of LIF signals. If we assume that the laser output is pulsed with a square wave output of intensity I and duration Υ then the population of level 2 at the moment the laser shuts off is:

$$N_2(t=\Upsilon) = \frac{(I/c)B_{12}\ N_T}{(I/c)(B_{21}+B_{12})+A_{21}+Q}\ (1-\exp\{-[(I/c)(B_{21}+B_{12})$$

$$+ A_{21} + Q]\}\Upsilon) \tag{9}$$

For short, non-saturating pulses, i.e., $((I/c)(B_{21} + B_{12}) + A_{21} + Q)\Upsilon \ll 1$, Eq. (9) becomes

$$N_2(t = \Upsilon) = (I/c)B_{12}\ N_T\Upsilon \tag{10}$$

and the fluorescence power density is:

$$P_F(t = \Upsilon) = (I/c)B_{12}\ N_T\Upsilon\ A_{21}\ h\nu \tag{11}$$

By using the relationship $B_{12} = A_{21}\ c^3/8\pi h\nu^3$ and putting in the time dependence of N_2 for $t > \Upsilon$ we find

$$P_F(t) = \frac{(I/c)\ c^3\ N_T\Upsilon\ A_{21}2\ h\nu}{8\pi h\nu^3}\ (\exp[-(A_{21} + Q)]t) \tag{12}$$

Two important features of pulsed LIF are evident from Eq. (12). First, the fluorescence is strongest just as the laser shuts off and decays at a rate $(A_{21} + Q)$. This phenomenon is utilized in determining radiative lifetimes and quenching rates as described later. Secondly, the fluorescence intensity is proportional to A_{21}^2. This clearly demonstrates the process of absorption and subsequent emission. It also shows that LIF signals for non-allowed (small A_{21}) transitions can be very weak.

One critical feature that we have neglected is the effect of linewidths on the LIF process. The coupling of radiation to an absorber is highly dependent upon the ratio of the source linewidth to the absorber's linewidth. The theory of line broadening is well established and beyond the scope or intent of this manuscript. An old but excellent treatment of this subject is given by Mitchell and Zemansky [1].

For many situations where total pressures in the fluorescence region are < 10 torr and where uv or visible wavelength lasers are used, the lineshape of the absorbing atom or molecule is adequately described by a Doppler profile. For infrared transitions or for atmospheric pressures collisional (Lorentz) broadening becomes important. For intermediate cases the linewidth is described by a Voigt function which is a convolution of Doppler and Lorentz lineshapes. Natural (lifetime) broadening is almost always negligible for all but the shortest lifetimes (<10 nsec).

For Doppler-broadened lines the absorption coefficient is given by [1]:

$$k_v = k_o \, e^{-(\omega^2)} \tag{13}$$

Where k_o is the absorption coefficient at line center and

$$\omega = 2(\ln 2)^{1/2}(v - v_o)/\Delta v_D \tag{14}$$

The Doppler width at temperature T for an emitter of mass M is:

$$\Delta v_D = \frac{2v_o(2R\ln 2)^{1/2}(T/M)^{1/2}}{c} \tag{15}$$

We observe from Eq. 13 that the absorption will be maximized for small ω or $(v - v_o) \leq \Delta v_D$. The reason that narrow band tunable lasers are so efficient at exciting atoms and molecules is that linewidths less than the Doppler width of the absorber can readily be produced and tuned to line center as described in the next section.

The above treatment of a two level system can be extended readily to multilevel systems such as molecules. Even with a multilevel absorber, the laser usually interacts with only two levels at any one time. Thus much of the two level treatment can be used.

3.   EQUIPMENT

Practically any experiment involving laser induced fluorescence one can envision will consist of three major components:

   a.   the laser source,

   b.   the sample chamber, and

   c.   the detection apparatus.

3.1 Laser Sources

Although some laser fluorescence experiments were possible
using fixed frequency discrete line lasers such as argon ion or
helium neon, the field really began to develop with the advent of
wavelength tunable dye lasers [2]. With dye lasers it was no
longer necessary to study molecules and atoms that had absorption
coincidences with fixed frequency lasers. Since the invention of
dye lasers the number of reports of laser induced fluorescence
has risen dramatically. Because of the influence dye lasers have
had in this explosion of research the salient features of these
devices are described below.

Dye lasers can be divided into two classes: a) flashlamp
pumped, and b) laser pumped. The first dye lasers were excited
using high intensity flashlamps and today several commercial companies offer a series of these laser systems. These devices are
inherently simple and offer the highest energy outputs. Energies
of >3 joules in a < 1.0 μsec pulse are available. They are very
useful for pumping weak transitions.

There are several pulsed lasers that have been successfully
utilized to pump dye lasers, the most important of which are:
a) 337 nm N_2 lasers, b) frequency doubled (532 nm) and tripled

(355 nm) Nd - YAG lasers, and c) excimer lasers. Typically the
laser pumped dye laser oscillator cavity is configured similar to
a design described by Hansch [3] and illustrated in Fig. 3.

Wavelength tuning in lasers such as illustrated in Fig. 3 is
obtained by using a diffraction grating as the rear reflector of
the cavity. The bandwidth is reduced further by inserting a beam
expander between the dye cell and the grating, thus filling more
grooves on the grating. To further narrow the linewidth an in-
tracavity etalon is often inserted in order to allow only one
longitudinal mode to oscillate. Linewidths of 10^{-2}Å can be

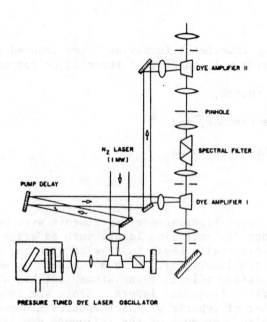

Fig. 3. Laser pumped dye laser of the Hansch design.

produced using this type of resonator. The output of the dye
oscillator can also be sent through an amplifier cell that is
pumped by the excitation laser as illustrated in Fig. 3.

The choice of pump laser depends largely on the requirements
of the particular experiment. Nitrogen laser pumped systems produce relatively low energy outputs (~ 10 μ J/pulse), but nitrogen
lasers are relatively simple devices. Both Nd/YAG and excimer
pumped systems produce much higher energies (~ 10 mJ/pulse)
largely because both these pump sources produce orders of magnitude more energy per pulse than does the nitrogen laser. All
three systems produce pulses of less than 10 nsec FWHM at repetition rates exceeding 20 Hz. Outputs ranging from 400-800 nm can
be obtained from all three systems by direct dye pumping, and
this range can be extended considerably on both ends by using
non-linear doubling and mixing techniques.

3.2 Sample Cell

The sample cell or reaction chamber in which the laser interacts with the fluorescing species can vary in complexity from
a glass cell to a crossed molecular beam. Some LIF experiments,
especially in combustion applications, are done in open air

flames. LIF can be utilized almost anywhere one can direct a
laser beam.

Most sample cells have the following characteristics: They
are constructed to be compatible with the environment of the experiment; they minimize scattered laser light; and they contain
viewing or collection optics. In our laboratory in which corro-
sive chemicals such as the halogens are used we construct stainless steel cells that can be used either as sealed cells or as
chemical reactor chambers. Light scattering which can severely
interfere with the LIF signal is minimized by using several conical baffles in the side arms where the laser enters and exits the
cell. Typically the laser is focused in the center of the cell
to increase the intensity and thus enhance the LIF signal.

A low f-number aspheric lens with its focus at the center of
the cell where the laser beam passes is satisfactory for collimating the fluorescence emission. Further lenses of selected f
numbers can be added to focus the emission onto the particular
detector being used.

The fluorescence power, S_F, actually reaching a detector is
related to the fluorescence power density, P_F.

$$S_F = \frac{P_F \Omega_L \eta V}{4 \pi} \qquad\qquad (16)$$

where V is the volume of the fluorescing region, Ω_L is the solid
angle of the collection lens, and η is the efficiency of the entire optical focusing system. We see at once why low f number
fast collection lenses are important. A schematic of a typical
reaction cell used in our laboratory is shown in Fig. 4.

3.3 Detection Systems and Signal Processors

The detection system used for most LIF experiments consists
of a light detector and a signal processor. For many applications a sensitive photomultiplier will suffice as a detector.
The GaAs tubes such as the RCA C31034 series are very useful because their spectral response is nearly flat over the spectral
region of 200 - 880 nm. Depending upon signal strength a wide-
band amplifier may be required to buffer the PMT from the signal
processor.

The choice of processor depends upon the signal strength and
should be given careful consideration. If the single pulse
signals are of sufficient strength then one can simply use a fast

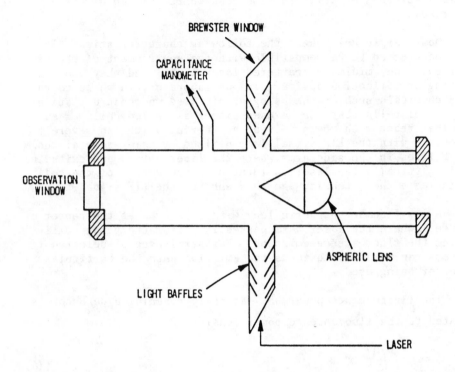

Fig. 4. Typical sample cell for LIF.

oscilloscope and a camera to record single events. If the signals are not strong enough for oscillographs then a form of signal averaging is required. One commonly used instrument is a
boxcar integrator. The boxcar derives its name from the fact
that it examines a small boxcar-shaped time increment or aperture
whose width is variable. One usually chooses the boxcar time
aperture to be much smaller than the period of the waveform being
analyzed. The position of the boxcar aperture with respect to
the beginning of the waveform is also variable. By utilizing an
integrating circuit of variable time constant, the portion of the
repetitive waveform under the boxcar increment is averaged with
the time constant selected, and a D.C. output voltage is obtained. By scanning the boxcar aperture over the entire waveform
with a ramp voltage the complete waveform can be averaged. Conversely, holding the aperture fixed at one position allows one to
average a single portion of the waveform. This is especially
useful in obtaining laser excitation spectra. The timing sequences of importance in a boxcar integrator are illustrated in
Fig. 5.

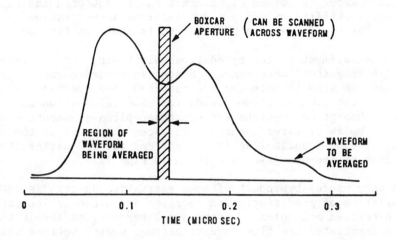

Fig. 5. Relation of boxcar aperture to the waveform being averaged.

Although boxcars can be used to analyze and average repetitive signals as short as a few nanoseconds, they do not effectively utilize the available input signal. Their effective duty factor is low because they only sample a small fraction of the waveform at any one time, and this can lead to times of several minutes to scan a waveform. In addition, long term drift in the laser power can adversely affect results obtained using a boxcar.

An alternative type of averaging utilizes a transient digitizer in combination with a signal averager. The transient digitizer divides the entire waveform into a large number of segments (typically 1024). Each segment is stored in a bin as a digital signal. Each time the laser fires the entire signal is added to the previously stored data. The averaging may have to be done in a microcomputer depending upon the type of digitizer being used. Transient digitizers are very efficient because each time the experiment is repeated the entire waveform is utilized in the averaging.

4.   APPLICATIONS OF LIF

4.1 Spectral Identification of Species

One of the most powerful uses of LIF is in species identification. The presence of a particular atom or molecule in the

reaction chamber or volume of interest can be detected and con-
clusively identified by using a dye laser as an excitation
source. The typical apparatus required is shown in Fig. 6.

The experiment is run by continuously tuning the dye laser
and monitoring the fluorescence. Since fluorescence can occur
only when the atom or molecule has absorbed some dye laser radiation the excitation spectrum yields the same information as a
detailed absorption spectrum and spectroscopic assignments can be
made. Examples of laser induced fluorescence spectra in the
molecule iodine monofluoride (B ← X system) for two different
bandwidths of the dye laser are shown in Fig. 7.

A very useful byproduct of some excitation spectra is the
determination of predissociation energies and dissociation energies in certain molecules. As the dye laser excites levels that
are predissociated the fluorescence becomes weaker because most

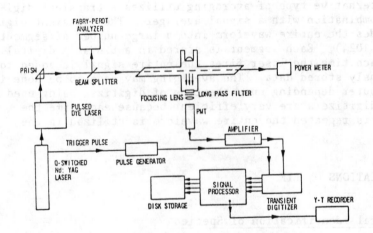

Fig. 6. Typical experimental setup for performing LIF experi-
         ments.

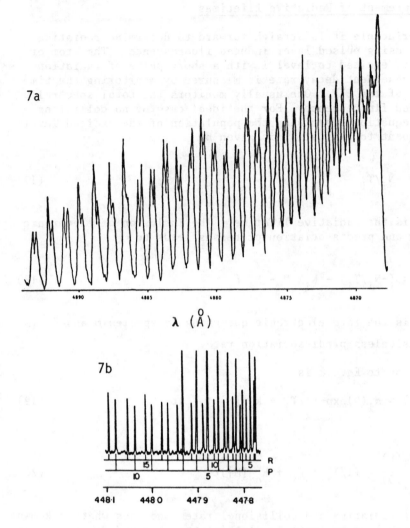

Fig. 7.   a) Excitation spectrum of IF (B← X) on the (3,0) band,
          ↓ Laser linewidth = 0.1Å. b) ↓ Excitation of (9,0)
          band, ↓ Laser linewidth = 0.01Å (from Ref [4]). In
          both 7a and 7b the laser pulse was 10 nsec long.

molecules dissociate before they can emit. Detailed work by
Clyne and co-workers [4] have produced precise measurements of
dissociation energies of several interhalogen molecules using
this approach. For example note that in Fig. 7b the decrease in
fluorescence intensity for J ≥ 7. All levels above this are
rapidly predissociated.

4.2 Measurement of Radiative Lifetimes

In principle it is straightforward to determine radiative lifetimes using pulsed laser induced fluorescence. The atom or molecule is excited to level i with a short pulse of radiation and the subsequent decay rate is measured by monitoring the time evolution of the LIF. One usually monitors the total spectrally undispersed fluorescence. For the ideal case of no collisions the rate equation describing the population of the excited level N_i subsequent to excitation is given by

$$\frac{N_i}{dt} = -N_i / \Upsilon_i \tag{17}$$

where Υ_i is the radiative lifetime. The inclusion of electronic quenching and predissociation processes leads to

$$\frac{dN_i}{dt} = (-N_i / \Upsilon_i) - N_i k_q M - N_i \Gamma_i \tag{18}$$

where k_q is the rate electronic quenching by species M and Γ_i is the collisionless predissociation rate.

The solution to Eq. 18 is

$$N_i(t) = N_i(0) \exp -(1/\Upsilon_i + k_q [M] + \Gamma_i) t \tag{19}$$

or

$$\ln \frac{N_i(t)}{N_i(0)} = (1/\Upsilon_i + \Gamma_i + k_q [M]) t \tag{20}$$

To extract radiative and collisional rates one uses what is known as a Stern-Volmer analysis by plotting the observed decay rate as a function of [M]. The slope yields k_q and the intercept gives $(1/\Upsilon_i + \Gamma_i)$.

Determination of the radiative lifetime of an electronic state is extremely important since fundamental properties such as the transition moment can be extracted. For diatomic molecules the lifetime is related to the transition moment | Re | by

$$1/\Upsilon = (64 \pi^4 / 3h) \Sigma_{v''} \quad Re^2 \quad q_{v'v''}^3 \tag{21}$$

where $q_{v' v''}$ is the Franck Condon Factor,

v is the frequency of the v' → v" transition

v' is the vibrational quantum number for the upper state

and

v" is the vibrational quantum number for the lower state.

For atoms oscillator strengths are determined in a similar manner. Quantities such as $|Re|$ and the oscillator strength are of theoretical interest because they serve as sensitive tests of calculated wavefunctions.

4.3   Energy Transfer Processes

An important area of study facilitated by LIF is the measurement of inelastic collisional processes such as vibration to translation (V-T) transfer. The experimental setup is similar to that used for measuring radiative lifetimes and quenching rates with the addition of a dispersing element to spectrally analyze the fluorescence. The dispersing element can be as simple as an interference filter or as complex as a monochromator.

The dye laser prepares the atom or molecule in a chosen excited v' level of an electronic state and the population of an adjacent v' level is monitored as a function of time. Because the fluorescence is dispersed, the signal strength is drastically reduced from the case in which total fluorescence is observed. Consequently signal averaging techniques become absolutely necessary. In Fig. 8 we illustrate V-T transfer in IF $B(^3\pi(0^+))$ with helium as the collision partner [5]. The dye laser populated the third vibrational level in IF(B) (V = 3) with an 8 nsec pulse and the time evolutions of V' = 4, 2, 1, and 0 are shown as an example.

This technique is extremely valuable because it allows one to study not only decay of the initially pumped level but also the appearance of population in adjacent levels caused by collisional transfer. Inelastic cross sections for a variety of collisional processes can be determined using the technique of temporally and spectrally resolved fluorescence.

4.4   Flowfield Visualization

One relatively recent and fruitful application of LIF is in the field of flow visualization which has proven to be an extremely valuable diagnostic for studying both mixing and combustion processes. LIF has been successfully used to detect

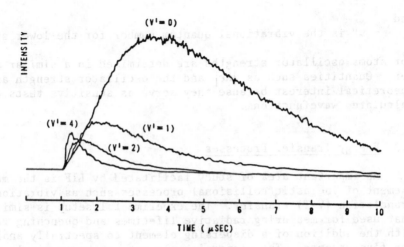

Fig. 8. Time evolutions of v' = 4, 2, 1, and 0 subsequent to
         v' = 3 being pumped by dye laser. Helium pressure was
         11 torr.

combustion reactants and products and to both temporally and
spatially resolve their concentrations. Excellent reviews of LIF
for uses in combustion research have appeared in the literature.

Flow visualization has provided fluid dynamicists with an
extremely powerful non-intrusive diagnostic. The degree of mixing of gas flows is critical to understanding and improving the
performance of devices such as chemical lasers and internal
combustion engines. Aerodynamic phenomena in both subsonic and
supersonic flows can also be studied using flow visualization.
The basic idea behind flow visualization is to seed a flow with a
molecule or atom and then pump the seed species with a laser
causing it to fluoresce. As an example of flow visualization we
describe some experiments on a supersonic mixing flowfield performed in our laboratory [6]. The apparatus is shown in Fig. 9.

Molecular iodine was seeded into the secondary flow of a
supersonic nozzle and the degree of mixing of the secondary and
primary flows was determined by causing the iodine to fluoresce
using an argon ion laser. The laser was focused to a line < 1 mm
in diameter. The fluorescence was intense enough to use photography as a diagnostic detector. One set of photos of the fluorescence is shown in Fig. 10. Only in the region where the I_2
has penetrated is fluorescence observed. For the data shown in
Fig. 10 the laser excited the I_2 at the exit plane of the nozzle.

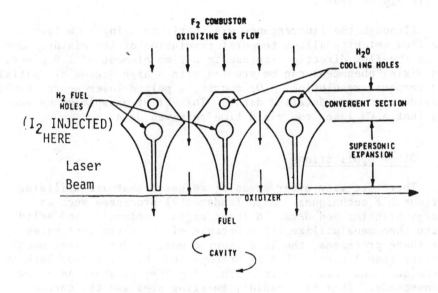

Fig. 9.   Detailed description of the CL-II nozzle: H_2 slit
width = .002 in.; F_2 throat width = .008 in.; F_2 exit
width = .151 in.; F_2 - F_2 nozzle width = .191 in.

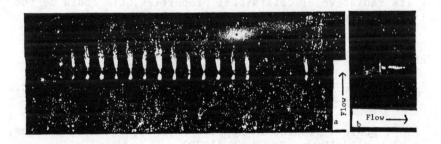

Fig. 10.   Visible fluorescence from I_2(B - X): a) viewed perpendicular to the optical axis; b) viewed along the optical axis.

The I_2 excited state lives ~10 μsec and the flow velocity is
~ 1 x 10^5 cm/sec, thus mixing downstream of the excitation point
is clearly observed.

Although the fluorescence was initiated using a CW laser,
the flow velocity allows temporal resolution of the mixing. One
mm in the flow direction represents a time element of 1.0 μ sec,
and mixing phenomena can be studied with a high degree of spatial
and temporal resolution. (Of course, a pulsed laser source would
provide even more detailed data.) This brief example shows one
way that a CW laser can yield time resolved data.

4.5 Other Applications

There are many other areas of research that are utilizing
various LIF techniques. Some fundamental processes such as
energy transfer and decay in large organic molecules and solid
state phenomena utilize LIF. Because of the ultra fast rates
for these processes, the laser source must produce pulses much
shorter than 1 nsec. This is accomplished by using mode locking
techniques that can produce trains of pulses as short as a few
picoseconds. This is a rapidly emerging area and the current
literature is rich with exciting reports.

5. SUMMARY

We have made a brief and incomplete survey of LIF and its
application to various areas of research. Because of its non-
intrusive nature and its relative simplicity, it has found many
uses and undoubtedly many more are yet to be discovered. When
one needs to search for the presence of a particular species or
study its radiative or collisional properties, LIF could be the
ideal diagnostic.

REFERENCES

1. Allan C. G. Mitchell and Mark W. Zemansky, "Resonance Radia-
    tion and Excited Atoms," Cambridge Press (1971).

2. F. P. Shafer, "Topics in Applied Physics, Vol 1," Springer,
    Berlin Heidelberg, New York (1973).

3. T. W. Hansch, Appl. Optics 11 895 (1972).

4. M. A. A. Clyne and J. S. McDermid, "Dynamics of the Excited State," John Wiley and Sons Ltd. (1982).

5. P. J. Wolf, R. F. Shea, and S. J. Davis (to be published).

6. N. L. Rapagnani and S. J. Davis, AIAA Journal 18 140 (1979).

MULTIPHOTON TECHNIQUES FOR THE DETECTION OF ATOMS

William K. Bischel

SRI International
Menlo Park, CA
USA

1. INTRODUCTION

Multiphoton processes have been studied since the pulsed
ruby laser was first invented in the early 1960's. However, it
is only recently that the development of lasers has progressed to
the point that quantitative multiphoton experiments can be made
with the confidence that the results can be reproduced in different laboratories. In particular, high-power dye lasers pumped
by either YAG or excimer lasers have opened the door to new techniques for the detection of atoms and molecules. Thus it has
only been in the last five years that multiphoton processes have
been seriously considered as a sensitive and quantitative detection technique for atoms and molecules. In that time we have
seen an explosion of new results, particularly in the area of
detecting atomic species.

The need for a sensitive detection technique for the atomic
radical species such as H, C, N, O, S, and F, particularily under
the nonideal experimental conditions found in discharge or combustion environments, has been well established for many years.
These atomic species have been unusually hard to detect since the
standard resonance fluorescence techniques require vuv wavelengths that do not propagate in the hostile environments.
Therefore, the use of multiphoton techniques (where the wavelengths required are usually in the near uv and therefore transmitted under most experimental conditions) can have a large
impact when applied to this detection application. I will thus
limit the discussion in this lecture to these atomic species,
using specific examples from our recent experiments on O atoms
[10, 9]. Application to molecular systems is straight forward

although the sensitivity limits derived for the detection technique will be at least two orders of magnitude lower than the
ones derived here for the atomic systems [27].

The study of multiphoton processes is a large field in which
new advances are being published at a rapid rate. Several reviews of multiphoton ionization processes, that are applicable to
the detection techniques discussed here, have recently been published [36; 21; 26]. In general, the transition rate to an excited state scales as I^n where I is the laser intensity and n is
the order of the process. Since the lowest order processes are
expected to give the largest excitation cross section, I will
restrict the discussion in this lecture to the case where n=2:
two-photon excitation.

There are two types of two-photon transitions that may be
useful for the detection of atoms. These are illustrated in
Fig. 1. The first is a <u>direct</u> transition that accesses final
states in the energy range of 65,000-100,000 cm^{-1}. The observation of probe beam absortion, or fluorescence and ionization from
the excited state, can then form the signal for this detection
technique. The last two detection techniques are call two-photon
excited fluorescence (TPEF) and two-photon resonant ionization
spectroscopy (TPRIS) [see, for example, 26]. The technique of
TPEF is generally useful for applications where a remote detection capability is required, since the fluorescence to be detected lies in the visible or near IR region of the spectrum.
This technique was first proposed for lidar detection of strato-
spheric 0 from a balloon platform [34]. However the first experimental demonstration of TPEF did not occur until 1981 [9].
There quickly followed proposals [45; 10] and demonstrations of
the techniques of TPEF and TPRIS for a number of the atomic radicals (see Table 1). We note here that all three of these techniques can determine absolute ground state number densities if
the two-photon absorption cross section, the excited state radiative lifetimes and collisional quenching rates, and excited state
photoionization cross sections are known. These quantities are
illustrated in Fig. 1.

The second type is a <u>Raman</u> transition where the final state
energies are on the order of 100-4000 cm^{-1}. Only atoms that have
their ground states split into its fine structure components can
be detected using this type of transition. Observation of gain
(or absorption) induced by the high power pump laser on a cw
probe laser would form the observed signal for this type of transition. This technique is called Stimulated Raman Gain Spectroscopy (SRGS) and has been applied to the spectroscopy of several
molecular systems [19]. We should note here that the technique
of coherent anti-Stokes Raman spectroscopy (CARS) and that of

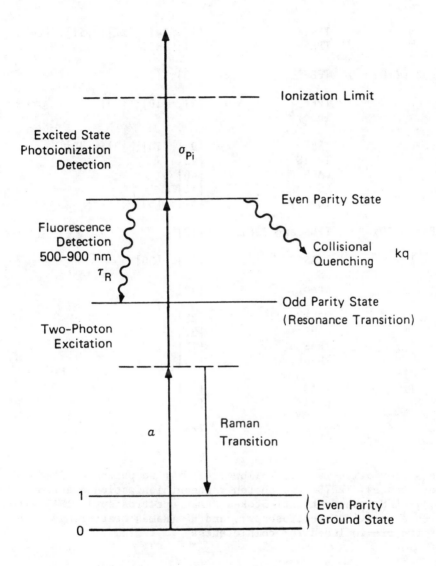

Fig. 1. Generalized level diagram for the detection of atoms using two-photon excitation.

TABLE 1
LASER DETECTION OF ATOMIC RADICAL SPECIES

ATOM	DETECTION METHOD*	REFERENCES
H	TPRIS TPEF	[1]; [4]; [22], [51], [24]; [7]; [32]
$C(^3P)$; $C(^1D)$	TPEF	[33]; [16]
N	TPEF Theory	[9]; [10]; [39]
O	TPEF TPRIS CARS RS Theory	[9]; [10]; [3]; [35] [23] [46] [14] [41]; [39]
$S(^3P)$; $S(^1D)$	TPEF and TPRIS	[12]
F	IRAS RS Theory	[44]; [30] [13] [42]
Cl	TPEF CARS IRAS Theory	[25] [37] [15]; [17] [39]
Br	IRAS CARS	[29] [43]
$I(^2P_{3/2}$; $^2P_{1/2})$	TPEF	[6]

*The abbreviations are as follows: TPEF (two-photon excited
fluorescence), TPRIS (two-photon resonant ionization spectro-
copy), CARS (coherent anti-Stokes Raman spectroscopy), IRAS (infrared absorption spectroscopy), and RS (Raman scattering). The
last two are included for completeness.

Infrared Absorption Spectroscopy (IRAS) [50 and references there-
in] could also be used to detect this Raman type transition. The
method of CARS is discussed later in this chapter. Examples of
atomic systems where the Raman type transition may be useful are
C, O, and F. Although the names of these two types of two-photon
transitions are different, and historically have been studied
separately, we shall see that the theoretical formalism is the
same. Thus they should be considered together when one is selecting a detection technique for a specific atomic system.

    Listed in Table 1 are the most interesting of the non-me-
talic atomic radical species and references to recent experimental and theoretical work relating to their detection using TPEF
and TPRIS. I have also included references to other optical detection techniques such as CARS, IRAS, and spontaneous Raman
Scattering (RS) for comparison and completeness.

    At this point, we should consider the question: when should
multiphoton processes be used as a detection technique? Since
the techniques of laser induced fluorescence (LIF) and spontaneous Raman scattering are relative simple to use and to analyze,
multiphoton excitation should only be used when these simpler
techniques are not applicable to the experimental system under
study. There are many circumstances when there is no other reasonable alternative, such as the detection of atomic radical species in hostile environments. For these applications, two-photon
excitation can be an extremely useful technique.

    Two-photon excitation (TPE) has several advantages when com-
pared to single photon methods such as LIF. First, direct TPE
can access energy levels that are two high for LIF. There are no
window tranmission problems, and the problem of propagating vuv
radiation is eliminated since the required lasers operate in the
wavelength region between 200-300 nm. Second, the Doppler width
can be minimized or eliminated for TPE processes thus allowing
more selectivity for the transition of interest. And thirdly,
the use of two lasers at different wavelengths and tight focal
volumes allows extremely good spatial resolution to be obtained
for cross beam geometries. The disadvantages must also be
weighted when considering TPE as a detection technique. These
include the fact that TPE has an inherently much lower sensitivity than LIF. In addition, absolute calibration of the atomic
number density depends on an accurate measurement of the laser
intensity in the focal volume. One final consideration is that
multiphoton absorption and dissociation by impurity molecules may
actually create the species one is trying to detect [35].

    With this brief introduction, we now move to a discussion of
the theory for two-photon excition processes in Section 2. Section 3 then describes our recent experiments on the detection of

0 atoms as an example of the experimental requirements for the
technique of TPEF. Section 4 then concludes with a comparison
of the sensitivity estimates for atom detection using TPE techniques in a low density environment.

2.   THEORY OF TWO-PHOTON EXCITATION

     Consider the system illustrated ·in Fig. 2. The generalized
two-photon cross section for the absorption (or gain) of a laser
with intensity I_2 with frequency v_2 induced by the presence of a
high intensity pump laser with intensity I_1 and frequency v_1 is
given by [BKR76 and references therein]

$$\alpha = \frac{\sigma}{I_1} = \frac{(2\pi)^3}{\hbar c^2} v_2 \left| P_{fg} \right|^2 g(v_1, v_2) \tag{1}$$

where P_{fg} is the second order matrix element given by

$$P_{fg} = \sum_i \mu_{fi} \, \mu_{ig} \left[\frac{1}{E_{ig} - hv_1} + \frac{1}{E_{if} + hv_1} \right] \tag{2}$$

and $E_{kn} = E_k - E_n$. Here the subscripts g and f denote the ground
and final states, respectively, of the two-photon transition.
Note that the energy demoninators in Eq. 2 can be rewritten to
reflect the type of two-photon transition (direct or Raman type)
by

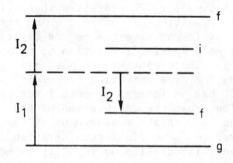

Fig. 2.   Level diagram for the notation used in the theory of
          two-photon excitation.

$$E_{if} + h\nu_1 = E_{ig} \pm h\nu_2 \tag{3}$$

where the − (+) indicates the direct (Raman) transition. In
Eq. 1 the term $g(\nu_1, \nu_2)$ is the lineshape function for the two-
photon transition, and the summation in Eq. 2 is taken over all
virtual states accessible by a one-photon transition from both
the g and f states. If we assume that there exists a single
intermediate state (still denoted by i) which forms the dominant
contribution to P_{fg}, assume a Doppler-limited lineshape function
(see below in Section 2.2), and include the effects of the finite
laser linewidth [MSA78], α in Eq. 1 can be written as

$$\alpha(J_f, J_g) = \frac{c\nu_2 \left|P_{fg}\right|^2}{\Delta E_{i\ell}^2 (\Delta\nu_D^2(\nu_1,\nu_2) + 2\Delta\nu_\ell^2)^{1/2}} \tag{4}$$

for a transition from a particular ground state angular momentum
component J_g to a particular excited state component J_f. Here
ΔE_{i1} is the difference in energy between that of the laser and
that of the intermediate state i, $\Delta\nu_D$ is the FWHM Doppler width,
and $\Delta\nu_\ell$ is the FWHM laser linewidth. In Eq. 4, if ΔE_{i1} is expressed in cm^{-1} and the matrix elements in debyes (10^{-18} esu), a
numerical value for the constant $C = 6.2 \times 10^{-25}$ yields α in
units of cm^4/W.

The selection rules for the two-photon transitions are determined by a consideration of the second order matrix element
P_{fg}. First, the states f and g must have the same parity with
$J_f - J_g = 0, \pm 1, \pm 2$ with transitions involving $0 \not\leftrightarrow 1$ being forbidden.
For additional selection rules we must examine the details of
calculating P_{fg} from Eq. 2. The matrix elements of P_{fg} can be
split into its radial and angular integrals:

$$\left|P_{fg}\right|^2 = \frac{1}{(\Delta E)^2} \left|< R_f^{nL}\|\mu\| R_i^{nL}> \; < R_i^{nL}\|\mu\| R_g^{nL}>\right|^2$$

$$\cdot \frac{1}{2J_g + 1} \sum_{M_g, M_f} \left|\sum_i < J_f \left|\cos\theta\right| J_i M_i> <> J_i M_i \left|\cos\theta\right| J_g M_g>\right|^2 \tag{5} \cdot$$

The angular matrix element here is expressed for linearly polarized light, whose polarization vector defines the axis of quantization. The radial integrals in Eq. 5 determine the absolute value of α, whereas the anglar integrals are responsible for the relative values of α among the fine structure components of the transition. We then expand the angular wave function in terms of their spin and orbital angular momentum components

$$\left| JM_J \right> = \sum_{M_L M_J} C_{M_L M_S M_J}^{LSJ} \left| LM_L \right> \left| SM_S \right> \qquad (6)$$

for each of the g, i, and f states. Since the laser is assumed to be linearly polarized, we obtain selection rules for P_{fg} of

$$\Delta M_J = 0$$

$$\Delta M_J = 0$$

$$\Delta M_J = 0 \; .$$

The relative intensities for the various fine-structure components of the two photon transition are then obtained by calculating the matrix elements using Eqs. 5 and 6, summing over the M_J values of states i and f and finally averaging over the M_J values of g. It is important to note at this point that the summation over all possible M_J components of the intermediate state i is inside the squared brackets in Eq. 5 while the summations over the M_J components of the f and g states are outside the squared brackets.

It should be emphasized that the theoretical analysis discussed above is for the <u>single</u> intermediate state approximation only. For many atomic systems the use of the resonance state as the intermediate state will give cross section values that are 50-90% of the total cross section calculated when all the intermediate states are included. Theoretical treatments that include all the relevant intermediate states can be found in references [41] [39] [47] [27].

2.1 Relative Fine-Structure Transition Probabilities

As an example of calculating the relative two-photon transition cross sections, let us consider the case of oxygen atoms. A partial energy level diagram is given in Fig. 3. The 0 atom

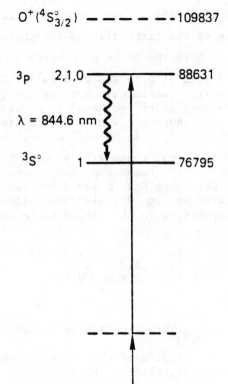

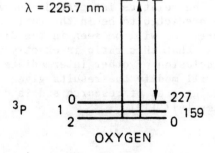

Fig. 3. Relevant energy levels for O atom detection using two-photon excitation processes.

possesses three ground state fine structure components $^3P_2$, $^3P_1$, and the $^3P_0$ at energies of 0, 158.5, and 226.5 cm^{-1}, respectively. The upper state of the first allowed two-photon transition is also a ^3P_J state where the three components are packed more closely with separations of 0.54 and 0.16 cm^{-1}. Consequently the laser can easily discriminate between the ground state J values but is in many cases too broad to resolve the upper state components. We therefore want to calculate the relative transition probabilities from Eq. 5 for each of the $J_g \rightarrow J_f$ transitions. We should note that these will also be the relative transition probabilities for the Raman type two-photon transition illustrated in Fig. 3. Combining Eqs. 5 and 6 for the specific case of 0 atoms illustrated in Fig. 3 yields the following formula for the relative intensities in the single state approximation [12].

$$\alpha(J_f J_g) \; \alpha \; (2J_g+1)^{-1} x \; \Sigma_{M_J} \left| \Sigma_{M_L} \; C^{11J}_{M_L M_S M_J} \; C^{11J_f}_{M_L M_S M_J} \right|^2 \tag{7}$$

where the coefficients $C^{11J}_{M_L M_S M_J}$ are the Clebsch-Gordon vector coupling coefficients. The results of this calculation are listed in Table 2. The actual relative intensities of the transitions are determined by a product of the transition intensity and the relative population of the lower level. The relative population of each of the fine-structure states are given in Table 2 at room temperature. In addition, since most laser linewidths are not narrow enough to resolve the upper state splitting, we need to sum over all the unresolved final ^3P_J states to obtain the experimental relative intensities. This is also shown in Table 2. The relative integrated experimental intensities are therefore predicted to be in the ratio of 1:0.28: 0.067 at room temperature. As will be seen in the discussion of the experimental results, that this ratio is exactly what we observe. Note that the inclusion of other intermediate states in the summation in Eq. 1 will modify the results given in Table 2. This question is an ongoing area of research and is discussed to some degree in references [41] [12].

TABLE 2
RELATIVE TWO-PHOTON INTENSITIES FOR OXYGEN

Ground $J_g \to J_f$	Intensity	Transition Temp.	Room State Population
$^3P_2$	$2 \to 2$	17/90	
	$2 \to 1$	1/10	0.742
	$2 \to 0$	4/90	
		1/3	
$^3P_1$	$1 \to 2$	1/6	
	$1 \to 1$	1/6	0.208
	$1 \to 0$	0	
		1/3	
$^3P_0$	$0 \to 0$	1/9	
	$0 \to 1$	0	0.050
	$0 \to 2$	2/9	
		1/3	

Intensity Ratio for Unresolved
Upper State 1:0.28:0.067

2.2   Absolute Two-Photon Absorption Cross Sections

As previously stated, absolute two-photon cross sections for atomic transitions of interest can be estimated if the single intermediate state approximation is made. This approximation is very reasonable for all the two-photon considered in this lecture since the intermediate state defined by the resonance transition has most of the oscillator strength. The two terms in α, that describing the radial wave functions (assumed to be constant for all fine structure states of the multiplet) and that describing the angular momentum factors described in the previous section, are written in Eq. 5. I estimate here absolute cross sections for the radial part of the matrix elements. The total cross section is then obtained by multiplying these results by the angular factors (see Table 2) which are on the order of 1/3.

We first must consider the lineshape factor in Eq. 1. There are two density regimes that experiments typically operate. The first is the low density limit where the two-photon lineshape is Doppler broadened. In this case, the peak value of the lineshape function is given by [8]

$$g(v_1,v_2)\Big|_{PEAK} = \frac{0.94}{\Delta v_D(v_1,v_2)} \qquad (8)$$

where $\Delta v_D(v_1,v_2)$ is the FWHM two-photon Doppler width. This Doppler width depends on the direction of propagation of the two laser beams and offers the possibility of significantly narrowing the Doppler lineshape. If the beams are copropagating (counter-propagating) for the direct (Raman) type of two-photon transition, the Doppler width can be calculated from the expression [8].

$$\Delta v_D = 2(v_1+v_2)(2\ell n2\ kT/M)^{1/2}$$
$$= 7.15 \times 10^{-7}\ (v_1 + v_2)\sqrt{T/M} \qquad (9)$$

where T is the temperature in degrees kelvin and the atomic mass M is in amu. However, if the beams are counterpropagating (a special circumstance not encountered in most experiments), the plus sign in Eq. 9 changes to a minus and the Doppler width decreases by a large factor. This effect occurs for both the direct and Raman type transition. It was first observed in 1975 [see 31 for references] and is called Doppler-free two-photon spectroscopy. In some cases, this effect can be very useful in the detection of atoms since the two-photon absorption cross section can be enhanced by several orders of magnitude (see [20] for example).

The second regime is the high density limit typical found in combustion experiments. For this situation, the lineshape is a collisionally broadened Lorentzian and the peak lineshape function is given by

$$g(v_1,v_2)\Big|_{PEAK} = \frac{2}{\pi\Delta v_L} \qquad (10)$$

where Δv_L is the FWHM linewidth. Note that in this case the linewidth is proportional to the density, a fact that is important to consider when estimating signal intensities.

We can now proceed to a calculation of the two-photon cross section assuming the low density limit with Dopper broadened

lines. We will also assume a single laser frequency $(v_1 = v_2)$
that is one half the two-photon transition energy. Using Eqs. 4
and 9 and the transition moment for the multiplets obtained from
Wiese et al [48], we can estimate the two-photon absorption cross
section for a number of atoms of interest. These calculations,
along with the relevant parameters for the two-photon transitions
considered, are given in Table 3. Note that all the listed

values of α are of the order of magnitude 10^{-27} cm^4/W. If similar calculations are carried out for molecules (see for example
H_2 in [27] [5]), the values for α are typically one to two orders

of magnitude smaller.

There are two theoretical calculations that include all the
relevant intermediate states available for comparison with these
estimates [41; 39]. Pindzola [41] has calculated for atomic
oxygen the cross section for the two-photon transition listed in
Table 3. Because he has used the natural linewidth (derived from

the radiative lifetime of $\tau=36$ nsec for the upper ^3P state) instead of the Doppler width in his formulation of $g(v)$ in his
version of Eq. 1, his value of the average cross section for the

multiplet transition $(\beta = 2.3 \times 10^{-43}$ $cm^4 sec)$ must be adjusted
for our line shape function. For consistency, we have used
Pindzola's line shape function (given in his Eq. 14) to obtain
his peak value for $g(v)$. The cross sections can now be compared
using the conversion equation

$$\alpha(cm^4/W) = \langle\beta_v^{HF}\rangle \frac{0.94\pi}{2\tau\Delta v_D \, hv_\ell} \tag{11}$$

This gives an ab initio value for the multiplet transition of
$\alpha=1.34 \times 10^{-27}$ cm^4/W in close agreement to our single-state estimate. It should be noted that the error bars on his calculation
are of the order ± 20%. Our value is also quite close to the
values of McIlrath et al [34]. This good agreement gives us confidence that the rest of the cross sections given in Table 3 have
a similar accuracy.

It is interesting to compare these cross sections for the
direct type two-photon transitions to cross sections calculated
for the Raman type transitions. If the polarized spontaneous
Raman scattering cross section is known $(\partial\sigma/\partial\Omega)$, then the second
order matrix element P_{fg} can be calculated from the expression

$$\left|P_{fg}\right|^2 = (\frac{c}{2\pi v_2})^4 (\frac{\partial\sigma}{\partial\Omega})_{\parallel} \quad . \tag{12}$$

TABLE. 3. ESTIMATES OF THE PARAMETERS FOR THE DETECTION OF SE-
LECTED ATOMS BY TWO-PHOTON EXCITATION.

		EXCITATION							FLUORESCENCE DETECTION	
Atom	Ground State (cm⁻¹)	Inter-mediate State (cm⁻¹)	Final State (cm⁻¹)	$\lvert\mu_{12}\rvert^2$ (D²)ᵇ	$\lvert\mu_{23}\rvert^2$ (D²)ᵇ	ΔE (cm⁻¹)	$\Delta\nu_D$ (cm⁻¹)	α(cm⁴/W) (× 10⁻²⁷)	Rad. Rate (× 10⁷ s⁻¹)	λ(nm) Obs'd.
H	^2S 0	$^2P^\circ$ 82258	^2D 97492	3.6	59	33511	1.2	0.5	6.6	656.4
C	$^3P^\circ_0$ 0	$^3P_1$ 60353	$^3P_2$ 71386	6.0	60	24701	0.25	5	2.5	908.8
N	$^4S^\circ$ 0	^4P 83366	$^4D^\circ$ 94883	1.08	104	35895	0.31	0.85	3.5	869.1
O	$^3P_2$ 0	$^3S^\circ_1$ 76795	$^3P_2$ 88631	2.6	54	32480	0.27	1.2	2.8	844.6
F	$^2P^\circ_{3/2}$ 0	$^2P_{3/2}$ 104732	$^2D^\circ_{5/2}$ 117624	3.7	52	45920	0.33	0.82	3.5	775.5

The differential scattering cross section has been measured or
estimated for several atomic systems. For the case of 0 atoms,
the polarized cross section for the $^3P_2 \to {}^3P_1$ Raman transition
at 158.265 cm^{-1} has been measured to be $(\partial\sigma/\partial\Omega)_\parallel = 6 \times 10^{-31}$
cm^2/sr at a pump wavelength of 532 nm [14]. Substituting this
value into Eq. 12, and assuming a density broadened Lorentzian
lineshape of 100 MHz gives $\alpha = 2 \times 10^{-29}$ cm^4/W, a value that is
approximately 6 times smaller than the corresponding direct two-
photon transition (combining values from Tables 2 and 3).

2.3 Signal Estimates for Two-Photon Excitation

All the potential detection techniques (absorption (gain),
fluorescence, and ionization) depend on a transfer of population
from the ground state to the excited state. In this section we
use a simple rate equation model to estimate the order of magnitude of the excited state population produced in the two-photon

excitation process. The excited state density N^* can be written as [5].

$$\frac{dN^*}{dt} = W_{fg}N_o - BN^* \tag{13}$$

where $W_{fg} = \alpha I^2/h\nu$ is the two-photon transition rate and B is the effective excited state lifetime equal to $B = A + Q + \sigma_{pi}I/h\nu_\ell$ (1/A is the radiative lifetime of the excited state, Q is the quenching frequency, and σ_{pi} is the excited state photoionization cross section). Here we have assumed that the ground state density is not depleted. Assuming a square laser pulse of length T_p, the solution to Eq. 13 is

$$N^* = \frac{\alpha I^2 N_o}{h\nu_\ell B} \left(1-e^{-Bt}\right) \qquad 0<t<T_p \; . \tag{14}$$

We can see that if $T_pB > 1$, Eq. 14 reduces to the steady state value of $N^* = \alpha I^2 N_o/h\nu_\ell B$. Note that N^* increases quadratically with laser intensity until

$$I > \frac{h\nu_\ell}{\sigma_{pi}T_p} \tag{15}$$

when the increase becomes linear. From Eq. 15 we can estimate the intensity required to saturate the excited state photoionization transition to be $I > 10^8$ W/cm^2 assuming $\sigma_{pi} = 10^{-18}$ cm^2 and $T_p = 10^{-8}$ sec. In order to efficiently channel the excited state energy into the fluorescence detection channel instead of the ionization channel, the photoionization transition must not be saturated. Using Eqs. 14 and 15, we can calculate a maximum N^* consistant with this requirement

$$N^* \cong \frac{\alpha h\nu_\ell N_o}{\sigma_{Pi}^2 T_p[1+T_p(A+Q)]} \tag{16}$$

Therefore, to obtain the largest N^* (consistant with the largest excitation volume), α must be as large as possible while σ_{pi}

must be as small as possible. If we take α to be approximately 10^{-28} cm^4/W (see Table 3) and the above laser pulse length and photoionization cross section, we calculate from Eq. 16 a value for N^* of $N^* = 10^{-2}N_o$. Note that if counter propagating beams are used (the Doppler-free case discussed previously), α can be increased by two orders of magnitude thereby funnelling most of the excited state energy into the fluorescence detection channel.

In order to calculate the absolute signal levels obtained in a particular experiment, we need to calculate the total number of excited states produced in the laser focal volume. To accomplish this, Eq. 14 must be integrated over the spatial distribution of the laser intensity within the observation volume. This integration is particularly simple if one assumes that the pump laser has a diffraction limited Gaussian spatial profile. If the observation volume is limited to one confocal parameter [28], the integration of Eq. 14 leads to [38]

$$N_T^* = \frac{\pi}{2} N_o \frac{\alpha}{hc} \frac{E_\ell^2}{T_P} \quad , \tag{17}$$

where E_ℓ is the total energy in the laser pulse. Note that this particularly simple expression is independent of the focusing conditions as long as the observation region is larger than one confocal parameter. In most cases of practical interest, the pump lasers will neither have a diffraction limited beam nor a Gaussian spatial distribution. To obtain accurate absolute signal levels, one must integrate Eq. 14 over the actual intensity distribution in the observation region. The result shown in Eq. 17 should only be considered an upper limit for the observed signals.

We should note here if the laser intensity in the focal volume exceeds the value calculated in Eq. 15, the focal volume should be made larger in order to observe fluorescence from the maximum number of excited states (thereby maximizing the sensitivity of the detection technique). If ions can be detected in a particular experiment (which will be more sensitive than detecting fluorescence), the intensity can be increased (by reducing the focal volume, raising the laser energy or reducing the laser pulse length) until the two-photon transition is saturated. We can calculate this saturation intensity from

$$I > \left(\frac{h\nu_\ell}{\alpha T_P}\right)^{1/2} \tag{18}$$

If the previous values for the parameters are substituted into Eq. 18, we find that $I > 10^9$ W/cm^2. At this intensity, 1/2 of the ground state atom density in the focal volume will be ionized.

At this point we can define a set of requirements for the pump laser to obtain the maximum sensitivity for the detection of atoms using two-photon excitation. First, the pulse length should be as short as possible for a given laser energy since the excited state density is proportional to $1/T_p$ (see Eq. 16). A laser pulse length of 10 nsec is adequate while a laser with 0.1-1 nsec pulse length would be ideal. Second, the laser should have as narrow a linewidth as possible. The linewidth should be less than the two-photon Doppler width of approximately 0.3 cm^{-1} (see Table 3) and time-bandwidth limited for those cases when it is useful to take advantage of the Doppler-free properties of counter propagating beams. Finally, the beam quality should be as close to diffraction limited as possible. This is particularly important if the laser energy is so small that intensities in the focal volume are less than 10^8 W/cm^2.

3.   EXPERIMENTAL APPARATUS AND ANALYSIS

In this section I describe the experimental apparatus necessary for the detection of atomic radicals using TPE methods. Although a variety of configurations are possible, I will present examples from our recent experiments on the detection of 0 and N atoms. Fig. 4 is a schematic diagram of the overall setup showing the laser wavelengths pertinent to the detection of 0 atoms. Many of the experimental details are described in [10]. I summarize below a few of the main points concerning this apparatus.

3.1 Atom Production

Before applying the TPE detection method to a specific experimental system, the apparatus should be breadboarded and debugged prior to utilization. This requires a laboratory source of atomic radicals. I describe below several different methods to obtain the required atomic densities.

In our experiments we produced the 0 and N atoms in a low pressure microwave discharge. A mixture of N_2 seeded in approximately 10 torr of He (with 5-10 mTorr SF_6 added to enhance the N atom production) was passed through a 2450-MHz discharge to

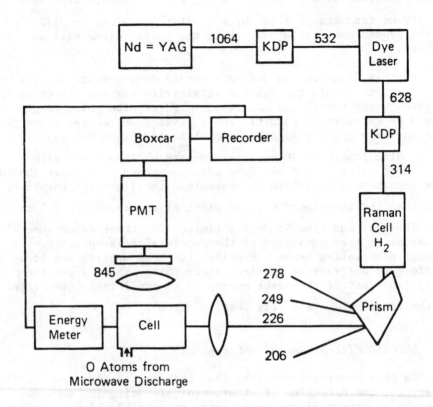

Fig. 4. Apparatus for the two-photon detection of atoms.

produce N atoms. Oxygen atoms were then produced by adding NO to
the flow a few centimeters down stream from the discharge. The
reaction NO + N → O + N_2 is used to produce the oxygen atoms

quantitatively, and it can be followed by chemiluminescence: a
red glow due to the N + N radiative recombination dominates below
the titration point (where the concentration of added NO equals
the initial N concentration); a very faint deep violet color due
to the NO near the titration point; and a bright greenish yellow

glow from the reaction O + NO → NO_2^* in the presence of excess

NO. From the value of the amount of added NO at the titration
point, we obtain the atom concentrations: most of the experiments

were conducted at atom concentrations of 3-10 x 10^{13} cm^{-3}. The

temperature in the observation region was determined to be room
temperature by scanning the near by γ bands of NO [10].

Other methods that have been used to produce atom concentrations include photodissociation of a parent molecule [12] [6]
[16] and chemical reactions in a high temperature flame [23] [3]
[14].

3.2 Frequency Conversion

TPE experiments usually require uv radiation in the region
of 200-300 nm. The wavelengths used in our experiments were generated using stimulated Raman shifting, also known as multiwave
parametric Raman scattering. This technique is based on many-
order anti-Stokes (AS) stimulated Raman scattering in molecular
hydrogen and has been recently studied experimentally in our
laboratory and elsewhere [49] [11] [40] as well as theoretically
[18] [2]. These experimental studies have demonstrated good conversion efficiencies in the uv for such a high-order nonlinear
process and indicate that the technique holds promise for the
routine production of vuv radiation. In this section I present
a brief description of the frequency conversion technique. The
details are described in [10].

A high power YAG pumped doubled dye laser (see Fig. 4) operating at 314 nm is focused into a cell containing hydrogen at
6-8 atm. The resulting nonlinear process produces a series of
lines separated from the pump wavelength by the fundamental
vibrational frequency of the H_2 (v_R = 4155 cm^{-1}). The conversion
efficiency into these sidebands can be very high, and in many
cases the pump energy is depleted by more than 50%. The light
emerging from the cell is recollimated, and the various Stokes/
anti-Stokes orders are separated for use in experiments using a
quartz Pellin-Broca prism.

Our experimental application uses the third anti-Stokes
order for the detection of both O and N atoms. For a pump laser
energy of 15-20 mJ at 314 nm, we typically obtain 50-100 micro-
joules at 226 nm. Additional details concerning this frequency
conversion technique can be found in [10]. Other methods for
obtaining coherent radiation at 226 include mixing the infrared
fundamental of the YAG laser with the doubled dye laser. This
allows one to obtain pump laser wavelengths down to 217 nm at
much higher energy. Energies over 1 mJ at 226 nm have been obtained using this process [33]. However, it is much more complicated than the anti-Stokes Raman technique (since it now requires
two crystals to be scanned instead of one) and the linewidth of
the resulting radiation is determined by the width of both the IR
and UV lasers. Since YAG lasers have typical linewidths on the

order of 1 cm^{-1}, the resulting linewidth will be much larger than the Doppler width of the two-photon transition (thereby reducing the two-photon cross section) unless the YAG laser linewidth is narrowed.

A YAG pumped doubled dye laser system that has either of these frequency conversion options is currently available commercially from several companies. Prices for such a laser system generally lie in the range of $100 K.

3.3 Fluorescence Measurements

The fluorescence at 845 nm (0 atoms) from the excitation volume was collected from the cell at right angles to the laser beam with a fast (f/2) Suprasil lens and focused through a filter onto a photomultiplier with high near-IR sensitivity (RCA C31034A). The signal passed through a fast gain-10 amplifier and into a boxcar integrator with narrow-gate capability (PAR model 165). Used in the gate scanning mode, this permitted lifetime determinations. Fig. 5 shows an excitation scan (60 pulse average) over each of the three fine structure components of the 0 atom transition. Very little background was observed (less than 1% of the peak signal); the noise arises from laser power fluctuations. The stick diagrams in Fig. 5 correspond to the expected positions of the upper state fine-structure components and are labeled by their J values. The frequency integrated signal strengths obtained from Fig. 5, when combined with the expected population distribution at room temperature, yield the relative intensity ratios for the three fine-structure transitions that are predicted by Table 2. If the upper state of the two-photon transition could be spectrally resolved, differences in the predictions between the single intermediate state approximation and the full theory [41] for these relative intensities could be determined.

We have also determined radiative lifetimes and collisional quenching rates for the excited state. One example of this kind of data is given in Fig. 6. Here we have plotted the decay frequencies as a function of pressure and fitted them to the equation.

$$k_d = k_r + k_q N \qquad (19)$$

A least-squares fit yields for the intercept $k_r = (2.56\pm0.09)\times10^7$ sec^{-1} or a radiative lifetime of 39 nsec. From the slope is obtained $k_q = (2.45\pm0.12)\times10^{-10}$ cm^3 sec^{-1}. This kind of data needs to

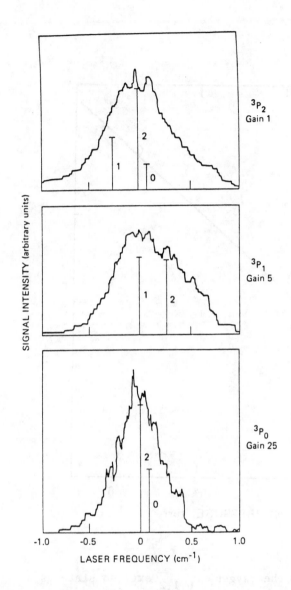

Fig. 5.   Examples of excitation scans through each of the 0^3P_J Transitions.   The stick diagrams correspond to the expected positions of the upper state components.   The laser frequency is in cm^{-1}; note that gain increases from top to bottom.

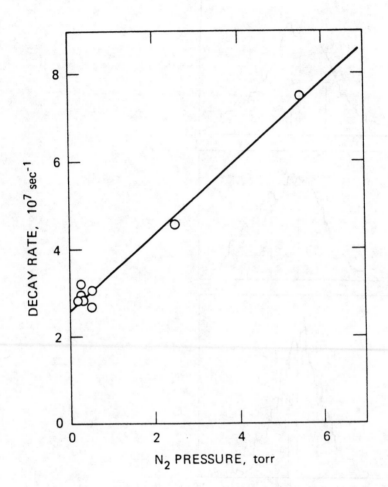

Fig. 6. Decay rate for the oxygen $^3P_{0,1,2}$ excited state as a
         function of N_2 buffer gas pressure determined using
         The zero pressure intercept gives a value for the life-
         time of the transition of T_R = 39 nanoseconds, while the
         slope gives a quenching rate of K_q = 2.5 x 10^{-10} cm^3/
         sec.

quantitative detection technique. Details of these measurements for both O and N atoms can be found in [10].

As stated previously, the other two signal acquisition methods following two-photon excitation are to detect the ions produced by the absorption of a third photon, and to detect the actual absorption (or gain in the Raman case). The ionization method has been sucessfully used by a number of authors in a variety of experimental circumstances (see Table 1). The absorption technique has primarily been used for Raman transitions in molecules and a good review of recent results has been recently given by Esherick and Owyoung [19]. All three of these methods should have application in a variety of experimental circumstances.

4. ESTIMATES OF SENSITIVITY LIMITS

In this section of the lecture we compare the expected sensitivity limits for the Raman gain, fluorescence and ionization detection techniques. I will present order of magnitude estimates based on reasonable values of the relevant parameters. For the most part, these sensitivity limits have not been demonstrated but instead are my best estimate of achievable limits with a properly designed detection system. In all the estimates presented below, I have assumed a diffraction limited near Gaussian pump laser that has a time-bandwidth limited linewidth. Such a laser system has been demonstrated in several laboratories but is not commercially available at this time. This will probably change in the near future.

4.1 Raman Gain Detection

The single pass Raman gain under focused conditions can be written as [19].

$$G = \frac{\pi \alpha P}{\lambda_P} N_o \tag{20}$$

where α is the Raman gain parameter calculated from Eq. 1, P is the pump power, and N_o is the population difference between the upper and lower states. If $P=10^6$ W, $\alpha = 2 \times 10^{-29}$ cm^4/W as calculated for O atoms in Eq. 12, and $\lambda_P=532$ nm, we calculate from Eq. 20 that $G=10^{-18} N_o$.

The sensitivity of detecting Raman gain is fundamentally limited by the shot noise of the probe laser. We have been able to measure Raman gains in our laboratory at a level of 10^{-5} of the incident cw laser intensity. Although other laboratories have demonstrated better detection limits [19], I will use this value as the minimum detectable gain. Substituting into Eq. 20, we determine a detection limit of $N_o = 10^{-13}$ cm^{-3} for the Raman gain (or absorption) detection method. If a crossed beam geometry is used to separate the pump and probe beams (as has been demonstrated in [19]), then this sensitivity limit must be increased by a factor of 5. A good comparison for the sensitivities of Raman gain spectroscopy, spontaneous Raman scattering and CARS is given in [19].

4.2 Fluorescence Detection

The total number of fluorescence photons arriving at the detector integrated over the radiative lifetime (1/A) of the excited state is given by

$$N_T = \eta_B \frac{\Omega}{4\pi} \eta_\phi N^* V \ , \tag{21}$$

where η_B is the branching ratio of the fluorescence into the detection bandwidth of the detector, $\Omega/4\pi$ is the solid angle collection efficiency, η_ϕ is the efficiency of the detection system including the photomultiplier efficiency and the filter transmission functions, N^* is the excited state density calculated from Eq. 13, and V is the observation or focal volume, whichever is smaller. In deriving Eq. 21 I have assumed no quenching of the excited state. If collisional quenching rate (Q) is significant, then the values calculated from Eq. 21 should be multiplied by the ratio of A/Q.

For the calculation we will assume a f/1.4 collection lens ($\Omega/4\pi = 3\times10^{-2}$), a system detection efficiency of 3×10^{-2} (as measured in our laboratory), a fluorescence branching ratio of 1 (e.g., the 0 atom transition at 845 nm), an excited state density calculated from Eq. 16 to be $10^{-2}N_o$ for 0 atoms with a pump laser intensity of 10^8 W/cm^2, and a conservative detection limit of 10 photons per laser pulse. We will assume that the observation volume is limited by apertures on the fluorescence detection system. The observation volume is defined as V=AL where A is the laser beam cross sectional area, and L is the observation length assumed to be 1 cm. If the laser intensity is $I=10^8$ W/cm^2 and the total energy is 1 mJ in a 10 nsec pulse, the beam area is approximately $A=10^{-3}$ cm^2. Combining all these factors, we calculate from Eq. 21 a detection sensitivy of approximately

$$N_o = 10^9 \text{ cm}^{-3}. \tag{22}$$

Although this sensitivity limit has yet to be achieved (estimates in [16] indicated $N\sim10^{10}$ has been demonstrated), this calculation illustrates the high sensitivity of this detection technique if diffraction limited, narrow-band lasers are used to excite the atom. The sensitivity limits for TPEF process using counter-pagating beams (Doppler-free) has been considered previously [20] and are consistent with those derived here.

4.3 Ionization Detection

To calculate the 0 atom sensitivity using the ionization detection method, we assume that the two-photon transition is saturated. We calculated from Eq. 18 that this would require an intensity of approximately 10^9 W/cm^2 in the case of 0 atom detection. At this intensity 1/2 the ground state density of 0 atoms will be ionized. The total number of ions observed at the detector (ion multiplier, multichannel plate or just a pair of parallel plates) will be $N_T=(1/2)N_oV$ where V is the laser excitation volume. If we assume the same laser parameters as for the fluorescence detection example ($E=1$ mJ, $T_P=10$ nsec) we calculate that an intensity of 10^9 W/cm^2 requires a beam area of $A=10^{-4}$ cm^2. Assuming an interaction length of $L=2$ cm and a signal to background detection limit of 10 ions per laser pulse (low density limit), we calculate an 0 atom detection limit of $N_o=10^5$ cm^{-3}. Note that this limit is 4 orders of magnitude smaller than the fluorescence detection limit calculated above for the same laser parameters. This example serves to illustrate the large advantage ionization detection has over any other method of atom detection. It has been successfully used in a combustion environment [22] as well as in low density applications (see Table 1). In addition to being the most sensitivity detection technique, it also has the advantage that the ground state atomic densities can be determined quantitatively if the two-photon transition is saturated. Unfortunately, there are many experiments where the ions can not be detected and for these cases, one of the other detection methods may be appropriate.

5.   CONCLUSION

In conclusion, I have tried in this lecture to give an introduction to the techniques of atom detection using two-photon
excitation. Atom detection using two-photon processes can have
good sensitivity, good spatical resolution, and in many cases the
sensitivity can be obtained with modest uv laser energies. The
sensitivity limits for any specific atomic system will depend
dramatically on the particular experimental conditions. However,
I have given order of magnitude estimates in the low density
limit that illustrate the atom detection sensitivities of the
three detection methods discussed in this lecture. The least
sensitive technique was the detection of gain or absorption
(SRGS). The sensitivity for the detection of fluorescence (TPEF)
from the upper state in the two-photon transition is approximately 4 orders of magnitude larger than that for absorption.
The detection of the ions created by the absorption of a third
photon (TPRIS) is the most sensitive detection considered in this
lecture, and is approximately 4 orders of magnitude larger than
that estimated for fluorescence detection. I expect that research during the next few years will reach and perhaps surpass
some of the limits estimated here for the detection of atoms
using these techniques. However, experiments demonstrating these
limits will require demanding specifications on the pump lasers
used in the experiments.

REFERENCES

1. C. P. Ausschnit, G. C. Bjorklund, and R. R. Freeman, Appl.
    Phys. Lett. 33 54 (1978).

2. J. R. Ackerhalt, Phys. Rev. Lett. 46, 922 (1981).

3. M. Alden, H. Edner, P. Grafstrom, and S. Svanberg, Opt.
    Comm. 42, 244 (1982).

4. G. C. Bjorklund, C. P. Ausschnitt, R. R. Freeman, and
    R. H. Storz, Appl. Phys. Lett. 33, 54 (1978).

5. W. K. Bischel, J. Bokor, D. J. Kligler, and C. K. Rhodes,
    IEEE J.Q.E. QE-15, 380 (1979).

6. P. Brewer, P. Das, G. Ondrey, and R. Bersahn, J. Chem. Phys.
    79, 720 (1983).

7. J. Bokor, R. R. Freeman, J. C. White, and R. H. Storz, Phys.
    Rev. A24, 6121 (1981).

8. W. K. Bischel, P. J. Kelly, and C. K. Rhodes, Phys. Rev. A13, 1817 (1976).

9. W. K. Bischel, B. E. Perry, and D. R. Crosley, Chem. Phys. Lett. 82, 85 (1981).

10. W. K. Bischel, B. E. Perry, and D. R. Crosley, Appl. Opt. 21, 1419 (1982).

11. D. J. Brink and D. Proch, Opt. Lett. 7 494 (1982).

12. P. Brewer, N. Van Veen, and R. Bersohn, Chem. Phys. Lett. 91, 126 (1982).

13. J. C. Cummings and D. P. Aeschliman, Opt. Comm. 31, 165 (1979).

14. C. J. Dasch and J. J. Bechtel, Opt. Lett. 6, 36 (1981).

15. M. Dagenais, J. W. C. Johns, and A. R. W. McKellar, Can. J. Phys. 54 1438 (1976).

16. P. Das, G. Ondrey, N. Van Veen, and R. Bersohn, J. Chem. Phys. 79, 724 (1983).

17. P. B. Davies and D. K. Russell, Chem. Phys. Lett. 67 440 (1979).

18. D. Eimerl, R. S. Hargrove, and J. A. Paisner, Phys. Rev. Lett. 46, 651 (1981).

19. P. Esherick and A. Owyoung, "High Resolution Stimulated Raman Gain Spectroscopy"; in Advances in Infrared and Raman Spectroscopy, Vol. 9 (Ed. R. J. H. Clark and R. E. Hester, Heyden and Sons, Ltd., London, 1982).

20. J. Gelbwachs, Appl. Opt. 15, 2654 (1976).

21. G. Grynberg, B. Cagnac, and F. Biraben, "Multiphoton Resonant Processes in Atoms," in Coherent Nonlinear Optics (ed. M. S. Feld and V. S. Letokhov, Springer-Verlag, Berlin, 1980).

22. J. E. M. Goldsmith, Opt. Lett. 7, 437 (1982).

23. J. E. M. Goldsmith, J. Chem. Phys. 78, 1610 (1983).

24. T. W. Hansch, S. A. Lee, R. Wallenstein, and C. Wieman, Phys. Rev. Lett. 34, 307 (1975).

25. M. Heaven, T. A. Miller, R. R. Freeman, J. C. White, and J. Bokor, J. Chem. Phys. Lett. $\underline{86}$, 458 (1982).

26. G. S. Hurst, M. G. Payne, S. D. Kramer, and J. P. Young, Rev. Mod. Phys. $\underline{51}$, 767 (1979).

27. W. Huo an R. L. Jaffe, "Ab Initio Calculation of the Two-Photon Absorption Cross Section of the $X^1\Sigma_g^+ \to (E_1F)^1\Sigma_g^+$ in H_2, Chem. Phys. Lett. (to be published, 1983).

28. H. Kogelnik and T. Li, Appl. Opt. $\underline{5}$, 1550 (1965).

29. J. V. V. Kasper, C. R. Pollock, R. F. Curl Jr., and F. K. Tittel, Chem. Phys. Lett. $\underline{77}$, 211 (1981).

30. G. A. Laguna and W. H. Beattie, Chem. Phys. Lett. $\underline{88}$, 439 (1982).

31. M. D. Levenson, <u>Introduction to Nonlinear Laser Spectros-copy</u> (<u>Academic</u> Press, 1982).

32. R. P. Lucht, J. T. Salmon, G. B. King, D. W. Sweeney, and N. M. Laurendeau, Opt. Lett. $\underline{7}$ 365 (1983).

33. C. H. Muller III, D. R. Eames, and K. H. Burrell, Bull. Am. Phys. Soc. $\underline{26}$, 1031 (1981).

34. T. J. McIlrath, R. Hudson, A. Aikin, and T. D. Wilkerson, Appl. Opt. 18, 316 (1979).

35. A. Miziolek, "Collisionally Induced Flame Emissions Following Oxygen Atom Two-Photon Excitation", paper WK5, 1983 Annual Meeting of the Optical Society of America, New Orleans, October 17-20, 1983.

36. J. Morelles, D. Normand, and G. Petite, "Nonresonant Multiphoton Ionization of Atoms," in Atomic and Molecular Physics, <u>Vol. 18</u> (Academic Press, 1982).

37. D. S. Moore, Chem. Phys. Lett. $\underline{89}$, 131 (1983).

38. B. R. Marx, J. Sumons, and L. Allen, J. Phys. B11, L273 (1978).

39. K. Omidvar, Phys. Rev. A22, 1576 (1980). (Relative intensities are wrong; multiplet matrix elements are probably correct.)

40. J. A. Paisner and R. S. Hargrove, in Digest of Conference on Laser Engineering and Applications (Optical Society of America, Washington D.C., 1981), postdeadline paper II-4.

41. M. S. Pindzola, Phys. Rev. A17, 1021 (1978).

42. H. Schlossberg, J.A.P. 47, 2044 (1976).

43. C. R. Quick and D. S. Moore, J. Chem. Phys. 79 759 (1983).

44. A. C. Staton and C. E. Kolb, J. Chem. Phys. 72, 6637 (1980).

45. K. Schofield and M. Steinberg, Opt. Eng. 20, 501 (1981).

46. R. E. Teets and J. H. Bechtel, Opt. Lett. 6, 458 (1981).

47. L. Vriens, Opt. Comm. 11 396 (1974).

48. W. L. Wiese, M. W. Smith, and B. M. Glennon, "Atomic Transition Probabilities," Nat. Stand. Ref. Data Ser. Nat. Bur. Stand. (1966).

49. V. Wilke and W. Schmidt, Appl. Phys. 16, 151 (1978); Appl. Phys. 18, 177 (1979).

50. J. Wormhoudt, A. C. Stanton, and J. Silver, "Techniques for Characterization of Gas Phase Species in Plasma Etching and Vapor Deposition Processes", to be published in Spectroscopic Characterization Techniques for Semiconductor Technology Vol. 452 (Proceeding of SPIE, 1983).

51. H. Zacharias, H. Rottke, J. Danon, and K. H. Welge, Opt. Comm. 37, 15 (1981).

COHERENT ANTI-STOKES RAMAN SCATTERING

James J. Valentini

Los Alamos National Laboratory
Los Alamos, NM
USA

1. INTRODUCTION

1.1 The Raman Effect

The Raman effect is an inelastic photon-scattering process, in which an atom or molecule absorbs one photon while simultaneously emitting another photon at a different frequency. The emitted photon is at a frequency, w_0, which can be greater or less than the frequency, w_1, of the absorbed photon:

$$w_0 = w_1 \pm \Delta w , \tag{1}$$

where $\Delta w > 0$.

The frequency shifts, Δw, are characteristic of the material responsible for the light scattering, and are found to correspond to the energy differences between rotational and vibrational levels in molecules (or less frequently electronic levels in atoms and molecules):

$$\Delta w = (E_r - E_g)/h = w_{rg} , \tag{2}$$

where r and g are labels identifying the states with energies E_r and E_g. Figures 1 and 2 show energy diagrams for Raman scattering. When $w_0 = w_1 - w_{rg}$ (Fig. 1) the process is referred to as Stokes scattering, and when $w_0 = w_1 + w_{rg}$ (Fig. 2) it is termed

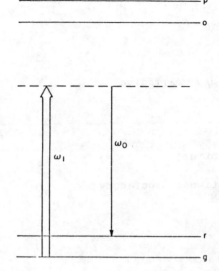

Fig. 1.   Energy level diagram for raman scattering (Stokes).

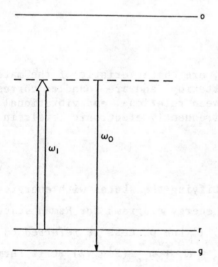

Fig. 2.   Energy level diagram for raman scattering (anti-Stokes).

anti-Stokes scattering. Because the frequency shifts of the
Raman-scattered light are the frequency differences between molecular states the Raman effect can be used as a spectroscopic
tool.

Unfortunately, the inelastic light scattering of the Raman
effect is extremely inefficient, and low intensity has remained a
problem in Raman spectroscopy, even with the use of laser sources
for sample illumination. This low efficiency means that sample
luminescence and laser-induced fluorescence can often mask the
Raman spectra, and Raman investigation of transient phenomena is
particularly difficult.

1.2   Coherent Anti-Stokes Raman Scattering

Coherent anti-Stokes Raman scattering (CARS) is a laser
technique which overcomes many of the problems connected with the
low efficiency of spontaneous Raman scattering and has expanded
the range of application of Raman spectroscopy in physics, chemistry, and biology. CARS is one of several coherent Raman spectroscopic techniques which are based on the existence of Raman
resonances in three-wave mixing. In three-wave mixing electric
fields (laser beams) at frequencies ω_1, ω_2, and ω_3 are mixed to
produce a fourth field at frequency ω_0. The amplitude of the
field at ω_0 is given by

$$\underline{E}(\omega_0) \propto \underline{\chi}^{(3)} \; \underline{E}(\omega_1) \; \underline{E}(\omega_2) \; \underline{E}(\omega_3) \; , \tag{3}$$

where $\underline{\chi}^{(3)}$ depends on the frequencies ω_0, ω_1, ω_2 and ω_3 as well
as the material in which the mixing takes place. The mixing is
coherent, since the field at ω_0 is in phase with the fields at
ω_1, ω_2, and ω_3.

For the process known as CARS the frequencies are related by

$$\omega_0 = \omega_1 - \omega_2 + \omega_3 \; , \tag{4}$$

where $\omega_1 > \omega_2$. When $\omega_1 - \omega_2 = \omega_{rg}$ where, as in Eq. 2, $h\omega_{rg}$ is
the energy difference between two molecular levels connected by a
Raman transition, this non-linear mixing is greatly enhanced.
These Raman resonances constitute a CARS spectrum, which is
scanned by varying ω_1 and/or ω_2 and detected via the coherently
generated signal at $\omega_0 = \omega_3 + \omega_{rg}$. In almost all CARS experiments $\omega_1 = \omega_3$ and ω_1 is held fixed in frequency while ω_2 is
scanned. Equation 4 thus becomes:

$$\omega_0 = 2\omega_1 - \omega_2 , \tag{5}$$

and at a Raman resonance

$$\omega_0 = \omega_1 + \omega_{rg} \tag{6}$$

and

$$\omega_2 = \omega_1 - \omega_{rg} . \tag{7}$$

Therefore, ω_1 is often referred to as the "pump" or "laser" frequency, ω_2 the "Stokes" frequency, and ω_0 the "anti-Stokes" frequency. An energy level schematic for the CARS process is shown in Fig. 3. Because the anti-Stokes signal beam at ω_0 is generated by a <u>coherent</u> mixing of ω_1 and ω_2 it is characterized by a high intensity, often several orders of magnitude greater than spontaneous Raman signals, and a small divergence, i.e., it is laser-like.

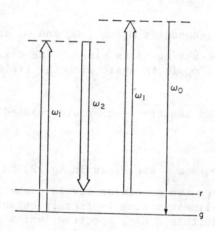

Fig. 3.   Energy level diagram for the CARS process. The "pump" frequency is ω_1, the "Stokes" frequency is ω_2, and ω_0 is the "anti-Stokes" frequency.

2.   PHENOMENOLOGY OF CARS

2.1 Electric Field Induced Polarization

The polarization, $\underline{P}$, is given by

$$\underline{P} = \chi^{(1)} \underline{E} + \chi^{(2)} \underline{E}^2 + \chi^{(3)} \underline{E}^3 + \ldots \,, \tag{8}$$

where $\chi^{(1)}$, $\chi^{(2)}$, $\chi^{(3)}$, . . . are the first-order, second-order, third-order, . . . dielectric susceptibility tensors, and $\underline{E}$ is the electric field. $\chi^{(1)}$ is responsible for classical linear optical phenomena, including refraction, Rayleigh scattering, and normal Raman scattering. $\chi^{(2)}$ gives rise to second harmonic generation, sum and difference frequency generation, hyper-Rayleigh scattering and other effects. It is $\chi^{(3)}$ which concerns us here, for it is responsible for CARS as well as other effects such as third harmonic generation. We limit our consideration then to the $\chi^{(3)}$ term in Eq. 8, and write:

$$\underline{P}^{(3)}(\underline{r},t) = \chi^{(3)} \underline{E}(\underline{r},t) \, \underline{E}(\underline{r},t) \, \underline{E}(\underline{r},t) \,, \tag{9}$$

where the spatial and temporal variation of the fields and induced polarization is made explicit.

We will assume that the electric field is made up of harmonic components which are plane waves travelling in the r direction:

$$\underline{E}(\underline{r},t) = 1/2 \sum_{j=1}^{3} [\underline{E}(\omega_j,\underline{r}) \exp i(k_j\underline{r}-\omega_j t) + c.c.], \tag{10}$$

where $k_j = |\underline{k}_j| = \omega_j n_j/c$, and n_j is the index of refraction at ω_j and c is the velocity of light. CARS and other three-wave mixing processes involve up to three distinct field frequencies j, so the summation in Eq. 10 is over j = 1 to 3. We see then that the polarization $\underline{P}^{(3)}$ will have frequency components at

$$\omega_0 = \pm \omega_k \pm \omega_\ell \pm \omega_m \,, \tag{11}$$

where k, ℓ, m = 1, 2, or 3. Thus $\underline{P}^{(3)}$ will have many frequency components, including $3w_1$, $3w_2$, and $3w_3$, which correspond to third harmonic generation, $w_1 + w_2 + w_3$, giving frequency summation, as well as a CARS component at $w_1 - w_2 + w_3$, where $w_1 > w_2$. For CARS we usually have only two field frequencies, w_1 and w_2, with $w_3 = w_1$, for which $w_0 = 2w_1 - w_2$.

If we take the driving fields to be plane light waves with parallel field vectors and with parallel propagation vectors aligned along the space-fixed z axis, we find that the vector direction of the source polarization $\underline{P}^{(3)}$ $(\underline{r},t)$ is also parallel to the vectors of the driving fields at w_1 and w_2. We can thus neglect the vector nature of the electric fields, as well as the tensor nature of $\underline{\chi}^{(3)}$, and the CARS polarization at w_0 becomes:

$$P^{(3)} (z,t) = 3/8 \, \chi^{(3)} (-w_0, \, w_1, \, -w_2, w_1)$$

$$\times \, E^2(w_1) \, E^*(w_2) \, \exp \, i[(2k_1 - k_2)z$$

$$- \, w_0 t] + c.c \qquad\qquad (12)$$

$\chi^{(3)} (-w_0, \, w_1, \, -w_2, \, w_1)$ represents the susceptibility for the process in which a polarization is induced at $w_0 = 2w_1 - w_2$.

2.2 Generation of an Electromagnetic Wave

The non-linear polarization, $P^{(3)}$ (z,t), at frequency w_0 acts as a source term in Maxwell's equations to produce an electromagnetic wave, $\underline{E}$ (r,t), at w_0. Taking our medium to be a gas occupying the space $z = 0$ to $z = L$, and neglecting dispersion we find that the generated wave at w_0 is also a plane wave traveling in the z direction:

$$\underline{E} \, (\underline{r},t) = E \, (z,t) = E \, (w_0,z) \, \exp \, i(k_0 z - w_0 t), \qquad (13)$$

with

$$E \, (w_0,z) = \frac{-3i\pi w_0}{2c} \, E^2 \, (w_1) \, E^* \, (w_2) \int_0^L \chi^{(3)} \, \exp \, (i\Delta kz) \, dz, \qquad (14)$$

where $\Delta k = 2k_1 - k_2 - k_0$, and for simplicity we have dropped the ω notation in $\chi^{(3)}$. Since we have neglected dispersion in our gas sample $\underline{k}_i = |k_i| = \omega_i n_i/c = \omega_i/c$, and we have for the exponential

$$\Delta k = 2k_1 - k_2 - k_0 = 2\omega_1 - \omega_2 - \omega_0 = 0 \tag{15}$$

When $\Delta k = 0$ the beams are referred to as phase matched. Therefore, we have:

$$E(\omega_0,L) = \frac{-3i\pi\omega_0}{2c} E^2(\omega_1) E^*(\omega_2) \chi^{(3)} L. \tag{16}$$

Now, using the fact that the intensity (power per unit area) is related to the field amplitude,

$$I(\omega) = (c/8\pi) E^2, \tag{17}$$

we find:

$$I_0(L) = \left(\frac{12\pi^2\omega_0^2}{c^2}\right)^2 I_1^2 I_2 |\chi^{(3)}|^2 L^2, \tag{18}$$

where I_i is the intensity of the beam at frequency ω_i.

2.3 Raman Resonance in $\chi^{(3)}$

The coherent anti-Stokes Raman scattering process which generates the anti-Stokes signal beam at $\omega_0 = \omega_1 - \omega_2 + \omega_3$ is greatly enhanced when $\omega_1 - \omega_2 = \omega_{rg}$, the frequency of a Raman allowed transition between molecular energy states g and r. The resonance behavior, which makes coherent anti-Stokes Raman scattering a useful spectroscopic tool, arises from Raman resonances in $\chi^{(3)}$.

This leads quite naturally to a separation of $\underline{\chi}^{(3)}$ into resonant (i.e., Raman resonant) and non-resonant contributions:

$$\chi^{(3)} \equiv \chi' + i\chi'' + \chi^{NR}, \tag{19}$$

where $\chi' + i\chi''$ is the resonant part of χ^3 and χ^{NR} the nonresonant part. Note that to avoid excessive superscripts the (3) identification is deleted from χ', χ'', and χ^{NR}, since only $\chi^{(3)}$ is important in CARS. In this separation of resonant and non-resonant contributions to $\chi^{(3)}$ we assume that χ^{NR} is real and independent of ω_0, ω_1, ω_2, and ω_3.

The amplitude of $\chi^{(3)}$ near the resonances is related to $\frac{d\sigma}{d\Omega}$, the spontaneous Raman scattering cross sections for the Raman transitions which cause these $\chi^{(3)}$ resonances:

$$\chi' + i\chi'' = \frac{N\Delta_{rg}c^4}{h\omega_2^4} \frac{d\sigma}{d\Omega} \frac{\omega_{rg}}{[\omega_{rg}^2 - (\omega_1 - \omega_2)^2 - i\Gamma_{rg}(\omega_1 - \omega_2)]} \, , \qquad (20)$$

where N is the number density of the sample, Δ_{rg} is the population difference between the energy levels r and g, and Γ_{rg} is the Raman transition linewidth.

At resonance $\omega_1 - \omega_2 = \omega_{rg}$, and

$$|\chi' + i\chi''|^2 = \left(\frac{N\Delta_{rg}c^4}{h\omega_2^4}\right)^2 \left(\frac{d\sigma}{d\Omega}\right)^2 \left(\frac{1}{\Gamma_{rg}}\right)^2 . \qquad (21)$$

2.4 Symmetry Properties of $\chi^{(3)}$

As a fourth-rank tensor, $\underline{\chi}^{(3)}$ has $3^4 = 81$ elements of the form $\chi_{\rho\sigma\tau\nu}^{(3)}(-\omega_0, \omega_1, -\omega_2, \omega_3)$, where ρ, σ, τ, ν = x, y, or z and $\omega_0 = \omega_1 - \omega_2 + \omega_3$. Symmetry restrictions however will limit the number of the elements which are independent and non-vanishing.

Since the vast majority of CARS experimental work has been done in isotropic media we will explicitly consider the symmetry properties of $\underline{\chi}^{(3)}$ only in isotropic materials. For such $\underline{\chi}^{(3)}$ has 21 non-zero elements;

$$\chi^{(3)}_{1111} = \chi^{(3)}_{xxxx} = \chi^{(3)}_{yyyy} = \chi^{(3)}_{zzzz}$$

$$\chi^{(3)}_{1122} = \chi^{(3)}_{xxyy} = \chi^{(3)}_{xxzz} = \chi^{(3)}_{yyxx} = \chi^{(3)}_{yyzz} = \chi^{(3)}_{zzxx} = \chi^{(3)}_{zzyy}$$

$$\chi^{(3)}_{1212} = \chi^{(3)}_{xyxy} = \chi^{(3)}_{xzxz} = \chi^{(3)}_{yxyx} = \chi^{(3)}_{yzyz} = \chi^{(3)}_{zxzx} = \chi^{(3)}_{zyzy}$$

$$\chi^{(3)}_{1221} = \chi^{(3)}_{xyyx} = \chi^{(3)}_{xzzx} = \chi^{(3)}_{yxxy} = w^{(3)}_{yzzy} = \chi^{(3)}_{zxxz} = w^{(3)}_{zyyz} \qquad (22)$$

Of these 21 non-vanishing elements only 3 are independent, since

$$\chi^{(3)}_{1111} = \chi^{(3)}_{1122} + \chi^{(3)}_{1212} + \chi^{(3)}_{1221} \qquad (23)$$

In almost all CARS experiments $w_1 = w_3$, so

$$\chi^{(3)}_{1122} = \chi^{(3)}_{1221}, \qquad (24)$$

and

$$\chi^{(3)}_{1111} = 2\chi^{(3)}_{1122} + \chi^{(3)}_{1212} \qquad (25)$$

From this one can see that $\chi^{(3)}$ in Eqs. (12), (16), and (18) to (21) is $\chi^{(3)}_{1111}$.

2.5 Selection Rules

Although the CARS signal is directly related to the ordinary Raman cross-section (cf. Eq. 21), the fact that it is a nonlinear three-wave mixing technique might lead one to expect different selection rules in CARS and spontaneous Raman scattering. However, an exact analysis shows that the selection rules for CARS and ordinary Raman scattering are the same.

2.6 Signal Intensity

The importance of CARS for Raman spectroscopy is principally a consequence of the fact that the CARS signal intensity can be several orders of magnitude greater than that produced in ordinary Raman scattering. This large signal enhancement arises because CARS is a coherent process.

From Eqs. (18) and (21) we can compute the magnitude of the CARS signal and compare it to the magnitude of the spontaneous

Raman signal obtained under similar conditions. We will consider
the Raman and CARS signals from hydrogen at 1 torr pressure. The
Q(1) vibrational transition of H_2 at ω_{rg} = 1.25 x 10^{13} sec^{-1}
(4161 cm^{-1}) has $\frac{d\sigma}{d\Omega}$ = 4.9 x 10^{-31} cm^2/sr at 532 nm and Γ_{rg} = 1.09
x 10^9 sec^{-1}. For a CARS experiment with ω_1 = 5.64 x 10^{14} sec^{-1}
(λ_1 = 532 nm) and ω_2 = 4.39 x 10^{14} sec^{-1} (λ_2 = 683 nm), we find
from Eqs. (18) and (21) that $I_0 \cong$ 0.01 W/cm^2 when I_1 = 10^6 W/cm^2
and I_2 = 10^5 W/cm^2, with a sample length, L, of 1 cm, and a population difference, Δ_{rg}, of 1. Taking all three beams to have
cross sectional areas of 0.30 cm^2 (0.62 cm diameter) we find that
$P_0 \cong$ 0.003 W. In spontaneous Raman scattering the signal power
at the Stokes or anti-Stokes shifted frequency, ω_0, is given by

$$P_0 = N \frac{d\sigma}{d\Omega} L P_1 \delta\Omega \quad , \tag{26}$$

where P_1 is the incident power and $\delta\Omega$ is the solid angle over
which the scattered light is collected. For the example above
this gives $P_0 \cong$ 2 x 10^{-8} W, assuming $\delta\Omega$ = 1 sr. The CARS signal
is thus 10^5 times larger than the spontaneous Raman signal.

For computing the CARS signal power we invoked the plane
wave result for the CARS intensity, Eq. (18), and assumed unfocused laser beams (beam diameter = 0.62 cm). However, unlike
the signal in ordinary Raman scattering, which depends only on
the incident laser power, the CARS signal power depends on the
intensity, i.e., power per unit area. Hence focusing of the pump
and Stokes beams at ω_1 and ω_2 will greatly enhance the CARS signal power. We can estimate the magnitude of this enhancement by
rewriting Eq. (18) as

$$P_0 = \left(\frac{12\pi^2\omega_0}{c^2}\right)^2 P_1^2 P_2 \frac{A_0}{A_1^2 A_2} |\chi^{(3)}|^2 L^2 \quad , \tag{27}$$

where A_i is the cross-sectional area of the beam at frequency ω_1.
Taking $A_0 = A_1 = A_2 = A$ we have

$$P_0 = \left(\frac{12\pi^2\omega_0}{c^2}\right)^2 P_1^2 P_2 \frac{1}{A^2} |\chi^{(3)}|^2 L^2 \quad , \tag{28}$$

and the CARS signal power will increase as the square of the decrease in area of the beams. The area can easily be reduced from
the 0.30 cm^2 value assumed in the computation above to 10^{-4} cm^2
over the 1 cm sample length by focusing with a 50 cm focal length
lens. This would yield an increase in P_0 of approximately 10^7,
and hence a CARS signal more than 10^{12} times greater than the
spontaneous Raman signal.

From Eq. (27) one might expect that the magnitude of CARS
signals would be restricted only by the power of available laser
sources. As with any other type of spectroscopy, however, there
are effects which limit the maximum signals which can be obtained. Probably most important among these is saturation, that
is saturation of the molecular transition from state g to state r
via the two-photon $w_1 - w_2$ transition (cf. Fig. 3). Other limiting mechanisms in CARS experiments include optical breakdown and
the dynamic Stark effect.

2.7 Spectral Lineshape

In Section 2.3 we expressed $\chi^{(3)}$ as a sum of resonant and
non-resonant components, with the resonant contribution having
real and imaginary parts:

$$\chi^{(3)} \equiv \chi' + i\chi'' + \chi^{NR} \tag{19}$$

Separating Eq. (20) into real and imaginary parts we find:

$$\chi' = \frac{N\Delta_{rg} c^4}{hw_2^4} \frac{d\sigma}{d\Omega} \frac{w_{rg}[w_{rg}^2 - (w_1 - w_2)^2]}{[w_{rg}^2 - (w_1 - w_2)^2]^2 + \Gamma_{rg}^2 (w_1 - w_2)^2} \tag{29}$$

and

$$\chi'' = \frac{N\Delta_{rg} c^4}{hw_2^4} \frac{d\sigma}{d\Omega} \frac{w_{rg} \Gamma_{rg} (w_1 - w_2)}{[w_{rg}^2 - (w_1 - w_2)^2]^2 + \Gamma_{rg}^2 (w_1 - w_2)^2} . \tag{30}$$

As before we assume χ^{NR} to be real and independent of w_0, w_1,
and w_2, so

$$\chi^{NR} = \text{constant.} \tag{31}$$

These contributions to $\chi^{(3)}$ are plotted in Fig. 4, while

$$|\chi^{(3)}|^2 = (\chi')^2 + (\chi'')^2 + (\chi^{NR})^2 + 2\chi'\chi^{NR} \quad , \tag{32}$$

is shown in Fig. 5. Since the frequency dependence of the CARS signal is determined by $|\chi^{(3)}|^2$, Fig. 5 gives the expected CARS lineshape, which is asymmetric for $\chi^{NR} \neq 0$.

Figure 5 gives the lineshape that would be observed for a Raman line far removed from other resonances. In general, in molecular CARS spectra one encounters several closely spaced Raman lines, for example the rotational lines in a vibrational Q-branch Raman transition. In this case adjacent lines can interfere with one another. Considering only two adjacent lines we can write:

$$\chi^{(3)} = \sum_{J=1}^{2} (\chi'_j + i\chi''_j) + \chi^{NR} \quad , \tag{33}$$

and

$$|\chi^{(3)}|^2 = (\chi'_1)^2 + (\chi'_2)^2 + (\chi''_1)^2 + (\chi''_2)^2 + (\chi^{NR})^2$$

$$+ 2\chi'_1\chi'_2 + 2\chi''_1\chi''_2 + 2\chi'_1\chi^{NR} + 2\chi'_2\chi^{NR} \tag{34}$$

Clearly, when there are several closely spaced lines the CARS lineshape function can become quite complicated. When the spacing between lines is not much greater than the linewidth, Γ_{rg}, numerical analysis of the CARS spectrum is often necessary if one wishes to obtain a quantum state population distribution or temperature from the observed spectrum.

3.   THE PRACTICE OF CARS

3.1 Experimental Apparatus

A typical CARS apparatus is shown in simplified form in Fig. 6. Basically the apparatus consists of two laser sources, one of which must be frequency tunable; laser beam steering optics for adjusting the beam overlap and crossing angle, θ, in the sample; focusing and recollimating lenses at the entrance and exit of the cell; spatial and spectral filters for discriminating

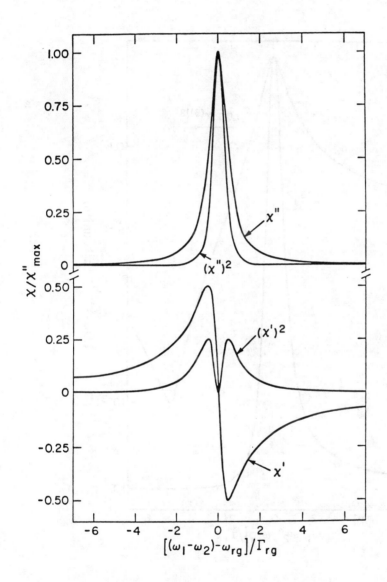

Fig. 4.   Plots of the real (χ') and imaginary (χ'') parts of the
third-order dielectric susceptibility.

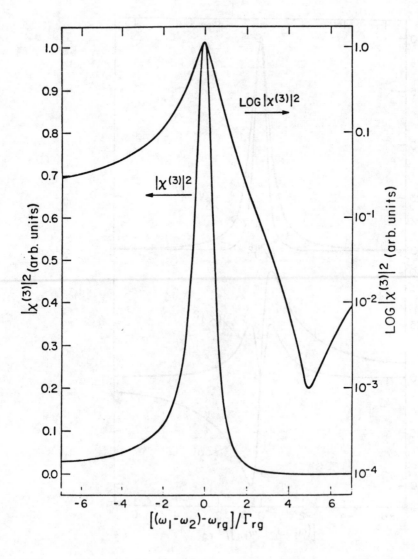

Fig. 5.   Lineshape that would be observed for a Raman line far
          removed from other resonances.

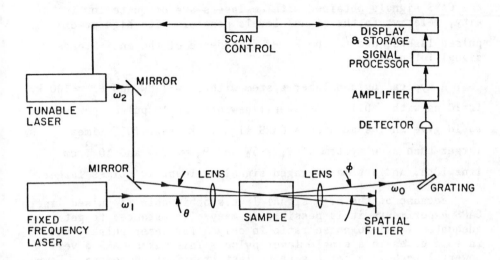

Fig. 6.   Typical CARS apparatus.

against scattered laser light and sample luminescence and fluorescence; a photodiode or photomultiplier to detect the signal;
and signal processing and recording electronics.

    The crossing angle, θ, is chosen so as to satisfy the phase-
matching condition:

$$\underline{k}_0 = 2\underline{k}_1 - \underline{k}_2 , \tag{35}$$

as discussed in Section 2.1. The magnitude of the wave vector
$\underline{k}_i$ is given by $\omega_i n_i/c$, where n_i is the index of refraction of
the medium at frequency ω_i and c is the velocity of light.   In
liquids $\theta \cong 1° - 3°$, while for low density gases, for which the
wavelength dispersion in the index of refraction is negligible,
$\theta = 0$ and therefore $\phi = 0$, and the pump and Stokes incident beams
and the anti-Stokes signal beam are collinear.   In this collinear
case spatial filtering of the signal beam from the input beams
can be accomplished only after some dispersive spectral filtering
via prisims or gratings.

    Several different kinds of laser systems have been used in
CARS spectroscopy.   In a few cases cw laser systems have been
employed, generally with an ion laser as the fixed-frequency
source and an ion-pumped dye laser as the tunable source.   Continuous wave sources have very low power outputs, and as a result

the CARS signals obtained with cw lasers are of quite low intensity, relative to the signal levels possible with high power pulsed lasers, due to the $P_1^2 P_2$ dependence of the anti-Stokes signal intensity.

A typical pulsed laser system with P_1 = 1 MW and P_2 = 100 kW in 10 ns with a 0.1 cm^{-1} laser linewidth and 10 pulses per second would give an instantaneous CARS signal at least 10^{10} times larger than a cw system of P_1 = 5W and P_2 = 0.5W and 10^{-3} cm^{-1} linewidth, and a time-averaged signal at least 10^3 times larger.

Because of the high signal levels achievable in pulsed laser CARS experiments it is possible in many circumstances to get adequate signal-to-noise ratio in only a few laser pulses, and in some cases in a single laser pulse. This makes CARS a very powerful technique for observing fast transient phenomena. However, the time required to scan the tunable laser source, w_2, through the Raman spectrum of interest is typically at least several minutes, so with a conventional scanning CARS apparatus the potential of CARS for studying transient phenomena cannot be fully realized.

An approach commonly referred to as multiplex CARS overcomes this difficulty. In multiplex CARS one operates the "tunable" laser in a broadband mode to produce output over a relatively wide spectral range. Carried out with a broadband Stokes laser source coherent anti-Stokes Raman scattering generates a signal at all frequencies within the range $w_0 = 2w_1 - (w_2' \pm \Delta w_2/2)$, where Δw_2 is the spectral width of the Stokes radiation centered at w_2', accessing all Raman resonances such that $w_{rg} = w_1 - (w_2' \pm \Delta w_2/2)$. A broadband dye laser, formed by replacing the laser grating or other spectrally dispersive elements by a mirror, can be made to oscillate over a bandwidth, Δw_2, of 100 cm^{-1} or more, a spectral range adequate to span an entire Raman band for most molecular species. With such a broadband source for the Stokes beam the CARS spectrum from an entire Raman band, e.g., the vibrational Q branch of a diatomic molecule, can be generated in each laser pulse, eliminating the need for spectral scanning. To record the spectrum one can use a spectrograph to disperse the anti-Stokes signal and an optical multichannel analyzer to collect the light.

Special CARS techniques are also necessary to take full advantage of another attractive feature of coherent anti-Stokes Raman scattering, namely the potential for highly spatially resolved spectroscopy. While the beam crossing required for phase-matching [Eq. (35)] in liquids and solids will generally permit high spatial resolution, the normal collinear phase-matched geometry in gas-phase experiments usually leads to very poor spatial resolution. For this reason several two-frequency, three-beam CARS techniques have been developed to permit crossed beam geometries in dispersionless media.

One such technique is termed BOXCARS because of the shape of the phase-matching diagrams. In BOXCARS the pump beam, ω_1, is split into two parts which are separately imaged into the sample at a crossing angle 2γ as shown in the simplified experimental schematic of Fig. 7. The angles θ and ϕ are then chosen to satisfy phase-matching, and are related by:

$$2 n_1 \omega_1 \cos \gamma = n_2 \omega_2 \cos \theta + n_0 \omega_0 \cos \phi \qquad (36)$$

and

$$n_2 \omega_2 \sin \theta = n_0 \omega_0 \sin \phi . \qquad (37)$$

Since phase-matching can be satisfied for any desired spatial resolution, down to a limit set by the beam waist at the focus, increasing resolution is accompanied only by the unavoidable loss in signal intensity due to decreasing sample volume. The interaction volume and spatial resolution are determined simply by beam overlap at the focus as in the case of only two crossed beams.

3.2 Background-Free Techniques

Due to the high intensity of CARS signals one seldom encounters in the application of CARS a situation where the signal per se is too small to detect. However, one frequently finds that although the resonant signal of interest is large the non-resonant background signal is equally large or larger, and the signal-to-noise ratio is limited not by the magnitude of the resonant signal, but by the ratio of the resonant signal magnitude to the non-resonant signal magnitude.

Several special CARS techniques have been developed to reduce such problems with non-resonant background interference by separating the resonant signal from the non-resonant background

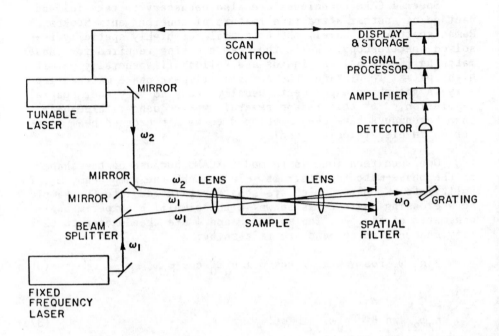

Fig. 7. Example of BOXCARS-technique apparatus in which the pump
        beam, ω_1, is split into two parts.

on the basis of the different polarization behavior of the
resonant and non-resonant third-order susceptibilities. Since
χ^{NR} and χ^{R} in general have different polarization behavior it is
possible, by appropriate choice of the direction of polarization
of ω_1 and ω_2, to generate a resonant signal with an electric
field component in a direction along which the non-resonant signal has no electric field component. By coupling this polarization orientation with polarization sensitive detection it is possible to eliminate the non-resonant contribution to the signal.
The polarization CARS techniques, CARE and ASTERISK, discriminate
against non-resonant contributions to $\chi^{(3)}$, and hence the non-
resonant signal, by exploiting the differences in the tensor
components of χ^{NR} and χ^{R}.

It is also possible to separate the resonant and non-resonant CARS signals on the basis of the different temporal behavior
of χ^{NR} and χ^{R}. The non-resonant contribution to $\chi^{(3)}$, which in

general is due to virtual electronic transitions, will have a "time constant" of the order of 10^{-14} seconds. In contrast, the resonant contribution, which is usually due to a real transition from one vibrational state to another, will have a lifetime determined by the vibrational dephasing time. The vibrational dephasing time can be as large as tens of picoseconds in condensed phase and hundreds of picoseconds in the gas phase.

3.3 Resonance CARS

Whenever any of the laser fields or the anti-Stokes signal field has a frequency equal to that of some one-photon allowed electronic transition, the coherent anti-Stokes Raman scattering will be enhanced. This effect is termed resonance CARS, and is analogous to the resonance Raman effect.

As they do in spontaneous Raman spectroscopy such one-photon resonances can lead to a very large signal enhancements in CARS spectroscopy. Increasing the signal in this way increases the ratio of the resonant CARS (i.e. Raman resonant) signal to the non-resonant background, and hence increases the signal-to-noise and improves the detectability for species in low relative concentration.

4.   APPLICATIONS OF CARS

4.1 General Applicability

In general CARS can be applied in all areas of research where spontaneous Raman spectroscopy can be used, as well as environments where spontaneous Raman is difficult or impractical to implement. CARS has been shown to be of particular usefulness in high-resolution spectroscopy, combustion diagnostics, time-resolved spectroscopy, spectroscopy of absorbing and/or fluorescing compounds, plasma diagnostics, and spatially-resolved spectroscopy. These applications of CARS are made possible by the large signal levels and high discrimination ($\sim 10^9$) against sample fluorescence and luminescence.

The high signal intensity often allows spectra to be obtained in only a few laser pulses, and even when this is not possible at much lower average powers than necessary for spontaneous Raman spectroscopy. This reduces problems associated with photolysis and degradation of absorbing samples in condensed phase. The high sensitivity of CARS, which facilitates detection

of species at densities of 10^{11} to 10^{12} cm^{-3} makes CARS suitable
for detection of photofragments, excited molecular states, and
molecular species in highly-ionized plasmas.

4.2   Combustion Diagnostics

Combustion research is probably the major application of
CARS spectroscopy. Almost all the desirable features of CARS
can be used to advantage in combustion studies. The sensitivity
facilitates single laser pulse measurements, particularly in the
multiplex configuration. The sensitivity and short pulse duration possible with pulsed lasers permits highly time-resolved
analysis of transient combustion phenomena. The coherent, dif-
fraction-limited character of the CARS signal beam makes remote
detection, which is necessary with many real combustors, simple.
The large discrimination against sample luminescence, made possible by the small signal beam divergence, eliminates many problems
encountered in doing spectroscopy in large-scale, high-tempera-
ture combustion environments. The high spatial resolution that
is possible with crossed-beam CARS techniques makes point density
and temperature measurements fairly routine. And the non-intrusive nature of CARS precludes any appreciable modification of the
sample being probed. Development of CARS for combustion diagnostics has proceeded far enough that applications to real combustors, e.g, jet engines and internal combustion engines, are now
being reported.

4.3   Plasma Diagnostics

The sensitivity and discrimination against sample luminescence inherent in CARS can be capitalized upon in making temperature and concentration measurements in electric discharges and
plasmas. Measurements of vibrational and rotational temperatures
have been made in discharges or plasmas containing nitrogen,
oxygen, hydrogen, and deuterium.

4.4   Time-Resolved Spectroscopy

There are many situations in the physical and biological
sciences in which the experimentalist needs to use a spectroscopic probe to investigate some transient species or temporally
varying phenomena. In order to be useful for any such investigation a spectroscopic method must have high sensitivity and high
temporal resolution. High sensitivity and temporal resolution
are two of the principal attributes of CARS, and as a result this

spectroscopic method is finding considerable use in the time-resolved spectroscopy. CARS is particularly useful in the investigation of time-dependent phenomena and transient chemical species because as a light scattering technique its temporal resolution is determined solely by the time width of the laser pulse used to effect it. This gives the experimentalist convenient control of the temporal resolution.

CARS experiments have been carried out with nanosecond and even subnanosecond temporal resolution to measure vibrational dephasing times, to obtain spectroscopic information on excited electronic states of molecules, to detect transient species, and to measure photofragment energy distributions.

4.5 Other Applications

CARS has been used for high-resolution molecular spectroscopy, as a non-perturbing diagnostic in free-jet expansions and molecular beams, and to study structure and dynamics in biological molecules.

THOMSON SCATTERING DIAGNOSTIC FOR INTENSE RELATIVISTIC ELECTRON BEAM EXPERIMENTS

Gary R. Allen, H. Parke Davis, and J. Chang

Sandia National Laboratories
Albuquerque, NM
USA

1. INTRODUCTION

Thomson scattering (TS) of ruby laser light by plasma electrons has become a widely used diagnostic for making non-perturbing, accurate measurements of electron temperature (T_e) and density (n_e) with good spatial and temporal resolution [1]. The application of TS reported here is unusual in its requirement for fast time resolution (~ 1 nsec), and is unique as regards the harsh experimental environment in which good signal-to-noise has been obtained [2]. The adverse environment includes high doses of X-rays and EMP, as well as high fill pressures (5-630 Torr) of air in which the plasma is created, and the attendant problems with luminescence of the air and laser breakdown of the air.

The diagnostic has been used to measure T_e and n_e in the plasma channel created by the propagation through 5 Torr of air by an intense relativistic electron beam (IREB) with 1.6 MV, 20 kA, and 70 ns pulsewidth. The IREB ionizes the air (Fig. 1) creating a cylindrical plasma channel (as in a lightning bolt) which is concentric with, and extends radically beyond the electron beam. Typical electron densities and temperatures of 10^{16} cm^{-3} and a few eV are produced, and the plasma conducts much of the return current of the electron beam.

To diagnose the plasma, a ruby laser beam, which is orthogonal to the IREB, is focussed onto the axis of the electron

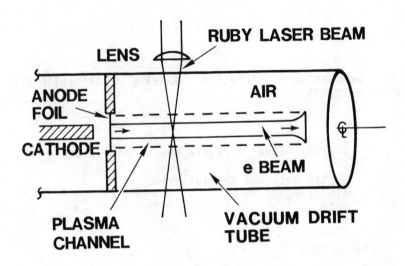

Fig. 1.   Schematic of an intense electron beam propagating
          through air, and the focussed ruby laser beam of the
          Thomson scattering diagnostic that measures the plasma
          channel parameters.

beam/plasma channel (Fig. 1). The incident, monochromatic (λ_o =
6943 Å) laser photons are scattered by the free electrons of the

plasma (Fig. 2a). In the scattering process, the laser photon is
Doppler-shifted in proportion to the velocity of the scattering
electron ($\Delta\lambda \sim \lambda_o\, v_e/c$). So, plasma electrons with some distri-

bution in velocity space will emit scattered photons with that
same distribution in wavelength space. The scattered light is
collected at an angle orthogonal to both the laser beam and the
IREB and transmitted by an optical system to a spectrometer
which disperses the wavelength distribution into a spatial distribution. An array of fiber optics bundles at the exit of the
spectrometer sorts the dispersed light into wavelength channels.
The light in each channel is detected by a photomultiplier
tube (PMT), and the electron velocity distribution function
$f(v_e)$ is determined from the signal amplitudes $S(\lambda)$ on the var-

ious wavelength channels.

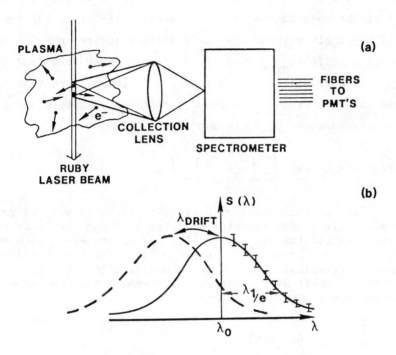

Fig. 2. Fundamentals of the Thomson scattering process.

a) Focussed ruby laser light is scatteded by plasma
electrons, dispersed by spectrometer, and detected
by an array of PMT's.

b) The solid curve shows the expected Thomson-scattered
spectrum from a Maxwellian plasma and the signals
which would be recorded on 8 PMT's, each looking at
a different wavelength channel. The ruby laser
wavelength is $\lambda_o = 6943$ Å; $\lambda_{1/e}$ is that wavelength
at which the signal is 1/e times that at λ_o. The
dashed curve represents a maxwellian that is shifted
(by an amount λ_{drift}) due to a fluid drift of the
plasma electrons.

Using several wavelength channels, as shown schematically in Fig. 2b, $f(v_e)$ is reconstructed and the following plasma electron parameters are determined: T_e from the Doppler-broadened width ($\lambda_{1/e}$) of the spectrum; n_e from the wavelength-integrated amplitude of the spectrum; and the fluid drift velocity, $v_{d,e}$, from the Doppler shift (λ_{drift}) of the entire spectrum away from λ_o. If $f(v_e)$ is Maxwellian and the geometry is chosen such that the Doppler shift due to $v_{d,e}$ is zero, then $f(v_e)$ is determined by only two independent parameters n_e and $\lambda_{1/e}$:

$$S(\lambda) = A\, n_e \, \exp\left[-\left(\frac{\lambda - \lambda_o}{\lambda_{1/e}}\right)^2\right]. \tag{1}$$

The constant, A, includes the TS cross-section, as well as geometrical, optical, and detector efficiencies, and the laser intensity. The constant is absolutely, and accurately, calibrated by substituting nitrogen for the plasma as a scattering medium (Rayleigh scattering) with a known cross-section. So, only two wavelength channels are necessary to determine the two independent parameters n_e and $\lambda_{1/e}$, where

$$\lambda_{1/e}[\overset{o}{A}] = 19.6(T_e[eV])^{1/2} \tag{2}$$

This paper reports the initial implementation of the diagnostic using only 2 PMT channels to minimize the complexity of the hardware with the following goals:

a. Establish the feasibility of getting good signal-to-noise in this uniquely harsh environment.

b. Obtain accurate measurements of T_e and n_e for comparison with theory. (Thomson scattering is the most accurate technique available for measuring these parameters.)

c. Assess the obstacles to using the diagnostic on larger accelerators at higher voltage, current, and air fill pressure, and with faster time resolution.

2.   THOMSON SCATTERING APPARATUS

The diagnostic is mounted on a 4' x 8' optical table which affords the portability and flexibility to implement is on

various experiments (Fig. 3a). The large, tapered cylinder
labeled MIMI is the transmission line of the IREB generator which
injects the electron beam into the small tube along its axis, to
the left. A tour of the apparatus, following the path of the
ruby laser, serves to highlight the features of the diagnostic.
A 5 mW HeNe laser beam, (1) aligned to the ruby beam, is used to
align all the components in the system. The ruby laser, consisting of an oscillator (2) and amplifier (3), produces a 6 mm diameter, 2 Joule beam with 2.5 mrad divergence in a 25 nsec FWHM
pulse (see Figures 3a, b for numbered component references). Part
of the laser beam is split off to a photodiode (4) to monitor the
timing of the laser relative to the electron beam. The principal
laser beam is steered by mirrors (5) (not all shown) to the
black-anodized TS section of the electron-beam drift tube (6) (see
Fig. 3b).

There, the laser beam is apertured (7) and focussed by a
150 mm focal length, A/R coated lens (8), which also serves as a
vacuum window, to a 400 μm diameter spot at the center line of
the drift tube. Beyond the focus, the laser beam is absorbed
in a beam dump (10) of Corning CS-4-72 blue glass mounted at
Brewster's angle. Thomson-scattered light is collected through
a vacuum window (11) by an objective lens (12). A set of baffles
(9) is located after the focussing lens and before the beam dump.
Each baffle tube consists of 5 apertures in series, with aperture
diameters chosen to be 30% larger than the laser beam diameter at
each point along the converging or diverging beam. The baffles
absorb any stray light which is scattered out of the laser beam
by the lens or beam dump. A viewing dump (13), identical to the
beam dump, provides a dark background against which the Thomson-
scattered light is viewed. The purpose of the baffles, viewing
and laser dumps, A/R coated lens, and black anodized surfaces
(inside the drift tube, baffles, and dumps) is to minimize the
amount of stray (non-Thomson-scattered) laser light which is collected by the objective lens.

The excellent stray light elimination allows the placement
of fiber optics wavelength channels only 7 $\overset{\circ}{A}$ away from the laser
line (at 6943 $\overset{\circ}{A}$) with zero stray light photons appearing as background noise on the TS signal, so that electron temperatures as
low as 1 eV (20 $\overset{\circ}{A}$ Doppler-broadened width) can be measured accurately.

The objective lens, along with field (14) and relay (15)
lenses images the TS volume (the intersection of the laser beam
with the acceptance cone of the objective lens) with 1:1 magnification onto the 400 μm wide entrance slit of the 2/3 meter, f/6
Instruments SA spectrometer (16). The light is dispersed by an

Fig. 3.   Photographs of the Thomson scattering diagnostic on the
          MIMI experiment.
          a) Overview of diagnostic and MIMI accelerator.

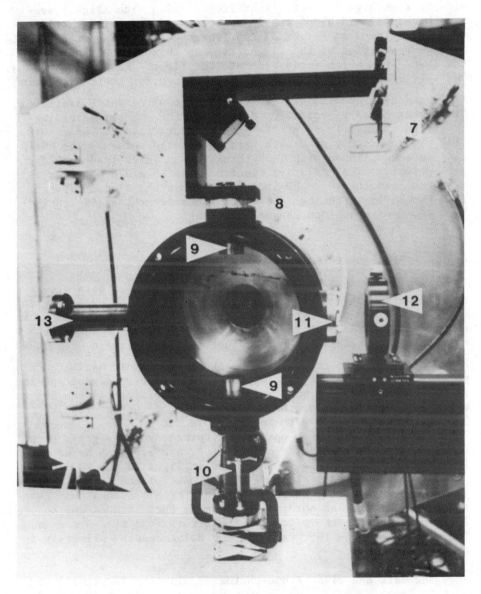

Fig. 3. Photographs of the Thomson scattering diagnostic on the
         MIMI experiment
         b) Close-up of the Thomson-scattering section of the
             8-inch diameter propagation tube for the electron
             beam.

1800 g/mm holographic grating to an inverse dispersion of 8 Å/mm at the exit of the spectrometer. There, the light is collected by bundles of plastic-clad silica fibers. Each bundle, 0.4 mm wide by 2 mm high collects light from a 3.2 Å wide slice (wavelength channel width) of the TS spectrum and transmits the light to a PMT. The 2 mm height of the fiber bundle defines the radial spatial resolution of the diagnostic, as this is the length of the laser beam which is imaged onto the fiber. Hamamatsu R928 PMT's (1 1/8" diameter, side-on) were chosen for their high gain (10^7) and high quantum efficiency (6% at 6943 Å) which assists in overcoming noise generated by X-rays and EMP. X-ray shielding sufficient to nearly eliminate X-ray luminescence in the fibers (17) and completely eliminate X-ray generated noise in the PMT's (18) is accomplished with 1/2" thick and 4" thick lead shields, respectively. The PMT's are mounted in RF-shielded housings inside the lead shield. EMP-induced noise in the signal cables from the PMT's to the oscilloscopes (in a screenbox) is reduced to ~ 10 mV with braided copper shielding.

3. EXPERIMENTAL DATA

In Fig. 4, dual-beam oscilloscope traces of TS data are shown from a shot on the MIMI accelerator at 1.6 MV, 21 kA, with a 70 nsec pulsewidth. The peak intensity of the ruby laser beam coincided with the peak current of the IREB (45 nsec into the electron beam pulse) at the TS volume. At this time n_e, T_e, and X-ray flux are also expected to be at or near their maxima. On each oscilloscope photo, the upper trace is the TS signal and the lower trace contains timing information. The width of the TS signals is determined by the 25 nsec width of the laser pulse. The small precursor to each TS signal is due to X-ray induced luminescence in the fiber optics. It preceeds the signal, even though the peak intensities of the laser and electron beams were synchronized, because the path of the TS light to the PMT's is circuitous and exceeds, in length, the path of the X-rays. This luminescence is easily shielded with additional lead around the fibers. There was no observable noise on these traces due to light emitted from the plasma or the air (5 Torr fill pressure). The ~ 20 mV hash on the traces in EMP noise induced primarily in the PMT's.

The left and right scope photos display the TS signals obtained from wavelength channels which are blue-shifted by 10 Å and 27 Å, respectively, from μ_o. The 10 Å channel signal corresponds to only 250 detected photons out of the 10^{19} photons which

$\Delta\lambda_1 = 10 \overset{\circ}{A}$ $\Delta\lambda_2 = 27 \overset{\circ}{A}$

$S_1 = 198 \pm 8$ mV $S_2 = 133 \pm 12$ mV

$$T_e = 4.24 \pm 0.85 \text{ eV } (\pm 20\%)$$

$$n_e = 1.86 \pm 0.22 \times 10^{15} \text{ cm}^{-3} (\pm 12\%)$$

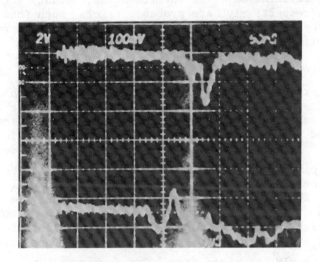

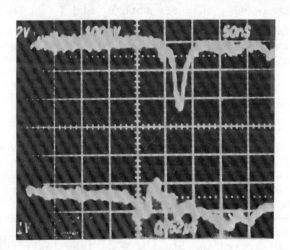

Fig. 4.   Oscilloscope records of the Thomson-scattered signals
           from the two wavelength channels on the MIMI experiment.

were focussed into the plasma. This emphasizes the necessity of
the care taken to eliminate stray laser light, as well as to reduce the X-ray and EMP induced noise levels.

Assuming that $f(v_e)$ is Maxwellian, T_e (= 4.24 $\pm$ 0.85 eV) is
determined from the ratio of amplitudes of the two signals, and
n_e (=1.86 $\pm$ 0.22 x 10^{15} cm^{-3}) is determined from the Rayleigh
scattering calibration. The uncertainties, amounting to $\pm$ 20%
and $\pm$ 12%, respectively, are probable errors, such that the true
value of the measured quantity lies within the stated range with
50% probability.

Measurements of T_e and n_e were similarly made on two other
MIMI accelerator shots but with the ruby laser synchronized at
times following the electron beam pulse. However, the unex-
pectedly large signals drove the PMT's beyond their linear response, and the data from these two later-time shots is highly
uncertain. The TS measurements of n_e and T_E from these three
shots are plotted vs. time (measured relative to the beginning of
IREB current pulse) in Figures 5a and 5b, respectively. The
error bars show the probable errors vertically and the 25 ns
pulsewidth of the laser horizontally. Also shown are the results
from the Air Propagation Code (APC) computer calculations of the
expected $n_e(t)$ and the $T_e(t)$ given the measured operating parameters of the MIMI electron beam. The agreement between theory
and experiment is very good for both n_e and T_e on the early-time
shot. In addition to the less accurate data on the two late-time
shots, the air chemistry calculations in the APC are not reliable
beyond 100 ns.

4.   IMPROVEMENTS AND PROBLEMS

In on-going experiments on more powerful beam accelerators
at higher electron beam voltages and currents and even shorter
pulse widths (20 ns rather than 70 ns), the emphasis is to obtain
faster time resolution (4 ns rather than 25 ns laser pulse) and
diagnose the plasma primarily during the 20 ns interval when the
electron beam is propagating. Furthermore, it is of interest to
measure these parameters during the 5 to 10 ns current rise time
of the electron beam with ~ 1 ns time resolution. It is relatively straightforward to obtain a 4 ns ruby laser pulse with an
existing laser pulse slicer driven by a laser-triggered spark gap
(LTSG). This capability is currently being developed on the TS
diagnostic. Furthermore, a 1 ns pulsewidth can be generated
using the fast LTSG-driven pulse slicer as described in the ac-
companying report on the subnanosecond holographic interferometry

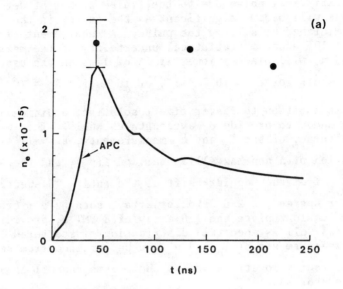

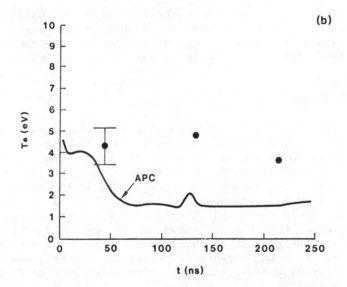

Fig. 5.   Thomson scattering measurements of electron density
          a) and temperature b) on three different MIMI shots at
          different times during and after the MIMI current pulse.
          The results at t = 45ns are from the data shown in
          Fig. 4.

diagnostic. The faster laser pulses will improve the signal-to-noise since the faster TS signal can be more easily discriminated against the slower changing noise signatures; however, the quantum statistical noise due to the finite number of detected TS photons will become significant as the energy in the laser pulse is diminished by slicing the pulse. A requirement of no worse than $\pm$ 10% quantum statistical uncertainty on the measurement of T_e and n_e will place a lower limit of 1 ns on the usable laser pulse width for $n_e = 10^{16}$ cm^{-3}, or 10 ns for $n_e = 10^{14}$ cm^{-3}.

In addition to faster time resolution, additional PMT's are being added to provide 8 wavelength channels. This will increase the accuracy of the T_e and n_e measurements, as well as permit observation of a non-Maxwellian tail on $f(v_e)$, and the measurement of $v_{d,e}$ from the Doppler-shift of the entire TS spectrum. As a further upgrade, a 2-D detector array, such as a micro-channel plate, could replace the 1-D array of 8 PMT's, providing a spatial (as well as spectral) distribution of scattered light so that measurements of T_e and n_e can be obtained from several ($\sim$ 10) radial locations across the plasma channel on a single accelerator shot.

Finally, the laser and detectors are being installed outside the concrete shielding walls, as opposed to the proximity installation shown in Fig. 4a. This will alleviate the need for overwhelming lead shielding on the fiber optics and PMT's.

5.   CONCLUSION

A Thomson scattering diagnostic employing 2 wavelength channels has been implemented in a harsh radiation environment, and accurate measurements of electron density and temperature have been made. The diagnostic is presently being installed on a more powerful electron beam accelerator (4 MV, 50 kA, 20 ns) where the electron beam will propagate through 100 Torr of air. On-going improvements to the diagnostic include faster time resolution, better immunity to the radiation environment, and additional data channels. The goal is to maintain or improve the accuracy of the measurements in the harsher radiation environment, and assess the scalability of the diagnostic to even more powerful electron beam accelerators.

REFERENCES

1. J. Sheffied, Plasma Scattering of Electromagnetic Radiation
    (Academic, New York, 1975.

2. G. R. Allen et al., Proceedings of the Fifth International
    Conference on High-Power Particle Beams, San Francisco
    (Sept. 1983), pp. 362-5, and references therein.

APPENDIX A

RANKING OF DIAGNOSTIC TECHNIQUES

TABLE A-1. ELECTRO-OPTICAL AND MAGNETO-OPTICAL MEASUREMENTS

Diagnostic	Characteristic Time	Applicability Range	Cost	Versatility	Comments
Kerr Effect					
$\bar{E}$	< 1 ns	10 kV/cm - 100 kV/cm	Moderate	Excellent	· Trade off between time response and sensitivity
V	< 1 ns	5 kV - 10 MV	Moderate	Excellent	· Can extrapolate over a wide range
ρ	100 ns	$10^{-8} - 10^{-4}$/cc	Moderate	Limited	· Δn quadratic in $\bar{E}$
Carrier Mobility	100 ns	$10^{-8} - 10^{-7}$ m^2/V-s			· Provides electrical isolation
					· Limited to reasonably birefringent liquids
					· Solutions are usually good solvents
					· Uncertainty less than 1 % but recording technique may be 5% or greater
					· Somewhat sensitive to field geometry
					· Conductivity and space charge effects .
Pockels Effect					
$\bar{E}$	0.001 - 0.1 ns	1 mV/cm - 100 kV/cm	Expensive	Excellent	· Solid media rather than liquid
V	0.001 - 0.1 ns	50 V - 1 MV	Expensive	Excellent	· Intrusive but electrically isolated
					· Δn linear in $\bar{E}$
					· Materials generally piezoelectric and sensitive to shock
					· Two linear regions of sensitivity
					· Long devices are sensitive but slow due to transient times

TABLE A-1. ELECTRO-OPTICAL AND MAGNETO-OPTICAL MEASUREMENTS (CONT'D)

Diagnostic	Characteristic Time	Applicability Range	Cost	Versatility	Comments
Pockels Effect					· Fastest devices are special installations
					· Uncertainty in measurement dominated by recording system ($\sim$5%)
Faraday Effect					· Linear effect
$\bar{H}, \bar{B}$	0.1 ns	1 G - 5 MG	Expensive	Excellent	· Electrical isolation
I	0.1 ns	50 A - 10 MA	Expensive	Excellent	· Small T dependence for diamagnetic materials, larger for paramagnetic
					· Trade off between time resolution and sensitivity
					· Sensitive to shock

TABLE A-2. CONVENTIONAL MEASUREMENT TECHNIQUES FOR VOLTAGE AND CURRENT

Diagnostic	Characteristic Time	Applicability Range	Cost	Versatility	Comments
Voltage					
Resistive Dividers	~ 100 ps	< 50 kV	Cheap	Limited	· 100-ps measurement requires TEM00 mode and considerable experimental techniques
	~ 1 ns	< 500 kV	Cheap	Excellent	· Connections to system ground
	~ 10 ns	> 500 kV	Cheap	Excellent	· Rise time geometry dependent
					· Intrusive
					· Usually requires secondary divider
					· Calibration not always at test conditions
					· Uncertainty on order of 5%
Capacitive Dividers	~ 1 ns	> 1 MV	Cheap	Excellent	· Intrusive measurement
	~ 50 ps	> 100 kV	Moderate	Limited	· Usually custom made
					· Calibration difficult
					· Low frequency cutoff ~ RC
					· 5 - 10% uncertainty in measurements
Current					
CVR/Shunt	50 ps	10 kA	Moderate	Moderate	· Energy limited
	> 500 ps	50 kA	Moderate	Moderate	· Thermal sensitive
	10 ns	1 MA	Moderate	Moderate	· d.c. coupled
					· Noise source
					· Skin depth factors

TABLE A-2. CONVENTIONAL MEASUREMENT TECHNIQUES FOR VOLTAGE AND CURRENT (CONT'D)

Diagnostic	Characteristic Time	Applicability Range	Cost	Versatility	Comments
Current (Cont'd)					
Trans. Line Current Sensor	> 500 ps	Milliamperes to Megaamperes	Moderate	Moderate	• Uncertainty on order of 5% • Requires azimuthal symmetry of magnetic field • Does not require integrator • Limited time range • Very high sensitivity 75 mA, ~ 1V/A
Rogowski Coil	0.5 - 1 ns	1 A - 5 MA	Cheap	Excellent	• Leaks in strong B field • Intrusive • Requires integrator • Difficult to obtain current standard for calibration • Position independent • Uncertainty on order of 5%
Current Transforms	10 ns	1A - 5 MA	Moderate	Excellent	• Does not require integrator • Volt-sec core limit • Intrusive • Uncertainty on order of 5%
Stripline Coil, di/dt	10 ps	< 10^{12} A/sec	Moderate	Moderate	• Intrusive • Can average azimuthal non-uniformities • Uncertainty on order of 5%

TABLE A-3. MISCELLANEOUS MEASUREMENT TECHNIQUES

Diagnostic	Characteristic Time	Applicability Range	Cost	Versatil- ity	Comments
Fields					
D-Dot Probe	~ 100 ps	Limited to modest fields, 50 - 100 kV/cm	Expensive	Moderate	· Rise time related to sensor size · Free-field measurement · Geometry sensitive · TEM00 mode for fast rise · Uncertainty 10% or greater · Requires primary standard
B-Dot Sensor	< 1 ns	500 V/m - 1 MV/m	Cheap	Moderate	· Speed related to size · Position dependent/ calibrate in place · Can average azimuthal nonuniformities · Requires primary standard · Uncertainty on order of 5%
Current					
Θ Probe	< 1 ns	1 - 10 kA	Cheap	Moderate	· Low frequency cutoff ~ L/R
Faraday Cup	5 ns	0.1 - 2.5 kA/cm^2	Moderate	Limited	· Susceptible to back-ground plasma · Calibration a function of J · Requires experimental finesse · About 5% uncertainty in measurements

TABLE A-4. SUMMARY OF OPTICAL TECHNIQUES FOR MEASURING GEOMETRY/VELOCITY

Diagnostic	Characteristic Time	Applicability Range	Cost	Versatility	Comments
Imaging Cameras					
Single Frame	3×10^{-5} ns	Extensive	Expensive	Excellent	· Calibrated using standard resolution chart
Rotating Mirror (Multiframe)	50 ns	Extensive	Expensive	Excellent	· Extremely short time resolution demands trade off with complexity and accuracy
Image Converter (Multiframe)	< 1 ns	Extensive	Expensive	Excellent	· Velocity measurement requires metric in image
Speckle	< 1 ns	Limited	Modest	Limited	· Multiframe system complex
					· May require intense light source
					· 1.6-ns frame rate commercially available
					· Used to measure changes in surface topology
					· Resolution technique specific, 10-10000 line pair/nm
					· Image converter technique approaches single photon sensitivity
					· Uncertainty is experiment specific - typically a few %
Streak Cameras					
Single Mirror	1 ns	Extensive	Expensive	Excellent	· Optically passive
Multimirror	0.1 ns	Extensive	Expensive	Limited	· Need intense light source
Image Converter	5×10^{-4} ns	Extensive	Expensive	Excellent	· Can be used to measure relative optical intensity
					· Interpretation sometimes difficult

TABLE A-5. REFRACTIVE INDEX MEASUREMENTS

Diagnostic	Characteristic Time	Applicability Range	Cost	Versatility	Comments
Index of Refraction, N					
Pulsed Laser Interferometer	< 0.1 ns	Can be employed over a wide range of parameters	Moderate	Excellent	· Nonperturbing · Accuracy depends upon symmetry · Uncertainty $\sim 10\%$ · Interpretation sometimes difficult · Sensitivity limits
Gradient N Schlieren	< 0.1 ns	Can be employed over a wide range of parameters	Moderate	Moderate	· Nonperturbing · Qualitative · Low accuracy in measurement · "Streaked" to obtain velocity · Interpretation · Sensitivity limits
Second Derivative of N	< 0.1 ns	Can be employed over a wide range of parameters	Moderate	Moderate	· Nonperturbing · Difficult to obtain quantitative results · Interpretation · Sensitivity limits
N, dN/dx, d^2N/dx^2	≤ 10 ps	10^{18}-10^{21}/cc 1 µm spatial resolution, velocity to 3×10^8 cm/sec	Moderate	Excellent	· Nonperturbing · Multiphenomena recorded on a single plate
Holographic Interferometry					· Highly stable and after-the-fact focusing · Interpretation sometimes difficult · Multiframe system is often a primary diagnostic · Uncertainty on order of 10%

TABLE A-6. X-RAY TECHNIQUES FOR FAST DIAGNOSTICS

Diagnostic	Characteristic Time	Applicability Range	Cost	Versatil- ity	Comments
Flash X-Ray					
All Imaging Techniques - Telescopes/ Microscopes	< 1 ns	Limited	Expensive	Limited	· Basic technique is simple. Experimental arrangement/application may lead to complex custom designs · Usually used to measure "shape" or "velocity" in an obscuring media · Limited to a few frames · Frame rate and number pulser-dependent · Frame rated up to 10^9/sec · Must use 1 pulser/frame at high rates
X-Ray Streak	0.02 ns	Limited	Expensive	Limited	· Custom design · Used as a diagnostic in fusion-pellet compression experiments

TABLE A-7. EMISSION SPECTROSCOPY DIAGNOSTICS

Diagnostic	Characteristic Time	Applicability Range	Cost	Versatility	Comments
State Ident					
Optical Region	< 1 ns	> 1 part in 10^{12}/cc	Inexpensive	Excellent	· Flexible · Standard technique · Large uncertainty in measurements
X-Ray Region	15 ps	1 part in 10^4/cc	Expensive	---	· Large uncertainty in measurements · Flexible technique · Custom designs
Ion/Neutral Density					
Optical Region	10 ns	$10^{12} - 10^{18}$/cc	Inexpensive	Good	· Must understand theoretical basis for multicomponent system for analysis · Large uncertainty
X-Ray Region	15 ps	$10^{15} - 10^{23}$/cc	Moderate	Excellent	· Total population inferred from excited population · Needs good spectral resolution
Ion/Neutral Temperature					
Optical Region	10 ns	0.2 - 10 ev	Inexpensive	Limited	· Complicated by line broadening by competing processes · Uncertainty on order of 20%

TABLE A-7. EMISSION SPECTROSCOPY DIAGNOSTICS (CONT'D)

Diagnostic	Characteristic Time	Applicability Range	Cost	Versatility	Comments
Ion/Neutral Temperature (Continued)					
X-Ray Region	15 ps	100 ev - 1 kev	Moderate	Limited	· Line broadening by competing processes · Uncertainty greater than 20%

TABLE A-8. SUMMARY OF ACTIVE OPTICAL TECHNIQUES

Diagnostic	Characteristic Time	Applicability Range	Cost	Versatility	Comments
Ion/Neutral Density					
Laser Induced Florescence (LIF)	~ 1 ps	10^{10} - 10^{19}/cc	Expensive	Good	· Requires good experimental technology · Signal averaging · Complex · Large uncertainty in measurements
Coherent Anti-Stokes Raman Scattering (CARS)	~ 1 ps	10^{12} - 10^{18}/cc	Moderate	Good	· About 10% relative uncertainty · Excellent in hostile environment
Elect Temperature					
Thomson Scattering	~ 1 ns	1 - 5 keV (10^{12} - 10^{19}/cc)	Moderate	Excellent	· Useful in harsh environment · Versatile technique · Adaptable to many geometries · Nonintrusive · 10 - 20% uncertainty
Species Indent					
LIF	1 ps	10^{10} - 10^{14}/cc	Expensive	Excellent	· Signal Averaging needed · Nonintrusive · Difficult experimentally · Laser limited

TABLE A-8. SUMMARY OF ACTIVE OPTICAL TECHNIQUES (CONT'D)

Diagnostic	Characteristic Time	Applicability Range	Cost	Versatility	Comments
Species Ident (Continued)					
CARS	1 ps	> 10^{12}/cc	Moderate	Excellent	· Excellent for time-dependence environment · Excellent in hostile environment · Useful for a wide range of transients · Laser limited · About 10% relative uncertainty · All species · 100 μm spatial resolution
Multiphoton	1 ps	Only for weakly ionized, low n_e, T_e (1 part in 10^9)	Expensive	Moderate	· Neutral detection 1 part in 10^9 · Absolute accuracy difficult · All atoms and molecules · Laser limited
State Ident					
CARS	1 ps	> 10^{12}/cc	Moderate to expensive	Excellent	· Good spatial resolution · All species · Insensitive to stray fields · Atomic or molecular · Laser limited

TABLE A-8. SUMMARY OF ACTIVE OPTICAL TECHNIQUES (CONT'D)

Diagnostic	Characteristic Time	Applicability Range	Cost	Versatility	Comments
State Ident (Continued)					
LIF	1 ps	> 10^{10}/cc, > 100 ppm	Expensive	Moderate	· Limited by proper laser choice · Atomic or molecular
Multiphoton	1 ps	---	Expensive	Moderate to good	· Single-state sensitive · All species · Neutrals
Fields					
(LIF) dE/dT	1 ns	E > 1 kV/cm	Moderate	Poor	· Restricted to plasmas with neutral He · Large uncertainty · Difficult experimentally · Sensitive to $\bar{E}$, not $\dot{E}$
$\bar{B}$ Zeeman Splitting	1 ns	---	Cheap	Poor	· Analysis difficult
Ion/Neutral Temperature					
LIF	ps	---	Expensive	Moderate	
Elect Energy Dist Fct					
Thomson Scattering	ns	---	Expensive	Limited	· Difficult measurement · Automatic data processing helpful · Small σ

TABLE A-8. SUMMARY OF ACTIVE OPTICAL TECHNIQUES (CONT'D)

Diagnostic	Characteristic Time	Applicability Range	Cost	Versatility	Comments
ELECT DENSITY					
Thomson Scattering	ps - ns	$10^{12} - 10^{19}$/cc	Expensive	Limited	• Trouble in weakly ionized plasmas • Small σ • Resolution related to laser power • Laser limited

PARTICIPANTS

Alff, J. J.
Swiss Federal Institute of
  Technology
High Voltage Laboratory
Physikstrasse 3
8092 Zurich
Switzerland

Baldo, G.
Padova University
6/A Gradenigo
3S100 Padova
Italy

Bauer, W.
Kernforschungszentrum
Karlsruhe, 1K
Postfacn 3640
D-7500 Karlsruhe
Federal Republic of Germany

Baum, C.
NTATT
Kirtland AFB, NM 87117
USA

Benenson, D. M.
State University of
  New York-Buffalo
Department of Electrical
  and Computer Engineering
4232 Ridge Lea Road
Amherst, New York 14226
USA

Bertazzi, A.
CESI
Via Rubattino, 54
20134 Milano
Italy

Biero, H.
CEA
Centre de Vaujours
BP7
93270 Sevran
France

Bischel, W. K.
Stanford Research Institute
Molecular Physics Laboratory
MS-PN-027
333 Ravenswood Avenue
Menlo Park, CA 94025
USA

Bradley, L. P.
Lawrence Livermore National
  Laboratory, L-490
P. O. Box 5508
Livermore, CA 94550
USA

Brandelik, A.
Kernforschungszentrum Gmbh
Postfach 3640
D-7500 Karlsruhe
Federal Republic of Germany

Busch, G. E.
KMS Fusion, Inc.
P. O. Box 1567
Ann Arbor, MI 48106
USA

Camarcat, N.
Commissariat a l'Energie
  Atomique
Centre d'Etudes de Valduc
B.P. 14-21/IS SUR TILLE
France

Capellos, C.
Energétic Materials Division
LCWSL, AARADCOM, EMD
Bldg. 3022, DRDAR-LCE-D
Dover, NJ 07801
USA

Chandler, G.
Los Alamos National
  Laboratory
MS-F639
P. O. Box 1663
Los Alamos, NM 87545
USA

Chang, J.
Sandia National Laboratories
Diagnostics Division 1234
Albuquerque, NM 87185
USA

Chatterton, P.
University of Liverpool
Department of Electrical
  Engineering and Electronics
Liverpool L69 3BX
United Kingdom

Clifford, J. R.
Advanced Technology Division
Air Force Weapons Laboratory
AFWL/NTY
Kirtland AFB, NM 87117
USA

Cooke, C. M.
Massachusetts Institute of
  Technology
High Voltage Research
  Laboratory
155 Massachusetts Avenue
Cambridge, MA 02139
USA

Countney-Pratt, J. S.
American Bell, Inc.
Crawfords Corner Road
Nolmdel, NJ 07733
USA

Davis, S.
Physical Sciences, Inc.
DASCOB Research Park
P. O. Box 3100
Andover, MA 01810

Day, R. H.
Los Alamos National Lab
MS-D410
P. O. Box 1663
Los Alamos, NM 87545
USA

Demske, D. L.
Naval Surface Weapons Center
White Oak, Detonation Physics
  Branch, Energetic Materials
  Division
Silver Spring, MD 20910
USA

DiCapua, M.
Physics International
2700 Merced Street
San Leandro, CA 94577
USA

Eastham, D. A.
Daresbury Laboratory
Daresbury, NR
Warrington WA4 4AD
Cheshire
United Kingdom

Favre, M.
Imperial College
Physics Department
London SW7 2BZ
United Kingdom

Fenneman, D. B.
Naval Surface Weapons Center
F-12
Dahlgren, VA 22448
USA

Fiorito, R.
Naval Surface Weapons Center
Code R-41
Silver Spring, MD 20910
USA

Flatau, C.
Tektronix Europe B.V.
P. O. Box 8271180AV
Amstelveen
The Netherlands

Forman, P. R.
CTR-8, MS-F639
Los Alamos National Laboratory
P. O. Box 1663
Los Alamos, NM 87545
USA

Germer, R.
Technical University of Berlin
Zieten Str. 2
D Berlin 30
Federal Republic of Germany

Graf, W.
SRI International
Location 40477
333 Ravenswood Ave.
Menlo Park, CA 94025

Griem, H.
Department of Physics
University of Maryland
College Park, MD 20742
USA

Gripshover, R.
Naval Surface Weapons Center
Code F-12
Dahlgren, VA 22447
USA

Guenther, A. H.
Chief Scientist
Air Force Weapons Lab/CA
Kirtland AFB, NM 87117
USA

Gundersen, M. A.
University of Southern
California
Department of Electrical
Engineering
SSC 420, MC-0484
University Park
Los Angeles, CA 90089
USA

Hansen, N. E.
Lawrence Livermore National
Laboratory
Mail Stop L-45
P. O. Box 808
Livermore, CA 94550
USA

Hebner, R.
Building 220, Room B-344
National Bureau of Standards
Washington, DC 20234
USA

Hoeft, L.
BDM Corporation
1801 Randolf Road, SE
Albuquerque, NM
USA

Hofer, W. W.
Lawrence Livermore National
Laboratory
P. O. Box 808
Livermore, CA 94550
USA

Hugenschmidt, M.
Deutsch-Franzosisches
Forchungsinstitut
12, rue de l'Industrie
68003 Saint Louis
Haut-Rhin
France

Humphries, Jr., S.
University of New Mexico
Department of Chemical and
   Nuclear Engineering
Farris Engineering Center
Albuquerque, NM   87131
USA

Hyder, Jr., A. K.
Auburn University
202 Samford Hall
Auburn, AL   36849
USA

Jamet, F.
Institut Franco-Allemand
de Recherches de Saint-Louis
Boite Postale No. 301
68301 St. Louis, Cedex
France

Kapetanakos, C. A.
Naval Research Laboratory
Department of the Navy
Code 4704
Washington, DC   20375
USA

Koehler, H. A.
Lawrence Livermore National
   Laboratory
University of California
P. O. Box 808
L-43
Livermore, CA   94550
USA

Krehl, P.
Ernst Mach Institut
Fraunhofer Institut fur
   Kurzzeitdynamik
Eckerstasse 4
D-7800 Freiburg
Federal Republic of Germany

Kristiansen, M.
Department of Electrical
   Engineering
Texas Tech. University
Lubbock, TX   79409
USA

Krompholz, H. G.
Texas Tech. University
Department of Electrical
   Engineering
P. O. Box 4439
Lubbock, TX   79409
USA

Kunhardt, E. E.
Polytechnic Institute of
   New York
Rt. 110
Farmingdale, NY   11735
USA

Kuscher, G. F.
Ernst Mach Institute of the
   Fraunhofer Company
Eckerstr. 4
7800 Freiburg
Federal Republic of Germany

Laghari, J. R.
State University of New York-
   Buffalo
Department of Electrical and
   Computer Engineering
4232 Ridge Lea Road
Amherst, NY   14226
USA

Leeper, R.
Sandia National Laboratories
Diagnostics Division 1234
Albuquerque, NM 87175
USA

Luessen, L. H.
Head, Directed Energy
  Branch (F12)
Naval Surface Weapons Center
Dahlgren, VA 22448
USA

Mentzoni, M.
University of Oslo
Institute of Physics
Box 1038
Blindern Oslo 3
Norway

Miles, H. T.
Department of Physics
University College of Swansea
Singelton Park
Swansea, SA2 8PP
United Kingdom

Molen, G. M.
Old Dominion University
Department of Electrical
  Engineering
Norfolk, VA 23508
USA

Mourou, G.
University of Rochester
Lab for Laser Energetics
250 East River Road
Rochester, NY 14623
USA

Nahman, N. S.
National Bureau of Standards
EM Fields Division
Fields Characterization
  Group, 723.03
325 Broadway
Boulder, CO 80303
USA

Nicholas, A.
Commissariat a l'Energie,
  Atomique
B.P. 14
21120 IS-SUR-TILLE
France

Niemeyer, L.
Brown Boveri Research Center
CH 5405 Baden
Switzerland

Nolting, E. E.
Naval Surface Weapons Center
White Oak Laboratory
Silver Spring, MD 20910
USA

Nudelman, A.
Princeton Plasma Physics
  Laboratory
P. O. Box 451, Bldg. 1-N
Princeton, NJ 08544
USA

Peacock, N. J.
Culham Laboratory
Division B UKAEA
Abingdon, Oxfordshire
OS 14 3DB
United Kingdom

Perryman, R. A. G.
The Welding Institute
Abington Hall
Abington, Cambridge CB1 6AL
United Kingdom

Pfeiffer, W.
Fachebiet Elektrische
Meβtechnik, Fachbereich 17
Technische Hochschule
  Darmstadt
Schloβgraben 1
6100 Damrstadt
Federal Republic of Germany

Popovics, C.
Laboratorie de Physique des
  Milieux Ionises
Ecole Polytechnique
91128 Palaiseau Cedex
France

Price, R. H.
Los Alamos National Lab
E1-MS E526
Los Alamos, NM 87545
USA

Proud, J. M.
GTE Laboratories
40 Sylvan Road
Waltham, MA 02254
USA

Pugh, Jr., H. L.
Air Force Office of Scientific
  Research, Building 410
Bolling AFB, DC 20332
USA

Qureshi, A. H.
Wayne State University
Electrical and Computer
  Engineering
Detroit, MI 48202
USA

Raleigh, M.
Naval Research Laboratory
Code 4763
Washington, DC 20375
USA

Rizzi, G.
CESI
Via Rubattino, 54
20134 - Milano
Italy

Rose, M. F.
Director, Space Power
  Institute
Auburn University
Auburn, AL 36849
USA

Rutgers, W. R.
N. V. KEMA
P. O. Box 9035
6800 ET Arnhem
The Netherlands

Schoenbach, K. H.
Texas Tech University
Department of Electrical
  Engineering
P. O. Box 4439
Lubbock, TX 79409
USA

Schon, K.
Physikalisch Technishe
  Bundesanstalt
Bundesallee 100
D-3300 Braunschweig
Federal Republic of Germany

Schwab, A. J.
High Voltage Research
  Laboratory
University of Karlsruhe
Karlsruhe
Federal Republic of Germany

Schweickart, D. L.
Air Force Wright Aeronautical
  Labs
Aero Propulsion Lab
AFWAL/POOS-2/Bldg. 450
Wright Patterson AFB
Dayton, OH 45433
USA

Schwirzke, F.
Naval Postgraduate School
Department of Physics
Monterey, CA 93940
USA

Seeboeck, R.
CERN, European Organization
  for Nuclear Research
PS Division
CH-1211 Geneva 23
Switzerland

Sethian, J. D.
Naval Research Laboratory
Code 4762
Washington, DC   20375
USA

Ksvarenina, T. L.
Air Force Institute of
   Technology
AFIT/ENG
Wright Patterson AFB
Dayton, OH   45443
USA

Stamper, J.
Naval Research Laboratory
Code 4771
Washington, DC   20375
USA

Streiff, S.
Nicolet Oscilloscope Division
5225 Verona Road
P. O. Box 4288
Madison, WI   53711
USA

Sudarshan, T. S.
University of South Carolina
Electrical & Computer
   Engineering
Columbia, SC   29208
USA

Teich, T. H.
FG Hochspannung
Federal Institute of
   Technology
ETHZ
Physikstrasse
3 CH 8092 Zurich
Switzerland

Thompson, J. E.
University of Texas-Arlington
Electrical Engineering
   Department
Arlington, TX   76019
USA

Tondello, G.
Padova University
6/A Gradenigo
3S100 Padova
Italy

Tuma, D. T.
Carnegie-Mellon University
Electrical Engineering
   Department
Schenley Park
Pittsburgh, PA   15213
USA

Valentini, J. J.
Los Alamos National
   Laboratory
P. O. Box 1663
Los Alamos, Nm   87545
USA

Vitkovitsky, I.
Head, Plasma Technology
   Branch
Code 6670
Naval Research Laboratory
Washington, DC   20375
USA

Vossenberg, E. B.
European Organization for
   Nuclear Research, CERN,
CH-1211, Geneva 23
Switzerland

Warren, F. T.
University of South Carolina
Electrical & Computer
   Engineering
Columbia, SC   29208
USA

Wiesinger, D. J.
Hochschule der Bundeswenr,
   Munchen
ET/WE 7 E1 Engergieversorgung
D-8014 Neubiberg
Federal Republic of Germany

Williams, F.
Texas Tech University
Department of Electrical
  Engineering
Lubbock, TX 79409
USA

Wolzak, G. G.
Eindhoven University of
  Technology
P. O. Box 513
6600 MB Eindhoven
The Netherlands

Zahn, M.
Massachusetts Institute of
  Technology
High Voltage Research Laboratory
Building N-10
Department of Electrical
  Engineering and Computer
  Science
Cambridge, MA 02139

Abel inversion 702, 704, 773, 793
Abel transform 905
Abelian integral equation 664
analog-to-digital converter (ADC) 121
anti-Stokes (AS) stimulated Raman scattering 989
anti-Stokes/Raman technique 989

B-dot sensors 82, 114
   8AL (8 axis loop) 104
   8AL-1A B-dot loops 104
   MGL-3 B-dot sensors 99
   MGL-3A B-dot sensor 101
   OML-1A(A) B-dot sensor 96
birefringence 44, 45, 47, 48, 425, 785
   distributed linear birefringence 44
BLT equation 467, 469-470, 511-512, 532,
           535-536, 538-539, 541, 544-545
bolometry 796, 798, 803
   foil bolometer 803
Briet-Wigner resonance reaction 275

calorimeter 796, 803
capacitive divider 151, 178, 182, 184, 445
   coaxial capacitive divider 150, 163, 164
   geometric capacitive divider 194

carbon composition resistors 175, 197, 204, 205, 207
Cerenkov angle 836
Cerenkov light 836
channeltron 864
coherrent anti-Stokes Raman spectroscopy (CARS) 972, 974, 1003
   CARS apparatus 1012, 1015-1016
   CARS spectroscopy 1015, 1019, 1020
   CARS techniques 1017-1018, 1020
continuous fourier transformation (CFT) 354
convolution (Duhamel's) integral 445

current viewing resistor (CVR) 241
current sensors, 110
  circular, parallel mutual-impedance sensor (CPM) 92
  I-dot, one-turn, insertion unit (I1I) 92
  flush moebius mutual inductance (FMM) 110, 116,
  FMM-1A J-dot sensor 112
  OMM-1A I-dot sensors 110
  Outside core I (OCI) 115
  Outside Moebius Mutual Inductance (OMN) 110, 115

D-dot sensors 82, 86
  ACD sensor 87
  ACD-5A(A) D-dot sensor 101
  asymptotic conical dipole (ACD) 87, 186
  flush plate dipole (FPD) 88, 116, 121, 187
  FPD-1A D-dot sensor 89
  hollow spherical dipole (HSD) sensor design 86
  HSD sensors 86-87
  HSD-2B(R)- HSD-4A(R) D-dot sensor 86, 87
diamatrices 470, 513, 518-519
differentiating sensor 184, 189, 228, 232
direct time domain deconvolution 357, 360
discrete fourier transformation (DFT) 354, 360, 414
  DFT domain 374-375, 377-378, 379, 384, 399
Doppler width 890, 894, 927, 982-983, 987, 990, 1026-1027
Doppler-broadening lines 906, 926, 956

eigenmodes 470, 493, 500, 502, 508-509, 516, 540-541, 543
eigensuper vectors 540
eignevalues 470, 495, 540, 542
electric field sensors 86, 106
  parallel mesh dipole (PMD) 114
  PPD sensor 92
  PPD-1A(R) E sensor 91
EMP measurements 75, 106, 469, 549, 551, 553, 1023

FABRY-PEROT etalon 654
Fabry-Perot-type resonator 646
Faraday effects 30, 32, 41, 52, 781
Faraday rotation 24, 44-45, 47-48, 692
            709-712, 778-781, 793, 786
fast fourier transformation 360
fiber optics 41, 419, 572, 835-837, 844, 1024
  active mode 836
  fiber-optic coupling 633
  passive fiber optic system 836
  single-mode fiber 42-43, 423, 425, 428

flash x-radiography (FXR) 863, 867
  flash x-ray cinematography 855-857
  flash x-ray system 847, 853, 859
  flash-radiography 820, 845, 852, 854, 858, 867
  FXR source 867-869, 871 873, 879
  FXR system 863, 867, 876, 879, 880

fluorescence detection 985, 994-996
  high-speed fluorescers 844
framing photography 754
Fresnel bi-prism 706
Fresnel coefficients 814
Fresnel diffraction 694, 707
Fresnel equations 814

grounding 467, 562, 565, 569, 572-573,
        580, 586, 795, 836
  ground loops 57, 573, 589

H-field sensors 114
holography 672, 677, 686, 789
  four-frame holographic interferometry
        technique 729, 731, 741
  holograms 675
  holographic grating 1030
  holographic interferometry 677-678,
        729-730, 732-733, 774, 789
  holographic techniques 675

I-dot probes 110, 115
image converters 1, 599, 662, 684
  image intensifier 612, 626, 628, 855-856
  image recording 627
    charge coupled device (CCD) 633
  imaging spectrograph 810-811
index of refraction 42, 44, 49, 53,
        758, 780, 835, 838, 1015
interferometry 667, 684, 692, 697,
        710, 744
  interferograms 697, 703, 753, 777
  laser interferogram 664
  laser interferometry 685, 743, 766
  Mach-Zehnder interferometer 44, 837-840
  Moire-fringes 791, 772
  Moire-pattern 771
  Moire-Schlieren 771, 777
  polarization optical time domain
        interferometry (POTDR) 53,
  Rochon prism 774-775

SAGNAC interferometer 53-54
  shearing interferogram 697-700
  shearing interferometer 703-704
  time-differential interferometry 773-774, 777
iterative deconvolution algorithm (IDA) 448-451
iterative time domain deconvolution 359

KDP crystal 58, 60, 63-66, 749
Kerr effects 28, 30-31, 53
  dc Kerr effect 8
  Kerr cell 28, 32-36, 38, 596
  Kerr fluids 57
  Kerr liquid 32
Kirchoff approximation 118, 471, 475

laser diode 38, 427, 644, 840
laser induced fluorescence (LIF) 951-953, 955-956, 961, 964-965,
    975, 1003
  LIF experiments 952, 958, 959, 962
  LIF signals 955, 959
Lorentzian profiles 982, 984
layered synthethic microstructures 812-816
  LSM manufacture 813
  LSM technology 815, 817

magnetic dipole sensors 79-80, 588
magnetic field sensors 79, 94-95, 108, 224, 229,
  03L-1A sensor 89, 99
  cavity sensor 231, 235, 236
  CML (Cylindrical Moebius Loop) sensors 108, 115
  discrete magnetic field sensors 228
  equivalent circuit of magnetic field sensor 225, 567
  moebius gaps 115
  moebius geometry 228
  Moebius loop geometry 230
  MTL-1A(A) sensor 97
  MTL-2 98
  multigap loop (MGL) 95, 228
  multiturn loop (MTL) 97, 115
  octahedral 3-axis loop (03L) sensor 98
  one-conductor, many-turn loop (OML) 96
magnetostrictive 44, 52
Marx generator 28, 33, 35-36, 820
  Marx bank 28, 32, 437, 613
  Marx's output voltage 849
  Marx-surge generators 848, 851, 852
  Marx-surge pulser supply 847, 849-850
microchannel plate (MCP) 619
  MCP photodiode 871, 879
  MCP-gated camera 871-873, 879

modal dispersion 423, 841
Monte Carlo calculations 327
Monte Carlo code 327
Monte Carlo Computer Simulation 327

nondispersive x-ray diagnostics 795, 828
nuclear detectors 274
nuclear techniques 263
numerical aperture, NA 42, 425
Nyquist plot 155, 156

optical data links 1, 44
optical fiber 42, 52, 60, 419-420, 422, 426, 440, 443
   chemical vapor deposition (CVD) 421
   double-crucible process 421, 422
optical path lengths 36

Paley-Wiener criterion 354
Parseval's Theorem 376-377
photo-multipliers 33, 1015, 1024
photomultiplier tube (PMT)
PIN diode nuclear detectors 325, 840
PIN photodiodes 427
pinhole x-ray camera 810
Pockels 30
   Pockel's coefficients 44
   Pockels-cell electro-optical q-switch 654

radiation-to-light converters 835, 844
Raman effect 658 1001, 1003, 1019
   Raman gain 993-994
   Raman line 1012, 1014
   Raman resonances 1003-1004, 1007, 1016, 1019
   Raman Scattering (RS) 989, 994, 1001-1003, 1005
   Raman shifting 692
   Raman transition 983-984, 993, 1007-1008, 1012
Rayleigh scattering 423, 1005, 1026, 1032
refractive index 31
   Gladstone Dale equation 664
resistive dividers 146, 184-185, 191,
          193, 201, 207
   geometric resistive divider 194
Rogowski coil 240-241, 243-244

Saha equation 893
schlieren 743
   schlieren imaging 744, 758, 766
   schlieren measurement 764, 771

schlieren photography 661, 733
schlieren studies 602
schlieren techniques 643, 684, 759, 761
shadowgraphy 659, 661, 663, 684, 743-744, 747,
         752, 754, 760, 766
  dark field shadowgraphy 717, 720
  dual-time shadowgrams 715-716

shunts 152-153, 155, 555
  coaxial disk shunts 152, 153
  coaxial shunts 156

spectroscopy 796, 1011, 982
  filtered detector spectrometer 803
  filtered detector system 806, 808,
                     828-830, 832
  Gaunt factors 893, 895, 900, 903
  high-resolution x-ray spectroscopy 796
  infrared absorption spectroscopy (IRAS) 974-975
  K-edge filters 809-810
  low-resolution spectroscopy applications 796, 812
  spectrograph 939, 940, 945
  spectrometer 798, 1025, 1030
  time-resolved low-resolution spectroscopy 796
  time-resolved spectroscopy 1020-1021
Stark effect 891, 907, 1001
streak camera 60-61, 67-68, 438, 440, 596,
         734, 738, 744, 748, 756, 764
streak photography limits 596
  electrooptical streak camera 438-439
  Schardin limit 596-597
streaked shadowgraphy 743-744, 756

Thomson scattering 1023-1026, 1028-1029, 1033-1034
  Thomson beam-splitting prism 48
  Thomson parabola analyzer 322
tomography 905
topology 124, 467, 485, 487-488, 545, 571-572
  circuit topology 471-472, 477
  electromagnetic topology 118, 467, 567-568,
                     573, 577, 582
  network topology 469-470
  node-branch matrix 472, 474-475
  supermatrices 470, 489, 513, 517-519, 523, 526-527, 532-533
  supervectors 470, 513, 527, 533, 535, 538-539, 542, 544
  wave-wave matrix 478, 482, 526
two photon excitation (TPE) techniques 976, 987

verdet constant 32, 41, 837
vidicons 439, 627, 633
voltage measurements 175
voltage measurements in transmission lines 176, 469-470
voltage measurements with resistive dividers 175, 192

x-ray diagnostics 3, 796, 827, 835
x-ray diffraction 812-813, 859
x-ray diode (XRD) 800-801
 flat-response XRD 801, 803
x-ray optics 814, 817, 833
x-ray sensitive streak camera 815
x-ray shadowgraphs 758, 767, 874, 876

Young's modulus 44

Zeeman splittings 908

Drawing

THE HEAD AND FIGURE